TROPICAL WETLANDS – INNOVATION IN MAPPING AND MANAGEMENT

PROCEEDINGS OF THE INTERNATIONAL WORKSHOP ON TROPICAL WETLANDS: INNOVATION IN MAPPING AND MANAGEMENT, OCTOBER 19-20, 2018, BANJARMASIN, INDONESIA

Tropical Wetlands – Innovation in Mapping and Management

Editors

Yiyi Sulaeman

Indonesian Center for Agricultural Land Resource Research and Development, Agency for Agricultural Research and Development, Ministry of Agriculture, Bogor, West Java, Indonesia

Laura Poggio

ISRIC – World Soil Information, Wageningen, The Netherlands

Budiman Minasny

Faculty of Science, School of Life and Environmental Sciences, University of Sydney, Sydney, NSW, Australia

Dedi Nursyamsi

Indonesian Center for Agricultural Land Resource Research and Development, Agency for Agricultural Research and Development, Ministry of Agriculture, Bogor, West Java, Indonesia

CRC Press
Taylor & Francis Group
Boca Raton London New York

CRC Press is an imprint of the
Taylor & Francis Group, an **informa** business
A BALKEMA BOOK

Published by:
CRC Press/Balkema
Schipholweg 107C, 2316 XC Leiden, The Netherlands

First issued in paperback 2023

© 2020 Taylor & Francis Group, London, UK

CRC Press/Balkema is an imprint of the Taylor & Francis Group, an informa business

No claim to original U.S. Government works

ISBN: 978-1-03-257077-8 (pbk)
ISBN: 978-0-367-20964-3 (hbk)
ISBN: 978-0-429-26446-7 (ebk)

Doi: https://doi.org/10.1201/9780429264467

Visit the Taylor & Francis Web site at
http://www.taylorandfrancis.com

and the CRC Press Web site at
http://www.crcpress.com

Typeset by Integra Software Services Pvt. Ltd., Pondicherry, India

Library of Congress Cataloging-in-Publication Data
Applied for

Table of Contents

Tropical Wetlands — Innovation in Mapping and Management — Sulaeman et al. (Eds)
© 2020 Taylor & Francis Group, London, ISBN 978-0-367-20964-3

Foreword

The world's population is increasing, and by 2050 it is projected to increase by more than 35 percent. We need to double our production to supply enough food, feed, fuel, and fiber to the growing population. This need is further challenged by climate change and extreme weather conditions. Agricultural land area is relatively constant and even tends to decrease. To feed our people, we need both options, i.e., optimizing the existing agricultural land and opening new agricultural land. Every country has different strategies.

Narrowing yield gap in the existing agricultural land has become the main effort worldwide. Innovative farming systems have been developed to make agriculture more efficient in maintaining high production. In Indonesia, our agroecosystem is divided into ricelands (irrigated and rainfed), drylands (dry and wet climate), and wetlands (tidal and in-land agricultural land). The shortage of agricultural land in Indonesia means wetlands need to be converted to agricultural use.

Considering the large potential of wetland for agriculture, wetland is the main focus for the future of agriculture in Indonesia. We need to develop it profitably and sustainably. Technologies to map and optimize wetland are available worldwide and ready to be scaled up.

Such best practices should be documented as a reference for wetland development.

This book reviews and discusses current progress on mapping and managing tropical wetlands for agricultural development. It also presents the latest technologies, techniques, challenges, ideas on wetland management, and suggestions for future research needs. I am confident that this reference will be a valuable resource for many years to come for scientists, students, and policy-makers.

Husnain
Director, Center for Agricultural Land Resource Research and Development
Indonesian Ministry of Agriculture, Agency for Agriculture Research and Development

Preface

Tropical Wetlands- Innovation in Mapping and Management contains papers presented at the Tropical Wetlands conference held on 19-20 October 2018 in Banjarmasin, Indonesia. The contributions demonstrate the latest development in mapping and management of tropical wetlands. To enrich the contribution from a global perspective, authors from selected countries were also invited to contribute.

The contributions in this book address a range of topics including mapping and characterization, use and management, conservation, and case studies regarding water management, peatlands, and tidal land.

We hope that *Tropical Wetlands- Innovation in Mapping and Management* will help and stimulate the sustainable management of wetland for agriculture in addressing food security, environmental sustainability, and farmers' prosperity.

Yiyi Sulaeman
Laura Poggio
Budiman Minasny
Dedi Nursyamsi

Bogor, October 2019

Tropical Wetlands — Innovation in Mapping and Management — Sulaeman et al. (Eds)
© *2020 Taylor & Francis Group, London, ISBN 978-0-367-20964-3*

Editorial Board

Part A. Mapping and characterization of wetlands

Open digital mapping for accurate assessment of tropical peatlands

B. Minasny
School of Life & Environmental Sciences, The University of Sydney, Australia

Rudiyanto
School of Food Science and Technology, Universiti Malaysia Terengganu, Malaysia

Y. Sulaeman
Indonesian Center for Agricultural Land Resource Research and Development, Bogor, Indonesia

B.I. Setiawan
Department of Civil and Environmental Engineering, Bogor Agricultural University, Indonesia

ABSTRACT: Peatlands are globally recognized for their high carbon stock. Worldwide estimates of soil carbon stocks have improved in the last decade due to the availability of large soil datasets and wide application of digital topsoil mapping. However, most mapping studies focus on mineral soils and do not include peatlands, leading to uncertainties in the global estimates. To reduce this knowledge gap, digital mapping techniques should be applied to peatlands. This study reviewed the controls of peatland distribution along with mapping in Indonesia. Open digital mapping techniques were introduced and demonstrated for charting peat thickness in Bengkalis Island in Indonesia. The approach makes use of the abundance of satellite data currently available in an open-access format. Additionally, it utilizes a range of machine learning algorithms in an open-source computing environment for mapping peatlands. It is upheld for transparency in the procedures, making it reputable and accountable. Due to its open nature, the method is scalable and appropriate for countries in tropical regions such as Indonesia. Field work and computer modelling can be executed in parallel for different regions and thus enhance the availability of high-quality peat information.

1 INTRODUCTION

Peatlands cover about 1-3 percent of the earth's land surface and hold probably 100 to 500 billion tons of carbon (C), an equivalent of 15-70% of atmospheric C, and 20-90% of global terrestrial C biomass (Köchy et al., 2015). Evidently, they have an important role in the global carbon cycle by storing a huge amount of C. A small loss of a peat from land clearance and mismanagement may contribute a large amount of CO_2 to the atmosphere. Indonesia has one of the largest peatlands in the tropics, estimated to be around 13 million hectares (BBSDLP, 2018).

The increasing climate and rapid land use changes have transformed peatlands into carbon source ecosystems. In addition, the warming environment causes a further release of carbon in peats. To achieve the Paris agreement target, restoration and good management of peatlands should be part of the climate change mitigation policy. Audit of carbon pools in the peatlands is important for national and international inventory reporting and reviewing carbon emissions. The report to the Intergovernmental Panel on Climate Change (IPCC) requires carbon estimates and corresponding uncertainties due to Land-use and Land-use Change and Forestry (LULUCF). Evidently, better maps for rapid assessment to support action and multi-stakeholder engagement in peatland is required (Crump, 2017).

Globally, peatlands are recognized for their high carbon stock, though their spatial distribution is still full of uncertainties. A good estimate of peat will be required for protecting and managing this rich carbon source. The Indonesian government use the thickness of the peat as a legislative measure for conservation purposes. This is because carbon loss and potential emissions reduction are highest in the areas with deep peats. Accurate peatland maps help to identify areas of threat, and the regions that need to be prioritized.

The traditional methods for mapping peatlands require a lot of field work and subjective interpretation. As well, analyzing field data and creating maps based on aerial photos or satellite imageries with expert opinion is always slow. The digital soil mapping (established twenty years ago) has revolutionized the way soil is mapped, though its application on peatland is limited. This study will demonstrate the use of digital soil mapping for improving tropical peatland inventories. To achieve this, an example of peatland mapping and C stock estimation in an area in Indonesia will be demonstrated.

2 CONTROLS OF THE SPATIAL DISTRIBUTION OF PEATLAND

To map peatland, an understanding of the controls of peat formation and distribution at different scales is needed (Limpens et al., 2008). At the global and continental scales, peat distribution is primarily driven by climate and vegetation while at a landscape the key determinants are hydrology and vegetation. At a local scale, it is determined by the biogeochemistry and plant-soil interactions (Figure 1). Therefore, to model and map peat at different spatial scales, relevant environmental factors possible to represent are needed.

Peat swamps in the tropics usually accumulate in mounds called domes. The peat domes often occur when organic materials mount up above the level of the surrounding two rivers. Their development is described by a conceptual model of water flow, an interplay between the water table and organic matter accumulation (Andriesse, 1988). Peat starts accumulating in an initial depression. In regions between 2 rivers, the accumulation of peat tends to canalize the main flow of water within the basin. The accumulation is then continued with vertical and horizontal growth of peat (as a dome), restricting the inflow of water, and thus rainfall the only source of supply. It was hypothesized that dome shape results mainly from the impeded drainage. While these conceptual models may describe peat formation qualitatively, they cannot be used for mapping purposes as the process occurring in different places may not be the same.

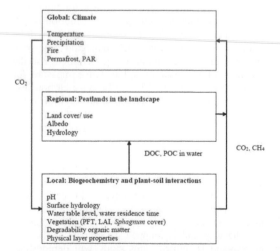

Figure 1. A conceptual model of the key biogeochemical and biophysical drivers of the peatland carbon balance at different spatial scales (from Limpens et al., 2011). PFT = plant functional type, LAI = leaf area index, DOC = dissolved organic carbon, POC = particulate organic carbon. (after Limpens et al., 2008, creative common license).

3 PEATLAND MAPPING IN INDONESIA

The latest estimate of peatlands area in Indonesia is about 13.2 million hectares (BBSDLP, 2018), mainly located in coastal and sub-coastal lowlands of the eastern coast of Sumatra, west and central Kalimantan, and Papua islands. Smaller areas are also be found in Aceh, West Sumatra, and Sulawesi (Polak, 1975).

Peat in Indonesia was recorded during the Dutch colonial period, where Koordes and Potonie found a large area of peat in Kampar, east of Sumatra in 1895 by debunking the myth against its occurrence in the tropics because of the high temperature (Joosten, 2016). Another pioneer Dutch scientist, Betje Polak, did a lot of work in this regard, and in 1952 he presented the first peatland map of Indonesia at a "rough" estimate of 16,350 million ha (Joosten, 2016). Subsequently, Andriesse (1988) gave an estimate of 17,000 million ha. In 1981, the Indonesian Soil Research Institute estimated an area of 27,000 million hectare. The latest estimate is 13,200 million hectares.

Coastal areas are favorable for peat formation, despite the high temperature experienced in the tropics. The low organic matter decomposition is mainly caused by: (1) oxygen deficiency due to constant inundation and (2) low pH and high levels of toxic chemicals such as Sulphur compounds, free aluminum, polyphenols and acidity (Bruenig, 1990). The other factors which may also contribute to peat development include suitable topography, high equatorial rainfall, and low silt content in rivers (Radjagukguk, 1997). For this reason, the rate of organic materials accumulation is larger than the rate of decomposition. The deposition of peats in the coastal area may have occurred quite recently. A measurement of the age of peat deposits at several coastal peatlands in the eastern coast of Sumatra found the peat-clay interface was about 4000-4500 years old while the surface peat ranged from 45 to 660 years (Brady, 1997).

Tropical peatland in Indonesia is often characterized by the occurrence of a convex-shaped dome in the relatively flat landscape, usually is laid down between two rivers. A transect survey along the dome revealed a thick peat layer in the center of the dome with thin edges. From a hydrological point of view, the dome shape can prevent flooding due to rapid radial drainage (Neuzil et al., 1993).

Before the 1990s, peatlands were considered marginal lands and heavily exploited. Rapid deforestation, excessive peatland drainage, and intensive fire have degraded them and increased carbon gas emissions to the atmosphere (Margono et al., 2014). With increasing awareness of climate change issues, peatland management has become an important factor for consideration in Indonesia. The government attempted to restore the degraded peatlands by issuing Government Regulation (PP) No. 71 Year 2014 and No. 57 Year 2016 on the conservation and management of peat ecosystems. From the provision, the peatland with a thickness greater than 3 meters must be conserved. Practically, one of the key impediments in peatland restoration and conservation is lack of a reliable high-resolution peat thickness map.

At a national scale (1:250,000), there are several versions of peat extent and thickness map. For instance, there is a vector format made available by an NGO, Wetlands International in 2004 (WI map) (Wahyunto and Subagjo, 2004). The WI map was digitized manually based on legacy source data and analysis of satellite imageries in 1990 and 2002.

The national peatland map (1:250,000) was afterwards updated by the Ministry of Agriculture in 2011 and lately in 2018 (MoA map). The MoA map was derived based on existing peatland charts with additional data and soil maps of Indonesia delineated with the help of SPOT5 images. Recently, it has become the official government map of peatlands in Indonesia.

The 1:250,000 chart is an indicative of peatland map. For effective spatial planning and policy making, local peat map with a resolution of 1:50,000 or spatial resolution of 30 m or finer is needed.

4 DIGITAL PEATLAND MAPPING

Traditional soil mapping requires many field observations and subjective interpretation by experts and this is labor intensive and costly. Many technologies for mapping peatlands have been tested in Indonesia, including flying sensors to detect peat from the above, the ground either proximally (near the ground) or remotely (high above the ground). However, they depended on cost and required spatial scale of mapping. Currently, the technologies such as GPR and electrical resistivity are only applicable along transects of 1-2 km, and may hardly be used for regional mapping. Similarly, flying Lidar to estimate surface elevation is expensive and not cost-effective. A single source of information provided by Lidar cannot be used to map peat extent and thickness.

Several studies tried to map peat distribution in the tropics based on satellite imageries (visible and infrared bands) by vegetation types (Wijedasa et al., 2012). Some used radar images to indicate wetness of the peat surface (Hoekman et al., 2010) while others tried to map thickness solely based on elevation (Jaenicke et al., 2008).

Digital soil mapping (DSM) was proposed by McBratney et al. (2003) as an empirical model and is based on the scorpan spatial prediction function approach:

$$P_x = f(s, c, o, r, p, a, n) + e \qquad (1)$$

where peat properties at spatial position x is a function f of soil factors (s), climate (c), organisms which include land use, human effects, and management (o), relief (r), parent materials (p), age or time (a), spatial position (n), and e is the spatially correlated errors.

Rudiyanto et al. (2018) simplified the model for peat mapping in Indonesia at a landscape scale, assuming the peat deposits (in an area) is at the same time and same parent materials:

$$P_x = f(s, o, r, n) + e \qquad (2)$$

The model assumes the factors are in steady-state, and the observations should cover the whole range variation in covariates for it to be extrapolated to the whole area. Spatial layers from remotely sensed images may be used to represent the soil, organism, relief, and spatial position factors. The form f may range from a simple linear to a more complex machine learning models such as neural networks or random forests. Digital soil mapping tries to automate the process of mapping. With a few lines of codes, a model can be generated by linking field observations with scorpan variables and be interpolated to the whole area of interest to make a digital map (Adhikari et al., 2014).

Digital soil mapping requires three components: the input in form of field and laboratory observational methods, the process used in terms of spatial and non-spatial soil inference systems, and the output which is spatial soil information systems, including outputs such as rasters along with the uncertainty of predictions.

5 OPEN DIGITAL MAPPING

Researchers from the University of Sydney, Australia and Institute Pertanian Bogor in Indonesia proposed an open digital mapping methodology (Rudiyanto et al., 2018). This would make use of open data in an open-source environment as a cost-effective and accurate method for mapping peat and its carbon stock for large areas in the country.

This open methodology combines field observations with factors known to influence peat depth distribution in space. These factors are represented by multi-source remotely sensed data derived from freely-available open statistics.

- Soil condition reflecting the moisture content of the peat may be detected using radar (ALOS PALSAR and Sentinel 1 data)
- An organism is represented by land use map optical images (Landsat TM).
- Relief is epitomized by a digital elevation model from the Shuttle Radar Topography Mission,
- Spatial position is denoted by distance to rivers and seas.

The method makes use of advanced machine learning models to predict peat depth every 30 m on the land surface. To make it transparent, accountable, and reputable, all mapping procedures were scripted in the R open source and free software. Since all the codes and data are stored digitally, if new observations are available, the whole map may be updated easily. Also, digital format offers interactive maps which allow users to view the area precisely in the field.

As the method is "open" in nature, it is scalable to a national extent for countries in tropical regions. Field work and computer modelling can be executed in parallel for different regions and thus enhance the availability of high-quality peat information.

6 CASE STUDY DIGITAL MAPPING OF PEAT THICKNESS OF BENGKALIS ISLAND

The study area was located in the eastern part of Bengkalis Island in Indonesia. The area about 50,000 hectares is located at N1.2502° to N1.5637° and E102.2658° to E102.5087°. The detail of this study may be found in Rudiyanto et al. (2018).

6.1 *DEM correction*

Peatlands in tropical regions, such as Indonesia, usually occur in the coast with a relatively flat area. Nevertheless, elevation appears to be an important predictor in peat thickness modelling. As discussed in the preceding section, peat often accumulate in a dome-shaped structure.

There are several sources of data which may be used to derive a digital elevation model (DEM). In this study we examined two freely-available DEMs: the SRTM and AW3D.

The SRTM (Space Shuttle Radar Topography Mission) was flown on board the Space Shuttle Endeavor mapped 80% of the Earth's land mass from 11 to 22 February of 2000. The freely-available one arc-second (around 30 m resolution at the equator) DEM from the Shuttle Radar Topography Mission (SRTM) proved to be an important covariate for mapping peats around the world. Due to the relatively short wavelength (C-band), the return electromagnetic energy was affected by vegetation canopy.

ALOS World 3D (AW3D) is a global digital elevation model (DEM) derived from the Panchromatic Remote-sensing Instrument for Stereo Mapping (PRISM) on board the Advanced Land Observing Satellite (ALOS). It was operated from 2006 to 2011and it was freely available at a resolution of 30 m DEM.

The DEM products are affected by vegetation. In lowland areas with rapid land use changes, there are challenges in the use of the raw DEM. Figure 2 shows two free DEM products (SRTM and AW3D) for the eastern part of Bengkalis Island in Indonesia. The difference in elevation was due to the tree height differences because of deforestation and afforestation. For this reason, there was need to correct such vegetation influences.

The availability of the 2 DEM products captured in different years in an area with a dynamic land use allows for the estimate of tree heights in the deforested area. Figure 3 shows the difference between SRTM and AW3D DEM with a time lag of 8 years. The elevation differences are due to trees lost as a result of deforestation. Therefore, tree height in some deforested areas may be calculated. The height data were intersected with a cloud-free Landsat 5 TM image close to the SRTM acquisition period. A neural network model was generated to predict vegetation height throughout the area based on Landsat bands. This process created a fine-scale tree height corrected DEM.

6.2 *Digital mapping of peat thickness*

Based on the tree height corrected DEM, two terrain attributes; multiresolution index of valley bottom flatness (MRVBF) and topographic wetness index (TWI) were derived. The MRVBF is a topographic index

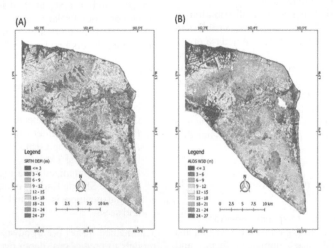

Figure 2. DEM of the eastern part of Bengkalis Island, Indonesia: (A) SRTM, (B) ALOS W3D.

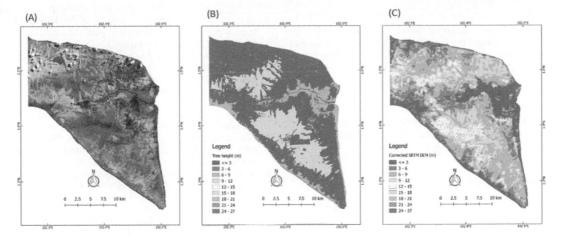

Figure 3. (A) Landsat image for the Bengkalis island, (B) modelled tree height estimated from the difference between AW3D DEM and SRTM, and (C) vegetation corrected SRTM.

designed to identify areas of deposition at a range of scales. The Topographic Wetness Index (TWI) combines local upslope contributing area and slope to quantify water flow paths. Besides, the Euclidean distance to rivers was also incorporated as covariates and radar images from Sentinel 1 were used.

The data of field peat thickness observation (n=159), and 15 selected environmental covariates described above were used in a Quantile Random Forest (QRF) model. The observed peat thickness ranged between 0 and 12 m with a median of

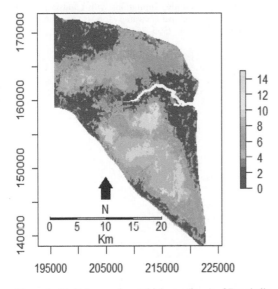

Figure 4. Digital map of peat thickness (in m) of Bengkalis Island.

6.0 m. From k-fold cross-validation, the QRF model has R2 = 0.97 and RMSE = 0.58 m, showing a good accuracy of the map.

The estimated cost and time required for map production (50,000 hectares at 30 m resolution) were 2 to 4 months and $0.3 and $0.5/ha respectively. There was a tradeoff between cost and accuracy.

This open digital mapping method is 15 times cheaper than LiDAR acquisition which costs about $5 to $15 per ha, depending on the remoteness, size, accessibility, and complexity of the area. LiDAR has been heavily endorsed as the method for mapping peatland in Indonesia without any clear scientific support or evidence. Compared to the conventional mapping, digital mapping is 3 to 4 times cheaper. As shown in many studies, the techniques used in digital mapping are more accurate and more cost-effective than conventional methods.

7 CONCLUSIONS

Digital soil mapping techniques have been successfully applied for nation-wide soil carbon mapping in various countries. This study has demonstrated the possibility of mapping peat thickness accurately using freely-available data, though it requires multiple sources of information.

The method is scalable to the whole country due to its open nature and may be carried out in parallel by many groups. With appropriate training, various groups in Indonesia may conduct mapping in a standard, and objective way. The authors envisage if carried out in parallel (both field work and computer modelling), utilizing existing observations and collecting new observations, a high-resolution peat map (estimate of peat thickness every 30 m) for Indonesia may be completed in 2 years.

It is possible for university students to participate in open digital mapping exercise. Students from universities taking the Soil Survey lectures may collect field data relating to peat thickness from pre-determined sites established using a random stratified sampling design. The sampling site may be determined using a hand-held GPS and measured thickness using a peat auger. The results may be used in a GIS class mapping peat thickness using kriging method. Further studies should employ advanced digital mapping techniques.

REFERENCES

Adhikari, K., Hartemink, A.E., Minasny, B., Bou Kheir, R., Greve, M.B. & Greve, M.H. 2014. Digital mapping of soil organic carbon contents and stocks in Denmark. *PLoS ONE*.0105519.

Andriesse, J. 1988. *Nature and Management Of Tropical Peat Soils*. Rome: Food & Agriculture Organisation.

BBSDLP. 2018. *Peta Lahan Gambut Indonesia* Skala 1: 250.000 (Indonesian Peatland map at the scale 1: 250.000). Indonesian Center for Agricultural Land Resources Research and Development, Bogor, Indonesia.

Brady, M.A. 1997. *Organic matter dynamics of coastal peat deposits in Sumatra, Indonesia*, PhD Thesis, University of British Columbia, Canada.

Bruenig, E. 1990. Oligotrophic forested wetlands in Borneo. In: A.E. Lugo, Brinson, M., Brown, S (eds), *Ecosystems of the World: Forested Wetlands*: 299-233. New York: Elsevier.

Crump, J. 2017. *Smoke on water–countering global threats from peatland loss and degradation*. A UNEP Rapid Response Assessment. United Nations Environment Programme and GRID-Arendal, Nairobi and Arendal.

Hoekman, D.H., Vissers, M.A. & Wielaard, N. 2010. PALSAR wide-area mapping of Borneo: methodology and map validation. *IEEE Journal of Selected Topics in Applied Earth Observations and Remote Sensing* 3: 605–617.

Jaenicke, J., Rieley, J.O., Mott, C., Kimman, P. & Siegert, F. 2008. Determination of the amount of carbon stored in Indonesian peatlands. *Geoderma* 147: 151–158.

Joosten, H. 2016. Changing paradigms in the history of tropical peatland research, tropical peatland ecosystems. In: *Tropical Peatland Ecosystem*: 33–48. Springer.

Köchy, M., Hiederer, R. & Freibauer, A. 2015. Global distribution of soil organic carbon–Part 1: Masses and frequency distributions of SOC stocks for the tropics, permafrost regions, wetlands, and the world. *Soil* 1: 351–365.

Limpens, J., Berendse, F., Blodau, C., Canadell, J., Freeman, C., Holden, J., Roulet, N., Rydin, H. & Schaepman-Strub, G. 2008. Peatlands and the carbon cycle: from local processes to global implications–a synthesis. *Biogeosciences* 5: 1475–1491.

Margono, B.A., Potapov, P.V., Turubanova, S., Stolle, F. & Hansen, M.C., 2014. Primary forest cover loss in Indonesia over 2000-2012. *Nature Climate Change* 4: 730.

McBratney, A., Mendonça Santos, M. & Minasny, B. 2003. On digital soil mapping. *Geoderma* 117: 3–52.

Neuzil, S.G., Cecil, C.B., Kane, J.S. & Soedjono, K. 1993. Inorganic geochemistry of domed peat in Indonesia and its implication for the origin of mineral matter in coal. *Special Papers-Geological Society of America*: 23–23.

Polak, B. 1975. Character and occurrence of peat deposits in the Malaysian tropics. *Modern Quaternary Research in Southeast Asia*: 71–81.

Radjagukguk, B. 1997. Peat soils of Indonesia: Location, classification, and problems for sustainability. *Biodiversity and Sustainability of Tropical Peatlands*: 45–54.

Rudiyanto, Minasny, B., Setiawan, B.I., Saptomo, S.K. & McBratney, A.B. 2018. Open digital mapping as a cost-effective method for mapping peat thickness and assessing the carbon stock of tropical peatlands. *Geoderma* 313: 25–40.

Wahyunto, R.S. & Subagjo, H. 2004. *Map of peatland distribution area and carbon content in Kalimantan*, 2000–2002. Wetlands International—Indonesia Programme & Wildlife Habitat Canada (WHC), Bogor, Indonesia.

Wijedasa, L.S., Sloan, S., Michelakis, D.G. & Clements, G.R. 2012. Overcoming limitations with Landsat imagery for mapping of peat swamp forests in Sundaland. *Remote Sensing* 4: 2595–2618.

Wetlands in Brazil

A. ten Caten
Federal University of Santa Catarina, Curitibanos, Santa Catarina, Brazil

ABSTRACT: It is often difficult to explain the importance of wetlands for economic development since these are places filled with water and, apparently, without immediate utility. In this paper, we will bring light to some aspects related to mangroves, salt marshes, and peatlands in Brazil. From the complexities involving wetlands definition and proper location all throughout the country, the worldwide importance of Brazilian Mangroves which could sum-up to nearly 8% of world area of this ecosystem, to the importance of salt marshes to carbon sequestration and its storage into the ocean's deep-waters and the peat soils (Organossolos) distributed in the entire continental area of the country, from low coastal areas (0 m) to mountains regions (2000 m), from northern-tropical to southern-temperate climates, we will seek to pinpoint some threats to these fragile environments, the controversy regarding their proper quantification and some possibilities for future research.

1 INTRODUCTION

The Brazilian territory, 8.5 million km², has a wide variety of wetland types that occupies an estimated 1.7 million km² (20%) of the land area (Junk et al., 2011). Wetlands systems can be found in the Amazon region, Savanna (Cerrado), and throughout the country, e.g. the Pantanal Matogrossense, flooded savannas of the Araguaia River, along with the Atlantic coast mangroves, where flooded wetlands are characteristic to the entire Brazilian coast (Table. 1) (Junk et al., 2014).

Table 1 outlines major wetlands in Brazil, which were made up of floodplains along the large rivers (Amazon), flooded savannas, coastal tidal wetlands. This paper will describe the general wetlands found in Brazil and will also discuss in detail mangroves, salt marshes, and peatlands in this country.

Table 1. Wetlands in Brazil.

Wetlands like large-river floodplains and the large periodically inundated savannas	Area (km²)
Main-stem Amazon River floodplain (várzea) covers	98,110
Flooded savannas in Roraima in Northern Amazonia (and Rupununi in Guiana)	16,500
Guapore River (into the Bolivian savannas of the Mamoré and Mortes Rivers)	92,100
The Pantanal (Brazil, Bolivia, and Paraguay)	109,590
Araguaia River (including Bananal Island)	58,600
Coastal tidal wetlands (mangroves and salt marshes Maranhão/Pará)	7,000

* According to (Junk et al. 2014) a total around 382,000 km².

1.1 *Brazilian wetlands and its challenges*

Brazilian wetlands are characterized by large fluctuations in the water level, also called 'flood-pulsing' since most of the country's regions face pronounced dry and wet seasons. This phenomenon causes a periodic sheet-flooding of large, flat interfluvial areas, with periodic filling-in of depressions with water, and the lateral inundation of large areas along streams and rivers. Besides, flood pulses are monomodal and predictable in these large floodplains. We can find multimodal, predictable tidal pulses in coastal wetlands with direct marine influence, and further inland, wetlands in coastal sand plains are subject to short, unpredictable, or monomodal pulses during the rainy season (Junk et al., 2014).

Brazilian wetlands are predominantly flood-pulsing wetlands with long terrestrial phases and do not exhibit accumulation of organic matter as periodic aeration facilitates the decomposition of organic matter. Organic matter accumulation as a result of permanent shallow inundation or long-term water-logging is a good indicator of permanent wet conditions (Junk et al., 2014).

According to Kandus et al. (2018), wetlands have their own identity and are often being treated only as a transitional space between terrestrial and aquatic ecosystems which causes difficulties to the better understanding of these complex wetlands ecosystems. Furthermore, physiognomies like swamps, marshes, floodplains, shallow lakes, lakes, lagoons, mangroves, peat bogs, and peatlands are all considered wetlands, and should all be studied and placed in an equivalent level to other ecosystems.

According to Junk et al. (2014), a wetland is a water table that fluctuates from dry to rainy season.

It is not the kind of environment that will accumulate much soil organic matter, at least until more country-wide surveys are carried out.

There are threats to wetlands, such as being exploited as solid-waste dumping sites, expansion of cities and housing areas, disruption by road construction, converting of wetlands into croplands, the leaching of fertilizers and pesticides as well as the run-off of untreated sewage from urban areas, and construction of hydroelectric reservoirs (Junk et al., 2014). Also, despite their importance, wetlands are rarely mentioned in federal legislation, state constitutions, or environmental legislation, and Brazil lacks a national policy that regulates their protection and management (Junk et al., 2014).

According to Kandus et al. (2018), South American countries do not have sufficient inventories of their wetlands. They lack detailed information on their areal extent, conservation status, or wetland type which compromises medium or long term monitoring plans.

In bibliometrics analyses carried out by Kandus et al. (2018), the authors explored how Remote Sensing have been reported in wetland-related studies in South America. They found a total of 153 papers, from 1960 to 2015, with 60% (90) of them focusing on macrosystems of South American mega-rivers Amazon River floodplain, Paraná River floodplain, and Pantanal at the Upper Paraguay Basin. In comparison to global publications, Remote Sensing of wetlands in South America represents, only 0.3% of the world literature. Although wetlands cover nearly 20% of the region, the lack of publications on wetlands is due to a higher attraction of research on the forest ecosystems (e.g. Amazon rainforest) among other causes.

Technological and economic difficulties hamper South American countries from carrying out long-term programs for environmental surveying and monitoring. Kandus et al. (2018) also believe that politicians in the region underestimate the importance of the wetlands ecosystems and their benefits for South America. Accordingly, no publications on peatlands were found in their bibliometric analysis of Remote Sensing application for wetland research in South America. The spatial resolution of the sensors is a strong limiting factor, especially if we take into consideration that, in many areas, wetlands are single patches or where the landscape is a mosaic of wetlands. This could be a strong limiting factor according to the authors since high-resolution imagery could be too expensive for institutions in South America (Kandus et al. (2018)).

1.2 Mangroves in Brazil

According to Magris & Barreto (2010), there is a total of 1,114,398 ha of mangrove in Brazil, which corresponds to 7.1% of these ecosystems throughout the world. The mangrove forests along the equatorial Brazilian coastline are among the largest on earth

(Kauffman et al. 2018). In Brazil, mangroves occur from the Amapá state (04°20′N) to the Santa Catarina one (28°30′S) (Schaeffer-Novelli et al., 2016).

The mean soil carbon stocks of salt marshes of the legal Amazon in the state of Pará, Brazil, are of 257 Mg C ha^{-1}, slightly higher than those used by the IPCC (Kauffman et al. 2018). Besides, the mean ecosystem carbon stocks of the Amazon mangroves was 511 Mg C ha^{-1} (a range of 362 Mg C ha^{-1} to 746 Mg C ha^{-1}). The authors have also found that mean soil carbon stocks in northern Brazil for mangroves to be 473 Mg C ha^{-1} which is below the mean value for global mangroves (885 Mg C ha^{-1}). This is due to the large tidal range in the area, linked to higher oxygen availability and, furthermore, the soils in Amazon mangroves are coarse-textured and carbon is susceptible to decomposition.

The mean carbon stock for mangroves in north-eastern Brazil, is 413 ± 94 Mg C ha^{-1}, with soils carbon stocking up to 80% of the total (Kauffman et al., 2018) with values among mangroves ranging from 129 Mg C ha^{-1} to 681 Mg C ha^{-1}. Land use changing from mangroves to shrimp ponds represented a 72% loss of the ecosystem carbon compared to nearby mangroves, with an after land use changing its value of 37 Mg C ha^{-1}. Up to 84% of the greenhouse gas emissions from mangrove conversion came from losses of soil C pools, with a soil bulk density increase and carbon concentration decrease throughout the soil profile even at depths >100 cm in areas of shrimp pond conversion. They highlighted the importance of sampling soil carbon deeper than 100 cm in estuaries.

There are extensive areas of shrimp farms in the northeastern coast of Brazil in areas of mangrove habitats (Magris & Barreto, 2010). The loss associated with the conversion of mangroves to shrimp ponds greatly exceed losses resulting from land cover change in upland tropical forests. The soil loss from the mangroves was 317 Mg C ha^{-1} which suggests that losses due to conversion are equivalent to 182 years accumulation (with a mean soil carbon accumulation rate for mangroves of 1.74 Mg C ha^{-1}) (Kauffman et al., 2018).

1.3 Salt marshes in Brazil

Mangroves and salt marshes capture large amounts of carbon from the atmosphere. Besides the hydrodynamic and geomorphic activities, they provide effective means for sequestering carbon in situ or exporting carbon offshore, where it enters the marine portion of the carbon cycle and storage in deep-water sediments (Schaeffer-Novelli et al., 2016). They have a great capturing capacity added to the long-term storage of carbon which makes those ecosystems essential for climate change adaptation strategies.

Salt marsh ecosystems occupy tropical, subtropical, and higher latitude coastal zones intermittently flooded with seawater. In Brazil, the most extensive salt marshes are located in the State of Rio Grande do Sul. Brazil's salt marshes occur all along the sheltered coast, either associated with mangroves or not (Schaeffer-Novelli et al., 2016).

1.4 Organossolos (Peatland or Histosols) in Brazil

In the Brazilian Soil Classification System (SiBCS), organossolos are considered as poorly evolved soils made up of organic material of black, very dark gray or brown coloration, resulting from the accumulation of vegetal remains, in varying degrees of decomposition, under conditions of restricted drainage (poorly drained environments), or at high altitudes with humid environments saturated with water for only a few days during the rainy season (Embrapa, 2013).

The deposition and accumulation of organic matter occur under conditions of excess water (H horizon) or free drainage (O horizon). Organossolos are soils composed of organic material (organic carbon content greater than or equal to 80 g kg^{-1} of air-dried fine earth (ADFE) and have a histic horizon (H or O) (Embrapa, 2013). The formation process of Organossolos in Brazil can occur due to the accumulation of litter in high altitude environments (mountainous regions) due to the continuous addition of vegetal material which presents a low rate of decomposition due to low temperatures and humid climate. Another important process for the formation of these organic soils is the organic materials deposited under hydromorphic conditions becoming humidified. In addition to humidification, there is also the process of decomposition - the transformation of original vegetation material into organic material - and the paludization process by the sustained and progressive colonization by plants of predominantly anaerobic environments (Perreira et al., 2005).

1.4.1 Occurrence of Organossolos
The real area of Organossolos in Brazil is controversial. In the 1:5,000,000 Solos do Brasil map, approximately 2200 km² of ORGANOSSOLOS HÁPLICOS Hêmicos are identified, corresponding to 0.03% of the territory. However, the Organossolos are also mapped in association with Podzols, Gleysols, Fluvisols, and Arenosols (Santos et al., 2011). Using legacy data from 129 profiles (Figure 1) of high organic carbon content Valladares (2003) estimated an area of 6,100 square kilometers corresponding to around 0.07% of the Brazilian territory. On the other hand, in a study by Perreira et al. (2005), it was reported that there are around 10,000 km² of Organossolos in Brazil, just over 0.1% of its territory. Considering that Brazil has more than 12,200 km² of Mangroves - distributed on more than 7000 km of coastline - where the presence of organic soils is remarkable, the real area of Organossolos is certainly unknown.

Among the difficulties to quantify more precisely the area of Organossolos in Brazil is their occurrence

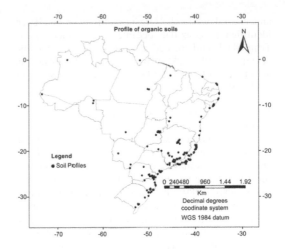

Figure 1. Distribution within Brazil of soils with high content of organic C (Beutler et al., 2017).

in the form of association and inclusion in complex areas, difficult to precisely categorize, as the areas of hydromorphic soils and in mangroves.

Also, Perreira et al. (2005) pointed out that due to the fact that only small scale soil surveys, in generalist and exploratory scales, are available for the entire country and do not allow the representation of Organossolos occurrence, which often happens in flooded or areas at high altitudes.

Among the legacy data examined by Valladares (2003), the author verified the presence of Organossolos from sea level (0 m), in poorly drained environments, up to 2000 m in low temperature and cool climate environments (mountainous regions). Soils with high organic carbon content can be found throughout the territory (Beutler et al., 2017).

The dataset classified by Valladares (2003) also showed the presence of a histic horizon, with around 18% of profiles having a thickness greater than 1.20 m, and there are even profiles with organic horizons greater than 3.00 m in the state of São Paulo. This legacy data collection demonstrated that there is a pattern showing that organic soil in mountainous regions develops less thick histic horizons (< 0.60 m). Furthermore, the same dataset showed that thicker histic horizons developed under hydromorphic conditions.

In the publication entitled Pedology: Soils of the Brazilian Biomes (Pedologia: Solos dos Biomas Brasileiros) (Curi et al., 2017), the occurrence of Organossolos Fólicos is mentioned within the Cerrado Biome in the Chapada dos Veadeiros regions. In the Atlantic Forest Biome, Organossolos are mentioned with frequent occurrence in the regions of altitude by the Araucárias Plateau (> 1500 m); and also in the Atlantic Forest Biome along the entire Brazilian coast in areas of Quaternary deposits of fluvial plains, mangroves, in areas of sandbanks (restingas) and coastal wetlands. Organossolos have also been identified in Brazilian oceanic islands such as the Trindade Island located at 1140 km in the Atlantic ocean (20°30'S / 29°19'W), these soils make up about 10% of the island's 13.5 km² area (Schaefer et al., 2017).

1.4.2 Agricultural use and consequences

Historically, there has been an agricultural interest and use of Organossolos in Brazil. Areas of these soils were already drained around the year 1729, the area of the nowadays city of Rio de Janeiro, to transform flooded areas into agricultural cultivation sites (Conceição, 1989). In 1984, the Brazilian Symposium on Organic Soils (Simpósio Brasileiro de Solos Orgânicos) was held, with several lectures with an agronomic approach, dealing with the possibility of exploring these areas for agricultural production mainly for horticulture, vegetable-growing within small agricultural family businesses (Andriesse, 1984). However, the exploration of Organossolos, and their adaptation to agricultural production resulted in draining and subsequent subsidence. Other side-effects can also be listed as a reduction of the organic volume, contraction, compaction, mineralization, combustion, wind erosion and formation of sulfuric acid with a decrease of pH in areas of Organassolos tiomórficos (Perreira et al., 2005).

2 CONCLUSIONS

Wetlands in Brazil are fragile ecosystems. They are vulnerable to environmental impacts and climate change. Therefore, research activities need to be strengthened in those areas. Brazil already has a National Committee on Wetlands to enforce actions related to the conservation and management of wetlands. Besides, important research institutions like The National Wetland Institute (INAU) and the Brazilian Agricultural Research Corporation – Embrapa Pantanal – are dedicated to the research in wetlands areas. With 27 sites and a total area of 26,794,454 ha, Brazil has the world larger extension of recognized areas of Wetlands of International Importance (Ramsar Sites). Future research should focus on multi-resolution, multi-temporal, and multi-sensor approaches to cope with the complexity of wetlands in South America.

The historical exploration of Organossolos for agricultural needs to be replaced by studies focusing on the environmental importance of these soils. Organossolos have a role in water retention, besides being chemical and physical filters, a place for maintenance of fauna and flora genetic reserves, the source of methane gas, carbon sequestration and delivering of ecosystem services. Areas of Organossolos which have not yet suffered from anthropogenic action should be mapped, protected, and conserved due to their intrinsic value and the

reduced prospects of sustainable agricultural use (Perreira et al., 2005).

Due to their occurrence in smaller mapping units, sometimes of only a few hectares, but with distribution throughout Brazilian territory, Organossolos will only be mapped and recorded in soil surveys of large scales, and geographically distributed throughout the entire country. Also, further studies are needed to better define the information for the 5th and 6th categorical levels for Organossolos within the Brazilian Soil Classification System.

REFERENCES

Andriesse, J. (ed.). 1984. *Simpósio Nacional de Solos Orgânicos*. Curitiba: MA/Provárzeas Nacional, Embrater, SEAG - Paraná e Acarpa/Emater.

Beutler, S.J., Pereira, M.G., Tassinari, W. de S., Menezes, M.D. de, Valladares, G.S. & Anjos, L.H.C. dos 2017. Bulk Density Prediction for Histosols and Soil Horizons with High Organic Matter Content. *Revista Brasileira de Ciência do Solo* 41: 1–13.

Conceição, M. 1989. *Natureza do húmus e caracterização de solos com elevado teor de matéria orgânica da região de Itaguaí - Santa Cruz, RJ*. Itaguaí: Universidade Federal Rural do Rio de Janeiro.

Curi, N., Ker, J.C., Novais, R.F., Vidal-Torrado, P. & Schaefer, C.E.G.R. (eds.) 2017. *Pedologia: solos dos biomas brasileiros*. Viçosa: Sociedade Brasileira de Ciência do Solo.

Embrapa. 2013 *SiBCS–Sistema brasileiro de classificação de solos*. Rio de Janeiro: Centro Nacional de Pesquisa de Solos - EMBRAPA.

Junk, W.J., Piedade, M.T.F., Schöngart, J., Cohn-Haft, M., Adeney, J.M. & Wittmann, F. 2011. A classification of major naturally-occurring amazonian lowland wetlands. *Wetlands* 31(4): 623–640.

Junk, W.J., Piedade, M.T.F., Lourival, R., Wittmann, F., Kandus, P., Lacerda, L.D. & Agostinho, A.A. 2014. Brazilian wetlands: Their definition, delineation, and classification for research, sustainable management, and protection. *Aquatic Conservation: Marine and Freshwater Ecosystems* 24(1): 5–22.

Kandus, P., Minotti, P.G., Morandeira, N.S., Grimson, R., Trilla, G.G., González, E.B. & Gayol, M.P. 2018. Remote sensing of wetlands in South America: Status and challenges. *International Journal of Remote Sensing* 39(4): 993–1016.

Kauffman, J.B., Bernardino, A.F., Ferreira, T.O., Bolton, N. W., Gomes, L.E. de O. & Nobrega, G.N. 2018. Shrimp ponds lead to massive loss of soil carbon and greenhouse gas emissions in northeastern Brazilian mangroves. *Ecology and Evolution* 8(11): 5530–5540.

Kauffman, J.B., Bernardino, A.F., Ferreira, T.O., Giovannoni, L.R., de O., Gomes, L.E., Romero, D.J. & Ruiz, F. 2018. Carbon stocks of mangroves and salt marshes of the Amazon region, Brazil. *Biology Letters* 14(9): 1–4.

Magris, R.A., & Barreto, R. 2010. Mapping and assessment of protection of mangrove habitats in Brazil. *Pan-American Journal of Aquatic Sciences* 5(4): 546–556.

Pereira, M.G., Anjos, L.H.C. & Valladares, G.S. 2005. Organossolos: Ocorrência, gênese, classificação, alterações pelo uso agrícola e manejo. In Torrado, P.V., Alleoni, L.R.F., Cooper, M., Silva, A.P., Cardoso, E.J. (eds.), *Tópicos em ciência do solo*: 233–276. Viçosa: Sociedade Brasileira de Ciência do Solo.

Santos, H.G. dos, Carvalho Junior, W. de, Dart, R. de O., Aglio, M.L.D., Sousa, J.S. de, Pares, J.G., Fontana, A., Martins, A. L. da S. & Oliveira, A.P.de. 2011. *O novo mapa de solos do Brasil: legenda atualizada*. Rio de Janeiro: Embrapa Solos.

Schaeffer-Novelli, Y., Soriano-Sierra, E.J., do Vale, C.C., Bernini, E., Rovai, A.S., Pinheiro, M.A.A. & Cintrón-Molero, G. 2016. Climate changes in mangrove forests and salt marshes. *Brazilian Journal of Oceanography* 64(2): 37–52.

Schaefer, C.E.G.R., Oliveira, F.S. de, Marques, F.D. & Solos Das Ilhas Oceânicas. 2017. In Curi, N.; Ker, J. C.; Novais, R.F.; Vidal-Torrado, P.; Schaefer, C.E.G. R. (eds.), *Pedologia: solos dos biomas brasileiros*: 546–597. Viçosa: Sociedade Brasileira de Ciência do Solo.

Tropical Wetlands — Innovation in Mapping and Management — Sulaeman et al. (Eds)
© 2020 Taylor & Francis Group, London, ISBN 978-0-367-20964-3

Peat and peatland mapping in Australia

B. Malone
A Agriculture and Food, Commonwealth Scientific and Industrial Research Organisation, CSIRO, Canberra, Australia

D. Kidd
Department of Primary Industries, Parks, Water and Environment (DPIPWE), Launceston, Australia

ABSTRACT: Peatland can be found throughout Australia from the wet tropics in the north to the temperate zone, the alpine regions in the south east to the coastal plains in the south west. However, the extent of peatland in Australia mainland is relatively small and receives little attention. The most studied peats in Australia are those found in Tasmania but there has been limited detailed mapping of peat formation in Tasmania, with field data scarce due to the access constraints inherent in the inhospitable south-west wilderness environments. Digital Soil Mapping of Organic Soil Carbon (SOC) throughout Tasmania has been done, and using 18% topsoil SOC content as a lower limit for peat soil, a map of peatland has been derived. Future work will improve the estimated coverage of carbon mapping and peat soils in Tasmania and Australia.

1 INTRODUCTION

Literature regarding the extent and nature of peatlands in Australia are well summarised in Pemberton (2005) and Grover (2006). Our summary of Australian peatlands mainly draws upon those resources and the literature therein.

Compared to the northern hemisphere, in general peatlands in the tropics and the southern hemisphere have been much less studied. There are obvious reasons for this which include there being less land at comparable latitudes, and much of the land is actually very dry. Additionally, in the tropics, high temperature and seasonal wet conditions also preclude the development of peats. Within the southern hemisphere, Australian peats have been much less extensively studied compared to those in New Zealand or Chile. Besides Tasmania, peatlands in Australia are not extensive, as the climate does not favour their formation (Cambell 1983). Because of their small extent, peats do not really appear on soil maps and there is no accurate estimate of their extent in Australia. According to the Australian Soil Classification (Isbell 1996), peats are considered as Organosols which are those soils that are not regularly inundated by saline water and either have more than 0.4m of organic materials within the upper 0.8 m, or have organic materials to a minimum depth of 0.1m if directly overlying rock or other hard layers. Organosols are then sub-divided on the basis of decomposition into fibric, hemic or sapric suborders and then sub-decided further according to chemistry and the underlying substratum (Isbell 1996).

As summarised in Grover (2006), wood peat, herbaceous peat and moss peat all occur in primarily in coastal and mountain areas of Australia. Coastal peats are mainly wood or herbaceous. Herbaceous and moss peats occur in mountain areas in the Eastern Highlands. Most of Australia's moss peats are within National Parks. Pemberton (2005) gives a good overview of specific areas and environments within Australia where peats occur. As described in Pemberton (2005), most Australian peatlands are late Pleistocene to Holocene in age with most having formed in the last 15,000 years following the last Pleistocene glaciation. In south eastern Australia this corresponded to the onset of more humid and maritime/temperate conditions following climatic change.

The most studied peats in Australia are those found in Tasmania and the moss peats found in the eastern highlands (Whinam and Hope 2005). Australia's body of research into peat includes general descriptions of specific peatland types and their distribution. Probably the biggest threats to Australian peats presently and in the past have been intensive animal grazing and fire. The slow recovery of peats subjected to these threats together with weed invasion and introduction of new plant species has seen number of research efforts directed towards active restoration of these environment. The other pervasive threat to Australian (and all global peatlands) is to do with climate change. In Australia the extent of these impacts have not been extensively studied, yet a study in Tasmania by Bridle et al. (2003) indicated that evaporation is greater than precipitation during drier months and that recent extended dry periods and lower water tables have been recorded for blanket bogs in the State, which could indicate the beginnings of a decline in peat accumulation there.

2 TASMANIA

Despite the ecological importance and considerable extent of peat soils in Tasmania's South West World Heritage wilderness area, estimated to be in excess of 1,000,000 ha (Pemberton, 2005), there is very little site data or mapping available to accurately map peat extent in Tasmania. Peat soils in Tasmania's south-west are generally shallow, averaging 30 cm in depth (Bridle, 1992), falling just below the classification range of 40 cm of Organosols under the Australian Soil Classification (ASC) for organic material overlying mineral soils, but meeting the requirements of Organosols having at least 10 cm of organic material over hard layers such as rock, weathered rock or gravels (Isbell, 2002).

As per the definitions of Organosols in Australia Soil Classification System (ASC), the type of peats generally found in Tasmania are fibrous, intermediate and 'muck', varying in humification, depth and moisture status (Bridle, 1992); these correspond to the ASC suborders of 'Fibric', 'Hemic' and 'Sapric' respectively (Isbell, 2002), and can exist together as different overlying peat layers depending upon topographic position.

There are many environmental factors that influence peat formation in Tasmania, including terrain (altitude, slope, and aspect), climate (rainfall, humidity, evaporation, and solar radiation), vegetation, parent material (type and weathering), groundwater and fire regime. The correlation between these environmental factors and peat occurrence, type, and depth is highly complex, with no single covariate alone considered the main driver of peat formation. Deeper, sapric peats generally occur at lower altitudes under buttongrass moorlands, while fibrous peats generally occur in relatively well drained woody and alpine vegetation regions (Bridle, 1992). Cool, wet conditions with low evaporation are considered favourable for peat formation in topographic hollows in alpine regions, leading to extensive peat formation in the Tasmanian Highlands, and lowland peats and blanket bogs forming on undulating low hills to sea-level on the Sate's west coast (Pemberton, 1989; Pemberton, 2005). Fire history, intensity and extent also strongly influence peat formation and loss in Tasmania, with variations in topographic position and fire interval leading to peat loss and erosion on steeper slopes, with faster post-fire peat recovery in topographic depressions (di Folco and Kirkpatrick, 2011).

3 PEATLAND MAPPING IN TASMANIA

There has been limited detailed mapping of peat formation in Tasmania, with field data scarce due to the access constraints inherent in the inhospitable south-west wilderness environments. Pemberton et al (1989) undertook considerable field work throughout the south-west region while mapping the Land System of Tasmania, producing 1:250,000 nominal-scaled conceptual land system component maps based on grouped similarities of rainfall, elevation, vegetation, topography and soils. These maps were later updated to include ASC estimates of each land system component (Cotching et al., 2009). Figure 1 shows the dominant soil orders map of Tasmania for Organosols (for each land systems component). Each component (up to 5 in total) correspond to a conceptual percentage of each land systems polygon. The approximate total area of peatland in Tasmania is 960,000 ha. However, it must be stressed that many of these components are conceptual, rather than field mapped, at a scale of 1:250,000.

The map (Figure 1) does not necessarily depict the absolute areas of Organosols, but the land systems containing Organosols (in described components or percentage estimates). It is therefore likely that these calculations are over-estimating the area of Organosols in Tasmania.

4 DIGITAL PEATLAND MAPPING

More recently, a Digital Soil Mapping (DSM) approach (McBratney et al., 2003) was applied by the Department of Primary Industries Parks Water and Environment (DPIPWE) in Tasmania as regional contributions to the Soil and Landscape Grid of Australia; part of this was to map soil organic carbon (SOC)

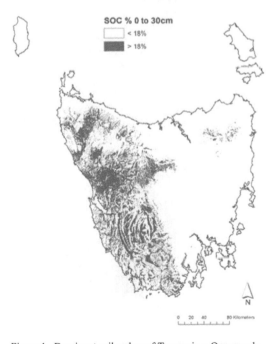

Figure 1. Dominant soil orders of Tasmania – Organosols.

content at 80 m resolution across the whole state for standard GlobaSoilMap (Arrouays et al., 2014) depths (Kidd et al., 2015). Good SOC predictions were obtained, with generally good validation diagnostics obtained across all depths to around 30 cm (the depth generally corresponding to maximum soil carbon development, with low SOC generally found in Tasmanian subsoils (Cotching, 2012).

Using these DSM surfaces for each depth to 30cm (corresponding to both better modelling diagnostics and the previously quoted average peat depth across Tasmania), a depth weighted mean of each layer was used to generate an estimated SOC content map across the state for 0 to 30cm. This was then split into areas of SOC < 18% as being non-peat soils, and SOC > 18% being considered as peat soils. The figure of 18% SOC is widely considered as constituting organic material, and therefore defined as peat soil (Isbell, 2002). Figure 2 shows the predicted extent of peat soils (SOC > 18%) using the Tasmanian depth-weighted DSM.

While generally showing similar spatial patterns to the land systems derived in the Figure 2 estimates (south-west peat predominance), the DSM products generally show more spatial detail and are better aligned with terrain. Average annual rainfall and terrain-based derivatives were found to be the most important predictors of SOC in Tasmania (high rainfall and lower slopes) (Kidd et al., 2015). The DSM peat-estimate product (0 to 30cm, SOC% > 18) corresponds to a total area of 1,147,820 ha. However, this mapping appears to be missing areas of coastal peat soils (many classifying as Podzols or Podosols in the Australian system (Isbell, 2002) around the far north-west, north-east, Flinders and King Islands, and was produced using limited site data in the South-West. This product should also be considered as a regional estimate of peat extent, due to the sparsity of calibration data used.

DPIPWE is currently re-running the DSM products at 30 m resolution, using recently acquired legacy data obtained from the Pemberton et al (1989) Land Systems mapping. However, both mineral (non-peat) and peat soils (defined as SOC% > 18) were both modelled together using the DSM approach described in Kidd et al (2015). Because peat soils are generally formed through different processes than many mineral soils, it would be advantageous to model peat formation, extent, depth and type separately. Consequently, DPIPWE has recently commenced a field program to identify non-peat and peat soils and depths in the Tasmanian World Heritage areas, with analysis of the carbon fractions to apply a DSM process to map peat thickness and extent. This is expected to be completed soon; it is hoped that the increased resolution, new datasets and remotely-sensed products (e.g. SENTINEL 1, Rudiyanto et al., 2018) will improve the estimated coverage of carbon mapping and peat soils in Tasmania.

5 CONCLUSIONS

Digital soil mapping techniques have been successfully applied for nation-wide soil carbon mapping in Australia. We need to extend such application to map peatland in Australia more accurately.

REFERENCES

Arrouays, D., Grundy, M.G., Hartemink, A.E., Hempel, J.W., Heuvelink, G.B.M., Hong, S.Y., Lagacherie, P., Lelyk, G., McBratney, A.B., McKenzie, N.J., Mendonca-Santos, M.D., Minasny, B., Montanarella, L., Odeh, I.O.A., Sanchez, P.A., Thompson, J.A., Zhang, G.L., 2014. GlobalSoilMap. Toward a Fine-Resolution Global Grid of Soil Properties, Advances in Agronomy, pp. 93–134.

Bridle, K., 1992. Organic Soils on Mt. Sprent, south west Tasmania: an analysis of correlations with local climate, microtopography and vegetation. Master of Environmental Studies thesis. University of Tasmania Hobart.

Cambell, E.O., 1983. Mires of Australasia. In: A.J.P. Gore (Ed.), Ecosystems for the World 4B Mires: Swamp, bog, fen and moor. Elsevier Scientific Publishing Company, Amsterdam.

Cotching, W., 2012. Carbon stocks in Tasmanian soils. Soil Research 50(2), 83–90.

Cotching, W.E., Lynch, S., Kidd, D.B., 2009. Dominant soil orders in Tasmania: distribution and selected properties. Soil Research 47(5), 537–548.

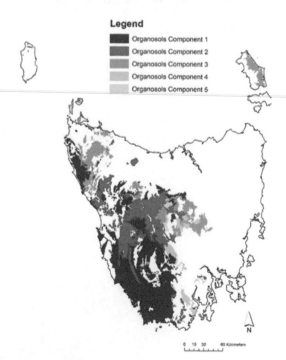

Legend

■ Organosols Component 1
■ Organosols Component 2
■ Organosols Component 3
■ Organosols Component 4
■ Organosols Component 5

N

0 15 30   60 Kilometers

Figure 2. DSM predicted peat soils - Tasmania.

di Folco, M.-B., Kirkpatrick, J.B., 2011. Topographic variation in burning-induced loss of carbon from organic soils in Tasmanian moorlands. Catena 87(2), 216–225.

Grover, S., 2006. Carbon and Water Dynamics of Peat Soils in the Australian Alps (PhD Thesis), La Trobe University, Bundoora, Victoria, 205 pp.

Isbell, R., 2002. The Australian Soil Classification - Revised Edition, Australian Soil and Land Survey Handbooks Series 4. CSIRO PUBLISHING, Australia.

Kidd, D., Webb, M., Malone, B., Minasny, B., McBratney, A., 2015. Eighty-metre resolution 3D soil-attribute maps for Tasmania, Australia. Soil Research 53(8), 932–955.

McBratney, A.B., Mendonça Santos, M.L., Minasny, B., 2003. On digital soil mapping. Geoderma 117(1-2), 3–52.

Pemberton, M., 1989. Land Systems of Tasmania. Region 7, South West. Department of Agriculture.

Pemberton, M., 2005. Australian peatlands: a brief consideration of their origin, distribution, natural values and threats. Journal of the Royal Society of Western Australia 88:81–89.

Rudiyanto, Minasny, B., Setiawan, B.I., Saptomo, S.K., McBratney, A.B., 2018. Open digital mapping as a cost-effective method for mapping peat thickness and assessing the carbon stock of tropical peatlands. Geoderma 313, 25–40.

Whinam, J., Hope, G., 2005. The Peatlands of the Australasian Region. In: G.M. Steiner (Ed.), Moore - von Sibirien bis Feuerland Mires - from Siberia to Tierra del Fuego. Biologiezentrum der Oberoesterreichischen, Linz, pp. 397–434.

Tropical Wetlands — Innovation in Mapping and Management — Sulaeman et al. (Eds)
© 2020 Taylor & Francis Group, London, ISBN 978-0-367-20964-3

Digital soil mapping for northern peatlands: Examples of mapping peats and their characteristics in Scotland

L. Poggio, R. Artz & A. Gimona
The James Hutton Institute Craigiebuckler, Aberdeen, Scotland (UK)

ABSTRACT: Peatlands contain a large amount of the carbon stocks of the biosphere. Mapping the extent, locations and conditions of peatland at the landscape scale has implications for carbon inventories, conserva- tion and ecosystem services assessments. Digital Soil Mapping provides a (geo)statistical framework to quantitatively model relationships between soil properties measured or observed on the ground and environmental covariates. The main aim of this paper was to present some successful examples of DSM for northern peatlands and their condition, in particular mapping and modelling peat presence and condition using a range of available earth observation sensors. Remote sensing data were also used to map the potential of the peat being degraded. The two results were combined to provide a probabilistic indication of where degraded peat can be found in Scotland. The results could be validated with good outcomes that visually agree with the expert evaluation and the common knowledge of the landscape. These probabilities and uncertainties need to be taken into account to improve further modelling such as earth surface modelling or carbon accounting.

1 INTRODUCTION

Peatlands contain a large amount of the carbon stocks of the biosphere, estimated between 113 and 543 Pg (10^5g) of carbon (Jackson et al., 2017; Kˇochy et al., 2015; Yu, 2012). Peat can range from moss peat in artic, sub-Arctic and boreal region to sedge and forest peat in temperate regions, to mangrove and swamp forest peat in humid tropics. 80% of the global peatlands occur in boreal regions of the northern emisphere (Bourgeau-Chavez et al., 2017).

Mapping the extent and locations of peatland at the landscape scale has implications for carbon inventories, conservation and ecosystem services assessments. Often peatlands were mapped as vegetation types (mainly wetlands) or as part of land cover (Mahdavi et al., 2017; Guo et al., 2017). Fewer examples exists of mapping peatland from a soil point of view, such as Montanarella et al. (2006); Bruneau and Johnson (2014). Despite the recent increase in Digital Soil Mapping (DSM) approaches (e.g. Keskin and Grunwald, 2018; Rossiter, 2018) for a large number of soil properties, there have been few efforts to map the locations and extent of peatland, see Minasny et al. (SUBM) for a recent review.

Digital Soil Mapping provides a (geo)statistical framework to quantitatively model relationships between soil properties measured or observed on the ground and environmental covariates that represent soil forming factors and to map the resulting predictions with uncertainty (McBratney et al., 2003;

Minasny and McBratney, 2016). DSM normally follows the scorpan approach, where the covariates are selected to represent the soil formation factors (Jenny, 1941):

$$S = f\left(s,\ c,\ o,\ r,\ p,\ a,\ n\right) + e$$

where the soil properties considered is function of s soil factors, c climate, o organisms (including land use and management), r relief, p parent material, a time, n spatial position and the spatially correlated error e.

Some recent examples of the use of DSM for mapping peatlands in northern regions are e.g. Sheng et al. (2004); Krankina et al. (2008); Rawlins et al. (2009); Kempen et al. (2009); Greve et al. (2014); Aitkenhead (2017). There were also efforts to map tropical peat using DSM approach such as (Margono et al., 2014; Rudiyanto et al., 2016). A variety of remote sensing sensors were used, both optical, radar and a combination of the two (Bourgeau-Chavez et al., 2017; Hribljan et al., 2017).

The main aim of this paper was to present some examples of DSM for northern peat and their condition, in particular mapping and modelling of peat presence and condition using a range of available remote sensing sensors.

1.1 Test area and data used

The models were run for the whole of Scotland (78, 000 km²), excluding Shetland islands because of the high cloud cover and low number of data available. Scotland has a fragmented distribution of different types of peat (Bruneau and Johnson, 2014; Joint Nature Conservation Committee, 2011).

The data sets defining presence or absence of peat in the soil were obtained from the Scottish Soil Profiles Data set. In total 7867 locations were available (Figure 1). The Scottish Soils Database contains information and data on soils from locations throughout Scotland. It contains the National Soil Inventory of Scotland (NSIS) profile samples collected on a regular 10 km grid of sampled locations (Lilly et al., 2010) and physical and chemical data from a large number of soil profiles taken to characterize the soil mapping units.

Data from the rolling six-year programme Site Condition Monitoring programme, based on Common Standards Monitoring (CSM) guidance (JNCC, 2004, 2009), for designated upland and lowland peatland sites in the period of interest were provided from the Site Condition Monitoring database by the Scottish Natural Heritage. In total 951 locations were available, 602 in favourable conditions and 349 in un- favourable conditions (Figure 2). This method assesses site condition on the basis of a number of criteria for each habitat type, in particular 1. feature extent 2. vegetation composition, i.e. frequency and cover of taxa which are indicators of favourable condition 3. vegetation structure, e.g. vegetation height, grazing, browsing, die-back of typical species 4. physical structure, e.g. excessive ground disturbance, burning, drainage or drying.

1.2 *Environmental covariates*

The covariates included are freely and globally available and were selected to describe, directly or

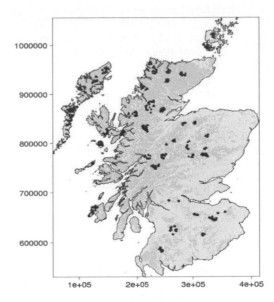

Figure 2. Locations of the points for peat conditions.

indirectly, the most important factors, namely topography, vegetation, climate and geographical position. The covariates were resampled to 100×100 m and the medians in each grid cell were used.

1.2.1 *Remote sensing*

1.2.1.1 Sentinel-1
The Sentinel-1 (S1) mission provides data from a dual-polarization C-band Synthetic Aperture Radar (SAR) instrument. This includes the S1 Ground Range Detected (GRD) scenes using the Interferometric Wide swath mode. The images were processed using the Sentinel-1 Toolbox to generate a calibrated, ortho-corrected product. Each scene was pre-processed with Sentinel-1 Toolbox using the following steps: Thermal noise removal, Radiometric calibration and terrain correction using SRTM 30 (Woodhouse, 2005). The final terrain corrected values were converted to decibels via log scaling log value = $10 \times \log10$(value) and quantized to 16-bits. The data were pre-processed, prepared, mosaiced and downloaded from Google Earth Engine (Gorelick et al., 2017).

VV and VH polarisations for the images available between 2015 and 2017 were used. VH is a vertically transmitted and horizontally received SAR (Synthetic Aperture Radar) backscatter signal from S1. VV is a vertically transmitted and received SAR backscatter signal. The polarisation ratio was calculated:

$$P \text{ olrt} = \frac{VH - VV}{VH + VV}$$

This ratio was proven useful for discriminating land covers, in particular with different canopy den- sities and it is an important parameter for soil moisture

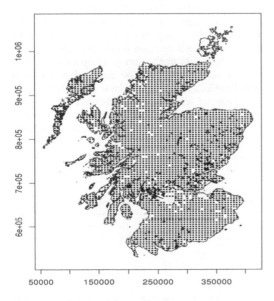

Figure 1. Locations of the profiles for peat presence.

detection (Becker and Choudhury, 1988; Patel et al., 2006; Gherboudj et al., 2011). The overall median (2015 to 2017) and monthly medians were calculated and used as covariates.

1.2.1.2 Sentinel-2
Sentinel-2 (S2) is a wide-swath, high-resolution, multi-spectral imaging mission supporting Copernicus Land Monitoring studies, including the monitoring of vegetation, soil and water cover, as well as observation of inland waterways and coastal areas. Each band represents TOA reflectance scaled by 10000. The data were mosaiced and downloaded from Google Earth Engine (Gorelick et al., 2017). The obtained mosaicked image (June to September 2016) showed some areas with snow and other smaller areas with some remaining clouds. The clouds and snow free areas allowed to fit the models and validated them.

The following band indices were calculated, where NIR is the near-infrared band, SWIR the short-wave infrared band with different wavelengths as indicated below:

1. NDVI
2. NDWI (Gao, 1996) using band SWIR between 2100-2300

1.2.2 *MODIS*
Eight-day composites from the MODIS satellite for mainland Scotland, the Western Isles and Orkney were downloaded for the period 2002-2011. Data for the Shetland Islands were excluded due to the much higher cloud coverage. The median of MODIS data for the 12 years were used, with cloud gaps filled using the method described in Poggio et al. (2012).

1.2.3 *Morphology*
The Digital Elevation Model (DEM) used as a covariate in the fitted models was SRTM (Shuttle Radar Topography Mission), further processed to fill in no-data voids (Jarvis et al., 2006; Rodriguez et al., 2006). SRTM has a spatial resolution of 90m with global coverage. The measures used were elevation, slope as the steepest slope angle, calculated using the D8 method (O'Callaghan and Mark, 1984), and topographic wetness index (TWI Sorensen et al., 2006).

2 METHODOLOGY

2.1 *Data sets preparation*

The soils were re-classified in three main categories of peat, peaty (or organo-mineral) and mineral, http://www.hutton.ac.uk/learning/soilshutton/soil-classification. The definition of peat in Scotland is based on the presence and thickness of the organic horizon (Bruneau and Johnson, 2014).

2.2 *Modelling approach*

An extension of the scorpan-kriging approach was used, i.e. hybrid geostatistical modelling, combining a trend model with kriging of the residuals. The approach features:

1. The fitting of a model to estimate the trend of the variable
2. Kriging of the residuals as spatial component to account for local details.

The trend was estimated with Generalized Additive Models (GAM; Wood, 2006) for binomial vari- ables. It produced the probability of presence for each prediction point (i.e. pixel). The GAM approach was used for the soil classification of peat as ordinal categorical variable. In case of the ordinal categorical regression, a linear predictor provides the expected value of a latent variable following a logistic distri- bution. The probability of this latent variable lying between certain cut-points provides the probability of the ordered categorical variable being of the corresponding category. The cut-points are estimated along side the model smoothing parameters, using the same criterion (Wood et al., 2016; Wood, 2006). The GAM implementation used relied on an internal cross-validation for model fitting (Wood, 2006). GAM was previously used for continuous soil properties (see e.g. Poggio et al., 2010, 2013; Poggio and Gimona, 2014, 2017a, b, for previous implementations).

The models produced residuals that were further modelled for spatial correlation using a kriging approach (Journel, 1996; Goovaerts, 1997). A variogram (Cressie, 1993) was fitted for the residuals. Exponential and spherical models (Deutsch and Journel, 1998; Goovaerts, 1997) were tested and the model providing the lowest AIC (Akaike information criterion) was retained. Anisotropy was also taken into account and the variograms were fitted accounting for the principal anisotropy axes (Goovaerts, 1997).

The sum of trend and corresponding kriged residual was calculated. The obtained probabilities were thresholded to obtain classes to compare with the validation set. The thresholds were calculated to minimise the mean of the error rate for positive observations and the error rate for negative observations (Wilson et al., 2004).

Table 1 presents a summary of the covariates used for each of the properties modelled in this study.

Table 1. Summary of models and covariates.

	Morphology	Modis	Sentinel 1	Sentinel 2
Peat presence	X		X	X
Peat degradation	X	X		

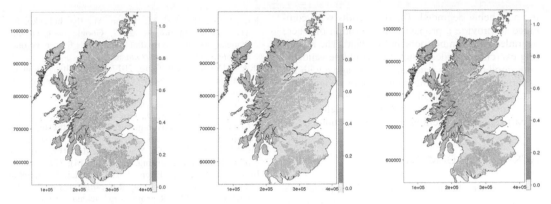

Figure 3. (a) Probability of being peat   (b) Probability of being in unfavourable conditions   (c) Combination.

2.3   Validation measures

The results were assessed using out-of-sample measures and compared for agreement between classes, spatial structure reproduction, computational load and covariates used. All data sets used were divided in calibration and validation sets (80:20) repeated 100 times using the same split was used for all combinations of covariates and methods. The median of each validation statistic is presented here. The measures calculated were:

1. Kappa statistic (Cohen, 1960): The Kappa statistic measures the proportion of correctly classified units after accounting for the probability of chance agreement.
2. Overall accuracy (Powers, 2011): it measures overall how often the predictions were correct
3. Spatial structured variance ratio (SSVR) (Vaysse and Lagacherie, 2015). The SSVR is defined as the complement to 1 of the nugget to sill ratio (Kerry and Oliver, 2008). The values range from zero to one. Values closer to one mean a higher proportion of the data explained by the spatial component.

2.4   Software used

The analyses were performed using open source software:

1. GRASS GIS (GRASS Development Team, 2017) for data management, preparation and visualization
2. The R software (R Core Team, 2017) for the statistical modelling. The following packages were used:
   a. amgcv for GAM (Wood, 2006)
   b. gstat (Pebesma, 2004) for fitting the variograms and kriging

c. raster (Hijmans and van Etten, 2013) for data management, preparation and visualisation
d. rgdal (Keitt et al., 2009) for data management
e. snow (Tierney et al., 2015) for parallelisation of the computations

3   RESULTS AND DISCUSSION

Figure 3a shows the probability of a pixel being defined as peat. this map was obtained with a model including Sentinel 1 and Sentinel 2 derived indices (Poggio et al., SUBM) with a kappa statistics of 0.82. The spatial pattern indicated a higher probability of peat along the Western coast and in the mountain areas.

Figure 3b shows the probability of a pixel to be in unfavourable conditions using MODIS indices as covariates. The most significant were the Soil Adjusted Vegetation Index, Normalised Water Difference Index and Land Surface Temperature at Night and the kappa statistics was 0.48 (Artz et al., SUBM).

Figure 3c shows the combined probabilities of a pixel being degraded peat (i.e. in unfavourable conditions). The highest values for being degraded can be found in the southern areas of the Borders and in mountain regions. Grazing and burning are common problems in these areas. The lowest values for being degraded are found in the Northern and Western regions where the largest extents of semi-natural peat can be found.

4   CONCLUSION

This paper showed some successful examples of DSM approach to map the extend of peat in a northern region using remote sensing data. Remote sensing data were also used to map the potential of the peat being degraded. The two results were combined to provide a probabilistic indication of where degraded peat can be found in Scotland. The modelling exercise provided results that could be validated with good outcomes and that visually agree with the expert evaluation and the common knowledge of the territory. These probabilities and uncertainties need to be taken into account for further modelling such as earth surface modelling or carbon accounting.

ACKNOWLEDGEMENT

This work was funded by the Scottish Government's Rural and Environment Science and Analytical Services division. Sentinel data were available from the European Space Agency and Copernicus Service. This study contains SNH information licensed under the Open Government Licence v3.0. Many thanks to the team that sampled and analysed the soils and to the team that set up the database.

REFERENCES.

Aitkenhead, M.J. 2017. Mapping peat in scotland with remote sensing and site characteristics. *European Journal of Soil Science* 68 (1): 28–38.

Artz, R.R., Johnson, S., Bruneau, P., Britton, A.J., Mitchell, R.J., Ross, L., Donaldson-Selby, G., Donnelly, D., Gimona, A. & Poggio, L. SUBM. The potential for modelling peatland habitat condition in scotland using long-term modis data. *Science of the Total Environment* 660: 429-442.

Becker, F. & Choudhury, B.J. 1988. Relative sensitivity of normalized difference vegetation index (ndvi) and microwave polarization difference index (mpdi) for vegetation and desertification monitoring. *Remote Sensing of Environment* 24 (2): 297-311.

Bourgeau-Chavez, L., Endres, S., Powell, R., Battaglia, M., Benscoter, B., Turetsky, M., Kasischke, E. & Banda, E. 2017. Mapping boreal peatland ecosystem types from multitemporal radar and optical satellite imagery. *Canadian Journal of Forest Research* 47 (4): 545-559.

Bruneau, P. & Johnson, S. 2014. Scotland's peatland - definiton and information resources.

Cohen, J. 1960. A coefficient of agreement for nominal scales. *Educational and Psychological Measurement* 20: 37-46.

Cressie, N. 1993. Statistics for Spatial Data. Wiley, New York.

Deutsch, C., Journel, A. 1998. GSLIB: Geostatistical Software Library and User's Guide, second edition Edition. Oxford University Press, New York.

Gao, B. 1996. NDWI - a normalized difference water index for remote sensing of vegetation liquid water from space. *Remote Sensing of Environment* 58: 257-266.

Gherboudj, I., Magagi, R., Berg, A.A. & Toth, B. 2011. Soil moisture retrieval over agricultural fields from multi-polarized and multi-angular radarsat-2 sar data. *Remote Sensing of Environment* 115 (1): 33-43.

Goovaerts, P. 1997. Geostatistics for Natural Resources Evaluation. Oxford University Press.

Gorelick, N., Hancher, M., Dixon, M., Ilyushchenko, S., Thau, D. & Moore, R. 2017. Google earth engine: Planetary-scale geospatial analysis for everyone. *Remote Sensing of Environment*.

GRASS Development Team, 2017. Geographic Resources Analysis Support System (GRASS GIS) Software, version 7.2.2. URL http://www.grass.osgeo.org

Greve, M., Christensen, O., Greve, M. & Kheir, R. 2014. Change in peat coverage in Danish cultivated soils during the past 35 Years. *Soil Science* 179: 250-257.

Guo, M., Li, J., Sheng, C., Xu, J. & Wu, L. 2017. A review of wetland remote sensing. *Sensors* 17 (4): 777.

Hijmans, R.J. & van Etten, J. 2013. raster: Geographic data analysis and modeling. *R package version* 2: 1-25. URL http://CRAN.R-project.org/package=raster

Hribljan, J. A., Suarez, E., BourgeauChavez, L., Endres, S., Lilleskov, E.A., Chimbolema, S., Wayson, C., Serocki, E. & Chimner, R.A., 2017. Multidate, multisensor remote sensing reveals high density of carbonrich mountain peatlands in the pramo of ecuador. *Global Change Biology* 23 (12): 5412-5425.

Jackson, R.B., Lajtha, K., Crow, S.E., Hugelius, G., Kramer, M. G. & Pieiro, G. 2017. The ecology of soil carbon: Pools, vulnerabilities, and biotic and abiotic controls. *Annual Review of Ecology, Evolution, and Systematics* 48 (1): 419-445.

Jarvis, A., Reuter, H., Nelson, A. & Guevara, E. 2006. Hole-filled seamless SRTM data V3. Tech. rep., International Centre for Tropical Agriculture (CIAT).

Jenny, H. 1941. Factors of Soil Formation. A System of Quantitative Pedology. McGraw-Hill.

Joint Nature Conservation Committee. 2011. Towards an assessment of the state of UK Peatlands. JNCC report No. 445. Tech. rep.

Journel, A. 1996. Modelling uncertainty and spatial dependence: Stochastic imaging. *International Journal of Geographical Information Systems* 10 (5): 517-522.

Keitt, T., Bivand, R., Pebesma, E. & Rowlingson, B. 2009. rgdal: Bindings for the Geospatial Data Abstraction Library. R package version 0.6-21. URL http://CRAN.R-project.org/package=rgdal

Kempen, B., Brus, D.J., Heuvelink, G.B. & Stoorvogel, J.J. 2009. Updating the 1:50,000 dutch soil map using legacy soil data: A multinomial logistic regression approach. *Geoderma* 151 (3): 311-326.

Kerry, R. & Oliver, M. 2008. Determining nugget: sillratios of standardized variograms from aerial photographs to krige sparse soil data. *Precision Agriculture* 9: 33-56.

Keskin, H. & Grunwald, S. 2018. Regression kriging as a workhorse in the digital soil mapper's toolbox. *Geoderma* 326: 22-41.

K¨ochy, M., Hiederer, R. & Freibauer, A. 2015. Global distribution of soil organic carbon part 1: Masses and frequency distributions of soc stocks for the tropics, permafrost regions, wetlands, and the world. *SOIL* 1 (1): 351-365.

Krankina, O.N., Pflugmacher, D., Friedl, M., Cohen, W.B., Nelson, P. & Baccini, A. May 2008. Meeting the challenge of mapping peatlands with remotely sensed data. *Biogeosciences Discussions* 5 (3): 2075-2101.

Lilly, A., Bell, J., Hudson, G., Nolan, A., Towers, W., 2010. National Soil Inventory of Scotland 1 (NSIS1): site location, sampling and profile description. (1978-1998). Tech. rep., Macaulay Institute.

Mahdavi, S., Salehi, B., Granger, J., Amani, M., Brisco, B. & Huang, W. 2017. Remote sensing for wetland classification: a comprehensive review. *GIScience & Remote Sensing* 0 (0): 1-36.

Margono, B.A., Bwangoy, J.-R.B., Potapov, P.V. & Hansen, M.C. 2014. Mapping wetlands in indonesia using landsat and palsar data-sets and derived topographical indices. *Geo-spatial Information Science* 17 (1): 60-71.

McBratney, A., Santos, M. & Minasny, B. 2003. On digital soil mapping. *Geoderma* 117: 3-52.

Minasny, B. & McBratney, A. 2016. Digital soil mapping: A brief history and some lessons. Geoderma 264, Part B, 301–3311, soil mapping, classification, and modelling: history and future directions.

Minasny, B., rjan Berglund, Connolly, J., Hedley, C., de Vries, F., Gimona, A., Kempen, B., Kidd, D., Lilja, H., Malone, B., McBratney, A., Roudier, P., ORourke, S., Rudiyanto, Padarian, J., Poggio, L., ten Caten, A., Thompson, D., Tuve, C. & Widyatmanti, W. SUBM. Digital mapping of peatlands a critical review. Earth Science Review.

Montanarella, L., Jones, R. & Hiederer, R. 2006. The distribution of peatland in Europe. *Mires and Peat* 1 (01).

O'Callaghan, J.F. & Mark, D.M. 1984. The extraction of drainage networks from digital elevation data. *Computer Vision Graphics Image Processing* 28: 323-344.

Patel, P., Srivastava, H.S., Panigrahy, S. & Parihar, J.S. 2006. Comparative evaluation of the sensitivity of multipolarized multifrequency sar backscatter to plant density. *International Journal of Remote Sensing* 27 (2): 293-305.

Pebesma, E. 2004. Multivariable geostatistics in S: the gstat package. *Computers & Geosciences* 30: 683-691.

Poggio, L. & Gimona, A. 2014. National scale 3D modelling of soil organic carbon stocks with uncertainty propagation - An example from Scotland. *Geoderma* 232-234: 284-299.

Poggio, L. & Gimona, A. 2017a. 3D mapping of soil texture in Scotland. *Geoderma Regional* 9: 5-16.

Poggio, L. & Gimona, A. 2017b. Assimilation of optical and radar remote sensing data in 3d mapping of soil properties over large areas. *The Science of the Total Environment* 579: 1094-1110.

Poggio, L., Gimona, A. & Brewer, M. 2013. Regional scale mapping of soil properties and their uncertainty with a large number of satellite-derived covariates. *Geoderma* 209-210: 1–14.

Poggio, L., Gimona, A. & Brown, I. 2012. Spatio-temporal MODIS EVI gap filling under cloud cover: an example in Scotland. *ISPRS Journal of Photogrammetry and Remote Sensing* 72: 56-72.

Poggio, L., Gimona, A., Brown, I. & Castellazzi, M. 2010. Soil available water capacity interpolation and spatial uncertainty modelling at multiple geographical extents. *Geoderma* 160: 175-188.

Poggio, L., Lassauce, A. & Gimona, A. SUBM. Modelling the extent of northern peat soil and its uncer- tainty with

sentinel: Scotland as example of highly cloudy region. *Geoderma*.

Powers, D.M.W. 2011. Evaluation: From Precision, Recall and F-Measure to ROC, Informedness, Markedness & Correlation. *Journal of Machine Learning Technologies* 2: 37-63.

R Core Team. 2017. R: A Language and Environment for Statistical Computing. R Foundation for Statistical Computing, Vienna, Austria, ISBN 3-900051-07-0. URL http://www.R-project.org/

Rawlins, B.G., Marchant, B.P., Smyth, D., Scheib, C., Lark, R. M. & Jordan, C. 2009. Airborne radiometric survey data and a dtm as covariates for regional scale mapping of soil organic carbon across northern ireland. *European Journal of Soil Science* 60 (1): 44-54.

Rodriguez, E., Morris, C., Belz, J., Chapin, E., Martin, J., Daffer, W. & Hensley, S. 2006. An assessment of the SRTM topographic products. Tech. Rep. JPL D-31639, NASA-Jet Propulsion Laboratory.

Rossiter, D.G. 2018. Past, present & future of information technology in pedometrics. *Geoderma* 324: 131-137.

Rudiyanto, Minasny, B., Setiawan, B.I., Arif, C., Saptomo, S. K. & Chadirin, Y. 2016. Digital mapping for cost-effective and accurate prediction of the depth and carbon stocks in indonesian peatlands. *Geoderma* 272: 20-31.

Sheng, Y., Smith, L.C., MacDonald, G.M., Kremenetski, K. V., Frey, K.E., Velichko, A.A., Lee, M., Beilman, D.W. & Dubinin, P. 2004. A highresolution gisbased inventory of the west siberian peat carbon pool. *Global Biogeochemical Cycles* 18 (3).

Sorensen, R., Zinko, U. & Seibert, J. 2006. On the calculation of the topographic wetness index: evaluation of different methods based on field observations. *Hydrology and Earth System Sciences* 10: 101-112.

Tierney, L., Rossini, A. J., Li, N.v& Sevcikova, H. 2015. snow: Simple Network of Workstations. R package version 0.4-1.

URL https://CRAN.R-project.org/package=snow

Vaysse, K. & Lagacherie, P. 2015. Evaluating digital soil mapping approaches for mapping globalsoilmap soil properties from legacy data in Languedoc-Roussillon (France). *Geoderma regional* 4: 20–30.

Wilson, K., Westphal, M., Possingham, H. & Elith, J. 2004. Sensitivity of conservation planning to different approaches to using predicted species distribution data. *Biological Conservation* 22: 99-112.

Wood, S. 2006. Generalized Additive Models: An Introduction with R. Chapman and Hall/CRC Press.

Wood, S. N., Pya, N. & Sfken, B. 2016. Smoothing parameter and model selection for general smooth models. *Journal of the American Statistical Association* 111 (516): 1548-1563.

Woodhouse, I. 2005. Introduction to Microwave Remote Sensing. CRC Press.

Yu, Z.C. 2012. Northern peatland carbon stocks and dynamics: a review. *Biogeosciences* 9 (10): 4071-4085.

Tropical Wetlands — Innovation in Mapping and Management — Sulaeman et al. (Eds)
© 2020 Taylor & Francis Group, London, ISBN 978-0-367-20964-3

Refining estimates of peat distribution, thickness and carbon stocks of Congolese peatlands: A brief review of knowledge gaps and research needs

Y.D. Botula
Faculty of Agronomy, Department of Natural Resources Management, University of Kinshasa, Kinshasa, Congo

E.T. Ngwamashi Mihaha
Réseau d'Actions des Ingénieurs Forestiers du Congo, Kinshasa, Congo

A. Lamulamu Kamukenge
Ministry of Environment and Sustainable Development, Congo

W. Kombe Ibey
Faculty of Natural & Renewable Resources Management, Water and Forests Department, University of Kisangani, Kisangani, Congo

ABSTRACT: A vast complex of peatlands was recently discovered in the Central Congo Basin ("Cuvette Centrale") in Africa and therefore remains understudied. Here, we briefly discuss five research needs in relation to pedometrics, a discipline of soil science that combines pedology and geostatistics: (1) cost-efficient data collection through appropriate sampling schemes; (2) digital mapping of peat distribution and thickness; (3) improvement of peat soil carbon stocks estimation; (4) description of the hydraulic properties of peatlands; and (5) contribution of hydropedology in understanding the hydrological functioning of the Congo Basin peat swamps. Overall, more scientific research is needed to provide up-to-date and relevant information for protection and sustainable management of the Congolese peatlands.

1 INTRODUCTION

Tropical peatlands cover approximately 440,000 km^2 or ca. 10% of the global peatland area (Page et al., 2011). Recently, vast amounts of peatlands were discovered by Dargie et al. (2017) in the Central Congo Basin (known as "Cuvette Centrale" in French), a region spanning the Democratic Republic of Congo (DRC) and the Republic of Congo (RoC). The discovery of large areas of peat soils in the swamp forests of the "Cuvette Centrale" in Central Africa has caught the world's attention. This region is known to be the home of three of the four African ape species: bonobo, western lowland gorilla, and chimpanzee, as well as the forest elephant and the dwarf crocodile (Miles et al., 2017; Dargie et al., 2018). The "Cuvette Centrale" peatland complex provides various ecosystem services and is considered the second largest tropical peatland in the world after Indonesia (Miles et al., 2017).

However, these tropical peatlands are highly vulnerable to potential land use changes and other threats (Dargie et al., 2018). Their drainage can lead to high emissions of greenhouse gases in the atmosphere and an increase in massive fire risks with serious consequences for the health and livelihood of people (Hayasaka et al., 2014). To protect these peatlands against natural and anthropogenic threats, we need to know where they are located and determine their thickness and extent. Forest inventory data collection in tropical peat swamp forests has long been a challenge, notably due to their large extent and inaccessibility combined with the difficult logistics of fieldwork posed by waterlogged conditions (Brown et al., 2018). To map the extent of "Cuvette Centrale" peatlands, a limited number of ground reference points were available from the DRC side, which contributes to considerable uncertainty in the estimated carbon stocks (Dargie et al., 2017, 2018). This large uncertainty in soil carbon stored in Congo Basin forested peatlands needs to be constrained with further investigations in this region (Fatoyinbo, 2017). Moreover, sustainable management of these peatlands requires the production of fine scale, accurate maps of peat distribution and its thickness. Traditional soil mapping techniques, particularly in developing countries, are too costly as they require many field observations, and the outputs can be too subjective (Rudiyanto et al., 2018). Due to limited resources and sparse data infrastructures in developing countries like DRC, cost-effective methods should be promoted to increase our knowledge of the newly discovered peatlands. Therefore, further studies about the extent, thickness, carbon content, and hydrological behavior of peat soils in DRC are needed.

Recently, pedometrics has emerged as a field of soil science that addresses soil problems by developing models to quantify soil variation over space and time using state-of-the-art data analysis techniques and analytical tools (Minasny et al., 2013). It may be defined as "the application of mathematical and statistical methods in the study of the distribution, the characterization and the genesis of soils" (Heuvelink, 2003). In summary, pedometrics combines pedology and geostatistics to provide: (1) faster and cheaper ways to collect data; (2) more efficient sampling schemes; (3) more accurate soil maps; and (4) an ability to predict soil properties and model soil behavior (McBratney et al., 2018).

In this article, we will explore research avenues on how pedometrics can contribute to the study of the Congolese peatlands. Five concerns will be briefly discussed: (1) cost-efficient data collection through efficient sampling schemes; (2) digital mapping of peat distribution and thickness; (3) improvement of soil carbon stock estimation; (4) description of the hydraulic properties of peatlands; and (5) contribution of hydropedology in understanding the hydrological behavior of Congo Basin peat swamps.

2 KNOWLEDGE GAPS AND RESEARCH NEEDS

2.1 Cost-efficient data collection through efficient sampling methodologies

Further research on recently discovered peat soils in the Congo Basin will essentially require more field observations, e.g., mapping their current distribution and estimating their carbon stocks and estimating the amounts of important greenhouse gases being exchanged with the atmosphere (Dargie et al., 2017, 2018). However, Congo peat swamp forests are difficult to access and waterlogged, which makes soil sampling particularly challenging (Fatoyinboh, 2017). Until now, a limited number of sampling campaigns have been conducted in the Congo Basin peatlands. Field sampling operations were conducted essentially in the RoC along nine previously identified transects, and the presence of peat was confirmed in eight of them (Dargie et al., 2017, 2018). Knowing the largest part of the total estimated peat area is located in the DRC (90,800 km^2 peat in DRC versus 54,700 km^2 in RoC), the small number of sampling points in DRC largely contributed to considerable uncertainty in the estimated carbon stock (between 6.3 and 46.8 Pg, 95% confidence intervals; Dargie et al., 2017, 2018). In developing countries, like DRC, the number of soil samples that need to be collected for large areas is constrained by the resources available. Minasny et al. (2008) defined better soil data as "data obtained more efficiently, so that a larger number of samples are analyzed at lower costs, in less time and with higher accuracy." To obtain better soil data, a cost-efficient sampling strategy should be implemented to ensure the variation at the most significant scale is captured (e.g.,

Minasny and McBratney, 2006). For example, Rudiyanto et al. (2016b) recommended the application of a sampling method for peat depth mapping using numerical stratification of elevation and the selection of sampling locations either by drawing transects that crossed all strata or by selecting random locations within each stratum. A detailed discussion on the statistical theory for sampling and monitoring in the environment has been provided by De Gruijter et al. (2006). Recently, Young et al. (2018a) provided guidelines to optimize soil carbon sampling efforts in coastal "blue carbon" ecosystems. This sampling methodology also needs to be modified to account for difficulty accessing the locations.

2.2 Digital mapping of peatland extent and thickness

The peatland probability map of the Central Congo Basin published by Dargie et al. (2017) was derived from remote sensing images based on a limited number of field observations; therefore, it is still uncertain. However, better management and conservation of tropical peatlands requires production of fine-scale maps of peatland extent and thickness (Rudiyanto et al., 2018). On the DRC side, only a portion of the region mapped as peatlands by Dargie et al. (2017) was surveyed for soil mapping purposes between 1958 and 1968 as shown in Figure 1 (Van Ranst et al., 2010). Mapping peatlands is a challenge, particularly in tropical regions where reliable information on peat extent, thickness, and carbon stock estimation is urgently needed (Lawson et al., 2015). Traditional soil mapping techniques are expensive and labor-intensive, as they require many field observations (Rudiyanto et al., 2018). The digital soil mapping (DSM) methodology proposed by McBratney et al. (2003) represents an interesting alternative and has been largely applied worldwide by soil scientists for mineral soils (Minasny and McBratney, 2016; Zhang et al., 2017).

In a recent study, Rudiyanto et al. (2016b) successfully applied the DSM framework to map peat thickness of two tropical peatlands in Indonesia. Young et al. (2018b) indicated there are few studies where DSM has been applied to predict the spatial distribution of peat soils and their properties. Rudiyanto et al. (2018) proposed an open-digital mapping approach as an evidence-based, cost-effective, and reliable method for mapping peat. This approach has the following advantages (Rudiyanto et al., 2018): it is derived from objective evidence of scientific research and uses open-source and free software with freely available data from multiple sources of information. It uses transparent algorithms and is therefore repeatable and accountable. The derived maps are dynamic products, as they can be easily updated with newly acquired data. Digital mapping produces accurate and high-resolution maps with confidence of the prediction. Moreover, this evidence-based method

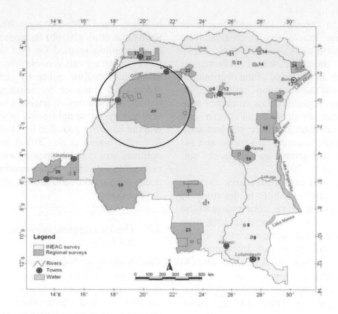

Figure 1. Soil survey coverage in the Democratic Republic of Congo (Source: Van Ranst et al., 2006). The black circle approximately represents the peatland region.

is cost-effective and timely and has the ability to use existing observations. Due to its open nature, it allows standardized and simultaneous actions that make it scalable to a national project.

Rudiyanto et al. (2018) estimated this open-digital mapping approach was 3 to 4 times cheaper than conventional mapping techniques. Recently, other methods have been used to map peatlands in tropical regions, such as ground penetrating radar (GPR) (e.g., Fyfe et al., 2014) and light detection and ranging (LiDAR) (e.g., Ballhorn et al., 2009). However, the LiDAR method requires expensive data acquisition, which will be too costly to apply to large areas like the "Cuvette Centrale" and has technical constraints. In addition, it requires ground observations as a model to predict the extent and thickness of peats.

Dargie et al. (2017) indicated that further measurements will be required to improve their first estimates of the area of peat within the Central Congo Basin, as these estimates essentially rely on peat-vegetation associations and remote sensing techniques. Various researchers showed the DSM framework could be successfully applied to countries with sparse data infrastructures (Minasny et al., 2008) and poorly-accessible regions (Cambule et al., 2013). We believe the open DSM methodology can be an attractive option to refine peatland maps of the Congo Basin, particularly in the DRC context, where very limited resources are allocated for scientific research. However, more localized studies with other methods, such as GPR or LiDAR, could be

conducted to provide additional information for specific cases or applications. Overall, Xu et al. (2018) stated that improving peatland mapping at regional and national scales represents an ongoing effort.

2.3 Improvement of peat soil carbon stocks estimation

Peatlands represent a major global carbon reservoir (528–600 Pg) with 10–30% of this carbon mass in tropical peatlands (Page et al., 2011; Hodgkins et al., 2018). Hartemink et al. (2014) considered that soils of ecologically-sensitive regions, like peatlands, are priority areas for soil organic carbon research. More detailed information on carbon storage in tropical peatlands remains missing, despite the widely recognized importance of peatlands for climate change mitigation (Lawson et al., 2015). This is particularly true for the Congo Basin, which is one of the most data-sparse continental regions (Creese and Washington, 2018). Dargie et al. (2017) estimated the Congo peat swamp forests store about 30.6 billion tons of carbon that represent almost 30% of all the soil organic carbon stored within tropical peatlands. However, the current estimate of carbon stocks in the Central Congo Basin peatlands has a high degree of uncertainty attached to it, particularly in DRC, where only a few field samples were collected compared to RoC (Dargie et al., 2017, 2018; Miles et al., 2017).

The high carbon density of tropical peatlands makes them an obvious candidate for emissions-management schemes, such as the UN-REDD+ (United Nations Reducing Emissions from Deforestation and Forest Degradation) program (Murdiyarso et al., 2010). In DRC, we need to know what carbon pools and stocks the peat swamp forests hold for a better implementation of the UN-REDD+ program in the Central Congo Basin. Accurate estimations of peat carbon stocks and fluxes in DRC peatlands are needed and must be added to the existing forest carbon stocks previously identified. This will provide more accurate baseline estimates against which projected reductions in carbon emissions can be measured, verified, and translated into carbon credits (Lawson et al., 2015).

However, laboratory analyses of soil data are expensive in DRC and RoC. Therefore, it would be necessary to develop predictive equations, known as pedotransfer functions (PTFs) (Bouma, 1989), for peat soil carbon estimation. For example, Warren et al. (2012) and Farmer et al. (2014) developed regression models that estimate carbon density (Cd, kg m^{-3}) of a tropical peat soil sample from its bulk density (BD, g cm^{-3}). Conversely, Rudiyanto et al. (2016) found that, when carbon content of a peat soil sample is greater than 0.5 g g^{-1} (i.e. 50%), there is no relationship between its carbon content and its bulk density. Thus, they proposed a simpler equation where Cd can be predicted from average carbon content Ĉc (0.5501 ± 0.0225 g g^{-1}) multiplied by the measured BD (i.e. Cd = Ĉc × BD = 0.5501 ± 0.0225 × BD). This PTF could represent an interesting alternative for peat carbon estimation in Congo swamp forests where peat soils are reported to have a carbon content greater than 0.5 g g^{-1} (i.e. 0.59 ± 0.03 g g^{-1}; Dargie et al., 2017). Therefore, it would be interesting to test PTFs based on datasets of peat soil samples from the Congo Basin and/or develop local PTFs with higher accuracy.

2.4 Determination of the hydraulic properties of peat soils

Knowledge about soil hydraulic properties is crucial for understanding water, energy, and carbon exchange processes between the land surface and the atmosphere (Montzka et al., 2017). Currently, few data exist on either hydraulic conductivity or water table dynamics in tropical peatlands (Lawson et al., 2014). Similarly, data derived from studies investigating the relationships between peat types and water holding capacity or flow rate are scarce (Leng et al., 2018). A deep understanding of the unique hydraulic properties of peats is necessary for reliable modeling of Congo Basin swamp forests hydrology.

However, obtaining direct measurements of hydraulic properties of peat soils can be time-consuming, labor-intensive, expensive, and therefore impractical for large-scale modeling, particularly for tropical peatlands that are often located in remote and permanently inundated regions. For Congolese peatlands, estimation of soil hydraulic properties (e.g., water retention capacity, hydraulic conductivity) using calibrated PTFs can provide satisfactory approximations of in situ conditions. This was demonstrated notably by Botula et al. (2013) for highly weathered soils in DRC and Nguyen et al. (2015) for Mekong Delta soils in Vietnam. In a comprehensive review of the challenges and perspectives on soil PTFs in Earth system science, Van Looy et al. (2017) noted PTFs need to be developed for volcanic ash soils and peat soils. Developing hydraulic PTFs for Congolese peatlands will contribute to this global effort.

2.5 Contribution of hydropedology in the understanding of peat hydrological behavior

Efforts should be devoted by the scientific community to explain the hydrological functioning of peatlands (University of Leeds Peat Club, 2017). The integrated approach proposed by hydropedology (i.e., the combination of hydrology and pedology; Lin et al., 2006) has been proposed as a response for rational management of tropical peatland ecosystems (Wösten et al., 2008) through addressing contemporary issues related to land uses affected by saturated soil conditions (Vepraskas et al., 2009). As "an emerging intertwined branch of soil science and hydrology that studies interactive pedologic and hydrologic processes and properties in the Earth's Critical Zone" (Lin et al., 2015), hydropedology can provide a better understanding of the links between Congo Basin peatland hydrology and the regional river system.

In their prospective study, Vepraskas et al. (2009) identified future directions for hydropedological investigation related to land use that are relevant for the Congo Basin peatlands. They proposed a framework that consists of six steps: (1) define land uses of interest; (2) define critical water table levels for each use; (3) estimate current water table levels in soils of interest; (4) extrapolate modeled water table data across broad geographic regions; (5) determine changes in water tables associated with climate change for future years of interest; and (6) create maps showing changes in potential land use patterns.

Moreover, Thompson et al. (2012) indicated that collaboration among DSM scientists and hydropedologists is expected to advance our understanding of the complexities of the soil–water continuum within and across soil landscapes. This was illustrated by Pennock et al. (2014), who applied hydropedology principles to predictive mapping of the distribution of wetland soils in the Canadian Prairie Pothole in Saskatchewan. We believe research on the Central Congo Basin peatlands dynamics will benefit from hydropedology to develop robust conceptual models of peat soil hydrology that characterize the hydrological functions of peatlands at various scales.

3 CONCLUSION

The recently discovered Central Congo Basin peatlands are important for climate change mitigation, as they store large quantities of belowground carbon. In this brief review, we identified knowledge gaps and research needs for Congolese peatlands in relation to data collection, mapping of peat extent and thickness, carbon stock estimation, peat hydraulic properties, and hydrological functioning. Presently, the Congo Basin peat swamp forests are relatively intact, but there are potential threats to their integrity that call for urgent actions to ensure their protection (Miles et al., 2017; Dargie et al., 2018). Moreover, Congolese peatland swamp forests are rich in biodiversity and knowledge of plant and animal species; particularly, fish and invertebrates are still incomplete (Dargie et al., 2018). Scientific research can significantly contribute to provide up-to-date and relevant information needed by various stakeholders to ensure conservation and drive sustainable management of the Congo Basin peatlands.

REFERENCES

Ballhorn, U., Siegert, F., Mason, M. & Limin, S. 2009. Derivation of burn scar depths and estimation of carbon emissions with LIDAR in Indonesian peatlands. *Proceedings of the National Academy of Sciences of the United States of America* 106: 21213-21218.

Brown, C., Doreen, S.B., Sjogersten, S., Clewly, D., Evers, S.L. & Aplin, P. 2018. Tropical peatland vegetation structure and biomass: optimal exploitation of airborne laser scanning. *Remote Sensing* 10: 671.

Botula, Y.-D., Nemes, A., Mafuka, P., Ranst, E. & Cornelis, W.M. 2013. Prediction of water retention of soils from the humid tropics by the nonparametric k-nearest neighbor approach. *Vadose Zone Journal* 12: 1-17.

Bouma, J. 1989. Using soil survey data for quantitative land evaluation. In B.A. Stewart (ed.), *Advances in soil science*: 177-213. New York: Springer Verlag.

Cambule, A.H., Rossiter, D.G. & Stoorvogel, J.J. 2013. A methodology for digital soil mapping in poorly-accessible areas. *Geoderma* 192: 341-353.

Creese, A. & Washington, R. 2018. A process-based assessment of CMIP5 rainfall in the Congo Basin: the September–November rainy season. *Journal of Climate* 31: 7417-7439.

Dargie, G.C., Lawson, I.T., Rayden, T., Miles, L., Mitchard, E.T.A., Page, S.E., Bocko, Y.E., Ifo, S.A. & Lewis, S.L. 2018. Congo Basin peatlands: threats and conservation priorities. *Mitigation and Adaptation Strategies for Global Change*. https://doi.org/10.1007/s11027-017-9774-8.

Dargie, G.C., Lewis, S.L., Lawson, I.T., Mitchard, E.T.A., Page, S.E., Bocko, Y.E. & Ifo, S.A. 2017. Age, extent and carbon storage of the central Congo Basin peatland complex. *Nature* 542: 86-90.

De Gruijter, J.J., Brus, D.J., Bierkens, M.F.P. & Knotters, M. 2006. *Sampling for Natural Resource*

Monitoring. New York: Springer.

Farmer, J., Matthews, R., Smith, P., Langan, C., Hergoualc'h, K., Verchot, L. & Smith J.U. 2014. Comparison of methods for quantifying soil carbon in tropical peats. *Geoderma* 214-215: 177-183.

Fatoyinbo, L. 2017. Vast peatlands found in the Congo Basin. *Nature* 542: 7639. doi:10.1038/542038b.

Fyfe, R.M., Coombe, R., Davies, H. & Parry, L. 2014. The importance of sub-peat carbon storage as shown by data from Dartmoor, UK. *Soil Use and Management* 30: 23-31.

Hartemink, A.E., Lal, R., Gerzabek, M.H., Jama, B., McBratney, A.B., Six, J. & Tornquist, C.G. 2014. Soil carbon research and global environmental challenges. *Peer J PrePrints* 2: e366v361. doi:http://dx.doi.org/10.7287/peerj. preprints.366v1.

Hayasaka, H., Noguchi, I., Putra, E. I., Yulianti, N. & Vadrevu, K. 2014. Peat-fire-related air pollution in Central Kalimantan, Indonesia. *Environmental Pollution* 195: 257-266.

Heuvelink, G. 2003. The definition of pedometrics. *Pedometron* 15: 11-12.

Hodgkins, S.B., Richardson, C.J., Dommain, R., Wang, H., Glaser, P.H., Verbeke, B., Winkler, B.R., Cobb, A.R., Rich, V.I., Missilmani, M., Flanagan, N., Ho, M., Hoyt, A.M., Harvey, C.F., Vining, S.R., Hough, M.A., Moore, T.R., Richard, P.J.H., De La Cruz, F.B., Toufaily, J., Hamdan, R., Cooper, W.T. & Chanton, J.P. 2018. Tropical peatland carbon storage linked to global latitudinal trends in peat recalcitrance. *Nature Communications* 9: 3640. doi: 10.1038/s41467-018-06050-2

Lawson, I.T., Jones, T.D., Kelly T.J., Honorio Coronado, E. N. & Roucoux, K.H. 2014. The geochemistry of Amazonian peats. *Wetlands* 34: 905-915.

Lawson, I.T., Kelly, T.J., Aplin, P., Boom, A., Dargie, G., Draper, F.C., Hassan, P.N.Z.B.P., Hoyos-Santillan, J., Kaduk, J., Large, D., Murphy, W., Page, S.E., Roucoux, K.H., Sjögersten, S., Tansey, K., Waldram, M., Wedeux, B.M.M. & Wheeler, J. 2015. Improving estimates of tropical peatland area, carbon storage, and greenhouse gas fluxes. *Wetlands Ecology and Management* 23: 327-346.

Leng, L.Y., Ahmed, O.H. & Jalloh, M.B. 2018. Brief review on climate change and tropical peatlands. *Geoscience Frontiers*, in press.

Lin, H.S., Bouma, J., Pachepsky, Y., Western, A., Thompson, J., van Genuchten, M.Th., Vogel, H. & Lilly, A. 2006. Hydropedology: Synergistic integration of pedology and hydrology. *Water Resources Research* 42: W05301. doi:10.1029/2005WR004085.

Lin, H., Patrick, D. & Green, T.R. 2015. Hydropedology: the last decade and the next decade. *Soil Science Society of America Journal* 79: 357-361.

McBratney, A., Mendonça Santos, M. & Minasny, B. 2003. On digital soil mapping. *Geoderma* 117: 3-52.

McBratney, A., Minasny, B. & Stockmann, U. 2018. *Pedometrics*. Cham: Springer.

Miles, L., Ravilious, C., García-Rangel, S., de Lamo, X., Dargie, G. & Lewis, S. 2017. Carbon, biodiversity and land-use in the Central Congo Basin peatlands. UN Environment World Conservation Monitoring Centre.

Minasny, B., Whelan, B.M., Triantafilis, J. & McBratney, A.B. 2013. Pedometrics research in the vadose zone – review and perspectives. *Vadose Zone Journal* 12: 1-20.

Minasny, B. & McBratney, A.B. 2006. A conditioned Latin hypercube method for sampling in the presence of ancillary information. *Computers & Geosciences* 32: 1378-1388.

Minasny, B., McBratney, A.B., & Lark, R.M. 2008. Digital soil mapping technologies for countries with sparse data infrastructures. In A.E. Hartemink, A.B. McBratney & M.I. Mendonça-Santos (eds.), *Digital soil mapping with limited data:* 15-30. London: Springer.

Minasny, B. & McBratney, A.B. 2016. Digital soil mapping: a brief history and some lessons. *Geoderma* 264: 301-311.

Montzka, C., Herbst, M., Weihermüller, L., Verhoef, A. & Vereecken, H. 2017. A global data set of soil hydraulic properties and sub-grid variability of soil water retention and hydraulic conductivity curves. *Earth System Science Data* 9: 529-543.

Murdiyarso, D., Hergoualc'h, K. & Verchot, L.V. 2010. Opportunities for reducing greenhouse gas emissions in tropical peatlands. *Proceedings of the National Academy of Sciences of the United States of America* 107: 19655-19660.

Nguyen, P.M., Khoa, V.L., Botula, Y.-D. & Cornelis, W.M. 2015. Evaluation of soil water retention pedotransfer functions for Vietnamese Mekong Delta soils. *Agricultural Water Management* 158: 126-138.

Page, S.E., Rieley, J.O. & Banks, C.J. 2011. Global and regional importance of the tropical peatland carbon pool. *Global Change Biology* 17: 798-818.

Pennock, D., Bedard-Haughn, A., Kiss, J. & van der Kamp, G. 2014. Application of hydropedology to predictive mapping of wetland soils in the Canadian prairie pothole region. *Geoderma* 235-236: 199-211.

Rudiyanto, Minasny, B. & Setiawan, B.I. 2016a. Further results on comparison of methods for quantifying soil carbon in tropical peats. *Geoderma* 269: 108-111.

Rudiyanto, Minasny, B., Setiawan, B.I., Arif, C., Saptomo, S. K. & Chadirin, Y. 2016b. Digital mapping for cost-effective and accurate prediction of the depth and carbon stocks in Indonesian peatlands. *Geoderma* 272: 20-31.

Rudiyanto, Minasny, B., Setiawan, B.I., Saptomo, S.K. & McBratney, A.B. 2018. Open digital mapping as a cost-effective method for mapping peat thickness and assessing the carbon stock of tropical peatlands. *Geoderma* 313: 25–40.

Thompson, J.A., Roecker, S., Grunwald, S. & Owens, P.R. 2012. Digital soil mapping: Interactions with and applications for hydropedology. In H. Lin (ed.), *Hydopedology. 1st edition*: 665-709. Amsterdam: Academic Press.

University of Leeds Peat Club: Bacon, K.L., Baird, A. J., Blundell, A., Bourgault, M-A., Chapman, P.J., Dargie, G., Dooling, G.P., Gee, C., Holden, J., Kelly, T., McKendrick-Smith, K.A., Morris, P.J., Noble, A., Palmer, S.M., Quillet, A., Swindles, G.T.,

Watson, E.J. & Young, D.M. 2017. Questioning ten common assumptions about peatlands. *Mires and Peat* 19: 1-23.

Van Looy, K., Bouma, J., Herbst, M., Koestel, J., Minasny, B., Mishra, U., Montzka, C., Nemes, A., Pachepsky, Y.A., Padarian, J., Schaap, M.G., Toth, B., Veroef, A., Vanderborght, J., van der Ploeg, M.J., Weihermüller, L., Zacharias, S., Zhang, Y. & Vereecken, H. 2017. Pedotransfer functions in Earth system science: challenges and perspectives. *Reviews of Geophysics* 55: 1199-1256.

Van Ranst, E., Baert, G., Mafuka Mbe-Mpie, P., Ngongo Luhembwe, M., Goyens, C., van de Wauw, J. & Verdoodt, A. 2006. Valorisation des données disponibles de ressources naturelles de la RD Congo pour une gestion rationnelle et durable des terres. Données cartographiques. *Gent overzee - Jonge onderzoekers delen hun passie mee.* Belgium: Royal Academy of Sciences.

Van Ranst, E., Verdoodt, A. & Baert, G. 2010. Soil mapping in Africa at the crossroads: work to make up for lost ground. *Bulletin des Séances de l'Académie Royale des Sciences d'Outre-Mer* 56: 147-163.

Vepraskas, M.J., Heitman, J.L., & Austin, R.E. 2009. Future directions for hydropedology: quantifying impacts of global change on land use. *Hydrology and Earth System Sciences* 13: 1427-1438.

Warren, M.W., Kauffman, J.B., Murdiyarso, D., Anshari, G., Hergoualc'h, K., Kurnianito, S., Purbopuspito, J., Gusmayanti, E., Afifudin, M., Rahajoe, J., Alhamd, L., Limin, S. & Iswandi, A. 2012. A cost-efficient method to assess carbon stocks in tropical peat soil. *Biogeosciences* 9: 4477-4485.

Wösten, J.H.M., Clymans, E., Page, S.E., Rieley, J.O. & Limin, S.H. 2007. Peat–water interrelationships in a tropical peatland ecosystem in Southeast Asia. *Catena* 73: 212-224.

Xu, J., Morris, P.J., Liu, J. & Holden, J. 2018. PEATMAP: Refining estimates of global peatland distribution based on a meta-analysis. *Catena* 160: 134-140.

Young, M.A., Macreadie, P.I., Duncan, C., Carnell, P.E., Nicholson, E., Serrano, O., Duarte, C.M., Shiell, G., Baldock, J. & Ierodiaconou, D. 2018a. Optimal soil carbon sampling designs to achieve cost-effectiveness: a case study in blue carbon ecosystems. *Biology Letters* 14 (9).pii: 20180416. doi: 10.1098/rsbl.2018.0416.

Young, D.M., Parry, L.E., Lee, D. & Ray, S. 2018b. Spatial models with covariates improve estimates of peat depth in blanket peatlands. *PLoS ONE* 13(9): e0202691. https://doi.org/10.1371/journal.pone.0202691.

Zhang, G., Liu, F. & Song, X. 2017. Recent progress and future prospect of digital soil mapping: a review. *Journal of Integrative Agriculture* 16: 2871-2885.

Mapping acid sulfate soil hydrogeomorphical unit on the peatland landscape using a hybrid remote sensing approach

W. Widyatmanti, D. Umarhadi, M.U.L. Ningam, Z. Sarah & K. Nugroho
Department of Geographic Information Science, Faculty of Geography, Universitas Gadjah Mada, Yogyakarta, Indonesia

Wahyunto & Y. Sulaeman
Indonesian Centre for Agricultural Land Resources Research and Development, Ministry of Agriculture, New Delhi, India

ABSTRACT: Peatland mapping is an important national agenda in Indonesia, due to human disturbances and other prominent issues related to conservation and environment. The exposure of mineral soils underlying these peatlands can further increase potential environmental problems when it contains acid sulfate soil (ASSoils - FeS_2), harmful mineral when exposed to oxygen. Remote sensing applications on peatland mapping are commonly utilized to identify its extent, but never examined its ASSoils potential area. This study aims: 1) to map the coastal ASSoil occurrence probability based on a hybrid remote sensing and hydrogeomorphical approaches and 2) to examine the ability of remote sensing in assisting the development of digital ASSoils mapping in tropical peatland. This study utilized the legacy peatland data of Kubu Raya and Ketapang areas provided by the Indonesian Ministry of Agriculture. The hybrid remote sensing image analysis approach consists of a combination of visual interpretation and digital image analysis. Multi-temporal satellite imageries (Landsat 8, Sentinel-2) representing rainy and dry seasons were processed to derive wetness, vegetation density indices and landcover maps. Soil characteristics data included the thickness of peat, underlying materials, and ASSoils distributions. Processed hydrological and topographical data produces maps of tidal, salinity, and sea water intrusion and micro-elevation accordingly. Visual remote sensing image interpretation was conducted to identify existing coastal landforms. Both intermediary maps are associated to the field survey data and are used to determine the peatland ASSoil hydrogeomorphology units. The development of this digital ASSoils mapping is important to underpin the strategic regional and national planning in the peatland environment.

1 INTRODUCTION

Tropical peatlands have an important role as a source of water storage that functions to stabilize water cycle, reduce runoff, and maintain the quality and volume of river flows, to prevent sea water intrusion (Osaki and Tsuji, 2016). Therefore, peatland drying, for plantation purposes as an example, can result in lowering the peat surface and triggering the oxidation causing loss of CO_2, nitrogen, other nutrients. Continuous oxidation and drainage could expose the underlying soil layer which could be mineral soil, quartz sand, or sediment containing pyrite. When the mineral layer is exposed to air, the pyrite-containing sediments can undergo an oxidation process and produce sulfuric acid (acid sulfate soil-ASSoils - FeS_2), a harmful mineral when exposed to oxygen, causing widespread acidification in limnology and soil ecosystems (Andriesse

and Sukardi, 1990; Sammut et al., 1996). When this acid contaminated environment is impacted by forest fires, it produces a greater amount of CO_2 in the atmosphere (Haraguchi, 2016). Therefore, peatland conservation and management planning, especially in sulfidic layer peatlands, are necessary to prevent the pyrite oxidation. Accurate peatland mapping is also compulsory as the best scientific knowledge for managing the peatland using comprehensive valuation.

Sustainable peatland management planning requires compulsory peat characteristics data, including peat extent, thickness, groundwater level, maturity level, and type of material that underlies the peatland. Until this time, those data have been spatially presented in the form of hardcopy and digital maps, using spatial statistics and detailed field survey data. The conventional mapping procedure is reliable for small and easily-accessed areas. However, it becomes inefficient when applied in a vast area because the field survey demands high costs, time, and effort. The work load increases when physical and chemical soil properties data is required to be collected from the field to complete the peat map data attributes (KLHK, 2017). Since advances in space technology have opened application possibilities of remote sensing application in soil mapping, especially in reducing manual fieldwork dramatically, its applications then become the main alternative for determining information and sampling in the peatland environment. Various methods have been developed intensively and carried out both from manual and digital methods (Cazals et al., 2016). However, remote sensing applications on peatland mapping are commonly utilized to identify its extent and thickness. Rarely peat mapping identified ASSoils potential area. In fact, it is possible to optimize the use of remote sensing using an ecological approach to identify the aforementioned peatland characteristics. Therefore, this current study aims to (1) map the potential occurrence of ASSoils on peatlands in coastal areas for the medium scale using the hydro-geomorphology unit approach and (2) assess the effectiveness of remote sensing data for the determination of hydro-geomorphological units in helping identify the potential presence of ASSoils on peatland using a variety of multi-temporal remote sensing spectral transformations. This study is conducted in West Kalimantan Province region, specifically in Kubu Raya and Ketapang Regencies (Figure 1), since the data of these areas are comprehensively available.

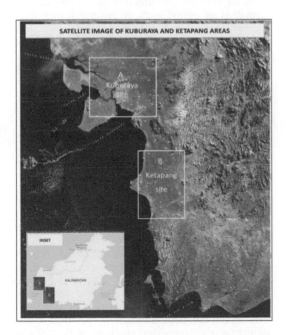

Figure 1. The study site of A. Kuburaya and B. Ketapang.

2 LITERATURE REVIEW

2.1 Peatland landscape with underlain pyrite sediment

Peatland consists of organic material that accumulated naturally in an anaerobic concave landscape. In the initial stage, the deposition process of this organic matter occurs in depressed areas or basins behind river embankments. With the presence of fresh water and brackish water that inundates these depressed areas, the process of decomposition of organic matter becomes very slow and inhibited. Subsequently, this organic matter accumulated and eventually formed peat deposits with varying thickness, depending on the topography of the mineral soil beneath this peat layer (Widjaja-Adhi et al., 2000; Subagjo, 2006).

According to Soil Taxonomy (Soil Survey Staff, 2014), peat soils are composed of organic materials with a minimum thickness of 40 or 60 cm, depending on specific gravity (Bulk Density - BD) and the level of organic matter decomposition. The mineral soils underlying the peat soil have three types, namely mineral soil with clay or sand texture, quartz sand, and soil/sediment containing pyrite or ASSoils.

ASSoils is a sediment containing pyrite (FeS_2), which is mostly formed under mangrove forests due to reduction of sulfate, which is mediated by sulfides bacterial in rich organic sediments (Dent, 1986; Fitzpatrick et al., 2008, 2012). Pyrite is inactive under waterlogged conditions, especially in coastal or lowland areas. However, when pyrite is exposed to oxygen, oxidation will cause a decrease in soil pH (below 4) due to the release of sulfuric acid and acidity of minerals associated with metal transformation. If the acid neutralizing capacity of the soil is exceeded, it will cause severe soil acidification. Infrastructure development related to agriculture, drainage, industry, urbanization, and fisheries developed in coastal areas can cause a decrease in soil and water pH if the presence of ASSoils is unknown (Ahern et al., 1998, Lin et al., 1995; Sammut et al., 1996). For this reason, regional development planning in peatland environment crucially requires the information of ASSoils occurrences, based on a representative field survey.

2.2 Remote sensing for peatland and ASSoil Mapping

Mapping peatlands has its own challenges, especially when it involves vast areas and difficult accessibilities to conduct field surveys. Unlike mapping peatlands in a local area or at a large scale, mapping peatland at a medium scale requires integrated understanding of landscape characteristics that can assist us in determining boundaries between peatland and non-peatland environments. Larger area mapping coverage is also an issue, since a larger number of samples is required to achieve a high accuracy. However, a combination of appropriate remote sensing data, landscape ecological knowledge, and measurements in the field can assist mapping proces for vast peatland area.

Tropical peatlands have different characteristics compared to non-peat environments, so mapping them using remote sensing data is a challenge. Peatland mapping has been done with various approaches to obtain information from remote sensing data, including the identification of vegetation type and structure parameters, unique micro topography, and high groundwater level. Figure 2 presents the example of surface features from aerial photographs and their comparison on the field based on a study conducted by Lawson et al. (2015), complimented by visual interpretation on high-resolution imageries from Digital Globe Archive Data and Planet Imageries from this current study. This comparison only represents one study area and using one-time acquisition of satellite image; however, it provides an understanding of the differences of tropical peatland environmental characteristics.

Figure 2. The example of different types of landcover on peatland environment from aerial photograph and field observation taken from Lawson et al. (2015) and the result of visual interpretation from high resolution satellite imageries (current study).

To date, studies on peatland mapping have been conducted utilizing various image resolutions and scales, from MODIS and Landsat-8 (Millard et al., 2018; Gumbricht, 2018), to Ikonos, QuickBird, WorldView-2 (Lawson et al., 2015; Connolly and Holden, 2017), Hyperspectral (Nordin, et al., 2018; Osaki and Tsuji, 2016; Haris et al., 2015), and Airborne Laser Scanning – ALS (Magdon et al., 2018). Sentinel-1 and Sentinel-2 became popular for landcover mapping applications and began to be intensively explored to map wetland and peatland Miranda and Meadows (2015); Mouche and Chapron (2015). Those aforementioned imageries were combined with Digital Elevation Model (DEM), such as SRTM, Alos-PALSAR, TanDEM-X, TerraSAR and LIDAR, are other remote sensing methods commonly applied to assist mapping on tropical peatland (Nunes et al., 2017; Schlund et al., 2015; Li and Chen, 2005). However, very little research has applied integrated remote sensing and GIS to associate the peatland landscape to the potential occurrence of the underlying layer of ASSoils. In fact, ASSoils potential landscape and landform can be differentiated from non-ASSoils environment, using Remote sensing and the GIS approach to identify the vegetation type and its distance to rivers and oceans (in terms of the water and soil salinity levels) (Widyatmanti and Sammut, 2017, Fanning et al., 2009, 2017; Beucher et al., 2017). Most of the studies utilised multi-resolution satellite imageries to derive biophysical variables information that influence the formation of pyrite, such as sediment with iron, organic matter, and sulfate contents, and low energy environment that allow bacteria to reduce sulfate to sulfides. Once these variables information was converted into spatial data, they were superimposed to create mapping units. These units then represent the hydrological and geomorphological characteristics of the associated landforms referred to as hydro-geomorphological units (Widyatmanti and Sammut, 2017).

Since the majority of peatland with ASSoils underlying layer developed under tidal condition, their landforms are commonly located in coastal areas, with mangrove or peat swamp vegetations (Hole and Campbell, 1985; Husson et al., 2000). These environmental signatures are possible to be extracted from remote sensing data. Hence, the combination of the basic understanding of remote sensing for landform identification and the signature landscape of peatland with ASSoils underlying layer become a proposed approach to develop the method to generate the hydro-geomorphic unit. These units will direct the determination of spectral transformation signatures for peatland with ASSoils environments.

3 MATERIAL AND METHOD

This study utilized legacy peatland data of Kubu Raya and Ketapang areas provided by ICALRRD, the soil research institute under the Ministry of Agriculture. The soil characteristics data included the depth of peat, underlying material, and ASSoils distributions. Processed hydrological and topographical data produces maps of tidal, salinity, and sea water intrusion and micro-elevation accordingly. The hybrid remote sensing image analysis approach consists of a combination of visual interpretation and digital image analysis. The mapping units produced from the interpretation results are designated as hydro-geomorphic units. These mapping units are the substance data to determine the soil sampling numbers and locations.

Multi-temporal satellite imageries (Landsat 8, Sentinel-2) representing rainy and dry seasons were processed to derive wetness, vegetation density in-dices, and land-cover maps. This study applied Modified and Soil-adjusted Vegetation Index (MSAVI and SAVI) to differentiate the vegetation characteristics on peat and non-

peat environments. Visual remote sensing image interpretation was conducted to identify the existing coastal landcover and landforms. These intermediary maps are substantiated to the field survey data and are used to determine the peatland ASSoil hydro-geomorphology units. Details of the method are provided on the flowchart below (Figure 3).

4 RESULTS AND DISCUSSION

4.1 The coastal peatland with ASSoils landscape from satellite imageries

Common peat soil classification in Indonesia comprises ombrogenous and topogenous peat. Based on peatsoil characteristics and maps, hydrogeomorphic units that have ASSoils content sediment are commonly topogenous peat. These types of peat with ASSoils are associated with a coastal environment, including its estuary and sometimes reaching the middle course of the river that own high levels of salinity caused by sea water intrusion. In this study, the salinity influence of peatland is assessed using remote sensing approach by identifying mangrove and nypa along the Kapuas Estuaries in Kubu Raya, and the Pawan Estuaries in Ketapang. This approach is only possible to apply in the pristine to low disturbance estuaries, where mangrove and nypa can still be identified through high-resolution satellite imagery. The distance information of salinity influence into the peatland is determined by field survey. However, topogenous peatlands mostly have underlying ASSoils, as long as the parent material of their catchment is of volcanic origin (Widyatmanti and Sammut, 2017).

Multi-temporal Landsat-8 and Sentinel-2 of Kuburaya and Ketapang regions image processing generated several maps, including maps of landcover, vegetation, and wetness indices. The true color image visual interpretation on Landsat 8 combined with SRTM, hydrology and geomorphological data results in a landform map using the hydrogeomorphical approach. Figure 4 presents the example part of visual interpretation maps that has been substantiated with the peatland and ASSoils Maps. Ombrogenous peat appears in a darker tone compared to tidal topogenous peat on both rainy and dry seasons (4A and 4B). The higher vegetation density on ombrogenous peat contrasted to the tidal topogenous one, added with its higher water table, causes low reflectance value on the satellite image. Likewise, when land use changes occupy the peatland environment, the dark tones during dry and rainy seasons still signify the ombrogenous peat in Kubu Raya (3C and 3D). A similar situation was observed in Ketapang (Figure 4).

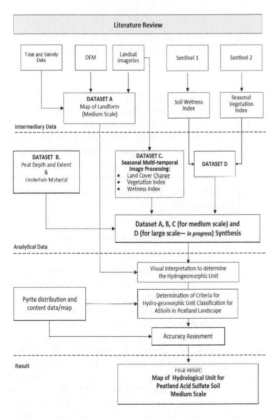

Figure 3. Flowchart diagram showing the work flow of Peatland ASSoils mapping using remote sensing and GIS.

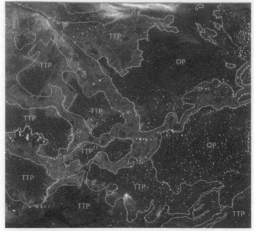

A. Landsat-5 Image, Kuburaya Area, April 1994, end of rainy season;

B. Landsat-8 Image, Kuburaya Area, August 2014, dry season

Figure 4. Multi-temporal landsat imageries showing the tidal topogenous peatland (TTP) and Ombrogenous Peatland (OP) and the land-cover changes in Kuburaya coastal area from 1994 to 2014.

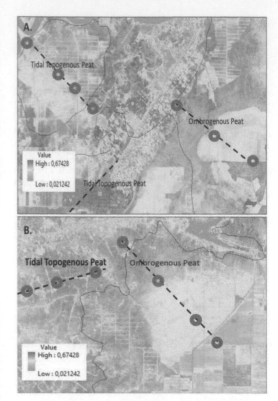

Figure 5. Spectral transformation results for vegetation and wetness indices for several sample/training areas.

4.2 The coastal peatland with ASSoils landscape from satellite imageries

This study proposes a sampling method for tropical coastal peatland and its underlying ASSoils. The sampling areas taken from the processed satellite imageries were distributed to represent peatland with tidal system, peatland with freshwater system, and non-peatland hydrogeomorphic units. Figure 5 portrays sample areas chosen to represent several hydrogeomorphic units combined with their vegetation and wetness indices. The hybrid approach assists the sampling sites determination within hydrogeomorphic unit based on the wetness index. The tidal topogenous peat associated with the hydrogeomorphic unit with ASSoils is presented by a lower index compared to ombrogenous peat, both on vegetation and wetness indices.

Plantation landcover that dominated the peat environment disrupts the consistency of the index of each hydro-geomorphic unit class. Therefore, the sampling areas have to be carefully selected to avoid bias range of the index among the classes. The on-going process of this study is to have more sampling areas, at least 50 for each class. Then, an accuracy assessment can be conducted.

4.3 Justification of hybrid remote sensing approach for coastal peatland with ASSoils

From this preliminary study, the hybrid remote sensing approach is potential to identify the landform unit associated to peatland with ASSoils underlying layers. The data integration using multi-temporal image interpretation, vegetation, and wetness index, collocated with data of the peatland location and its characteristics is applied to generate the hydrogeomorphology unit on peatland. The first identification of peatland landscapes must consider the distance of the peat hydrogeomorphology unit to the sea, river mouth, and river, which are still affected by tides. The units include alluvial plain, closed basin, estuary plain, backswamp, swalle/lagoon, and tidal/brackish water topogenous peat (Table 1). This ongoing study

Table 1. List Hydro-geomorphological unit that commonly exist in tropical peat environment and its possibility of ASSoils occurrences.

Hydrogeomorphic unit	Possibilities found on peatlands	The possibility of Acid Sulfate Soil Occurrences (ASSoil)
Alluvial plain	minor: 10-25%	Medium
Closed basin	dominant: 50-75%	Medium
Estuary plain	minor: 10-25%	High
Back swamp	fair-dominant: 25-75%	Medium
Swalle/lagoon	fair-dominant: 25-75%	High
Beach/shore	minor: 10-25%	High
Fresh water topogenous peat	predominant: > 75%	low/NA
Tidal/brackish water topogenous peat	predominant: > 75%	High
Ombrogenous peat	predominant: > 75%	Low
Peat dome	predominant: >75%	Low

will provide the range index values from the most applicable number for the sampling area and appropriate spatial modelling to achieve a high level of accuracy. Based on a literature review and this preliminary hybrid method, some ASSoils hydrogeomorphic units that refer to specific landforms are recognized (Table 1).

5 CONCLUSION

Based on the results described above, there are many significant points that can be concluded in this ongoing study, including:

a. The mapping of the ASSoil hydro-geomorphology unit on peatlands requires a better scientific understanding of the process of peatland formation and ASSoil, because this research is more complex than mapping non-peatlands.
b. The visual image interpretation approach cannot only use one original color image but must be supported by information of vegetation density and soil moisture, topographic conditions, and most importantly the distance of hydro-geomorphic units, which also represents landforms from the sea, estuary, and river with salinity level above 10 ppm.

c. The best spectral index transformation that can be used to map peatlands is SAVI and MSAVI, with a multi-temporal approach to obtain information on pure peatlands and those that have changed function.
d. The use of remote sensing data and GIS in this study provides an important contribution to provide a multi-scale, multi-resolution, and multi-spatial-temporal description of peatland conditions.
e. The more advanced contribution of remote sensing technology in the future will add value to research in the area of peatlands and acid sulfate soils to assist decision makers to contribute to sustainable peatland and coastal planning.
f. Future work: accuracy assessment method on ASSoils peatland mapping is in progress by collecting more samples/training areas in different locations that have similar characteristics of tropical coastal peatland.

ACKNOWLEDGEMENT

This work was supported by the Research Grant from Faculty of Geography, Universitas Gadjah Mada. We also thank Indonesian Center for Agricultural Land Resources Research and Development (ICALRRD), Agency for Agricultural Research and Development, Ministry of Agriculture for providing peat soil data; NASA for providing free Landsat and SRTM data; and European Space Agency (ESA) for Sentinel 1 and 2 Imageries.

REFERENCES

Ahern, C.R., Stone, Y. & Blunden, B. 1998. *Acid Sulfate Soils Assessment Guidelines*. Wollongbar, NSW, Australia: The Acid Sulfate Soil Management Advisory Committee.
Andriesse, W. & Sukardi, M. 1990. Survey Component: Introductions, Objective and Outline Papers Workshop on Acid Sulphate Soils in the Humid Tropics. AARD-LAWOO. Jakarta.
Beucher, A., Adhikari, K., Breuning-Madsen, H., Greve, M. B., Österholm, P., Fröjdö, S. & Greve, M.H. 2017. Mapping potential acid sulfate soils in Denmark using legacy data and LiDAR-based derivatives. *Geoderma* 308: 363–372.
Cazals, C., Rapinel, S., Frison, P.L., Bonis, A., Mercier, G., Mallet, C., Corgne, S. & Rudant, J.P. 2016. Mapping and characterization of hydrological dynamics in a coastal marsh using high temporal resolution sentinel-1A images. *Remote Sensing* 8(7): 570.
Connolly, J. & Holden, N.M. 2017. Detecting peatland drains with Object based Image Analysis and Geoeye-1 imagery. *Carbon balance and management* 12(1): 7.

Dent, D. 1986. *Acid Sulphate Soils: a Baseline for Research and Development*. Wageningen, The Netherlands: International Institute for Land Reclamation and Improvement ILRI.

Fanning, D.S., Rabenhorst, M.C., Balduff, D.M., Wagner, D. P., Orr, R.S. & Zurheide, P.K. 2009. An acid sulfate perspective on landscape/seascape soil mineralogy in the U.S. Mid-Atlantic Region. Geoderma Elsevier Science, Ltd.

Fanning, D.S., Rabenhorst, M.C., & Fitzpatrick, R.W. 2017. Historical developments in the understanding of acid sulfate soils. *Geoderma* 308: 191–206.

Fitzpatrick, R.W., Powell, B. & Marvanek, S. 2008. Atlas of Australian acid sulfate soils. In: Fitzpatrick, R.W., Shand, Paul (eds.), *Thematic Volume: Island Acid Sulfate Soil Systems across Australia, Perth, Australia* (chapter 2: 12 pp).

Fitzpatrick, R.W., Baker, A.K.M., Shand, P., Merry, R.H., Grealish, G. & Mosley, L.M. 2012. A modern soil-lansdcape characterization approach to reconstructing and predicting pedogenic pathways of inland acid sulfate soils. In: *Proceeding 7th International Acid Sulfate Soil Conference*. Finland: Vaasa.

Gumbricht, T. 2018. Detecting trends in wetland extent from MODIS derived soil moisture estimates. *Remote Sensing* 10(4): 611.

Haraguchi, A. 2016. Discharged Sulfuric Acid from Peatland to River System. In *Tropical Peatland Ecosystems*: 297–311. Tokyo: Springer.

Hole, F.D. & Campbell, J.B. 1985. *Soil Landscape Analysis*. London, England: Routlege & Kegan Paul plc.

Harris, A., Charnock, R. & Lucas, R.M. 2015. Hyperspectral remote sensing of peatland floristic gradients. *Remote Sensing of Environment* 162: 99–111.

Husson, O., Verburg, P.H., Phung, M.T. & Van Mensvoort, M.E.F. 2000. Spatial variability of acid sulphate soils in the plain of reeds, Mekong Delta, Vietnam. *Geoderma* 97: 1–19.

KLHK. 2017. Peraturan Menteri Lingkungan Hidup dan Kehutanan Republik Indonesia, (Ministry of Environment and Forestry, Republic of Indonesia), Nomor P.14/MENLHK/SETJEN/KUM.1/2/2017, Tata Cara Inventarisasi dan Penetapan Fungsi Ekosistem Gambut.

Lawson, I.T., Kelly, T.J., Aplin, P., Boom, A., Dargie, G., Draper, F.C.H. & Murphy, W. 2015. Improving estimates of tropical peatland area, carbon storage, and greenhouse gas fluxes. *Wetlands ecology and management* 23(3): 327–346.

Li, J. & Chen, W. 2005. A rule-based method for mapping Canada's wetlands using optical, radar and DEM data. *International Journal of Remote Sensing* 26(22): 5051–5069.

Lin, C., Melville, M.D. & Hafer, S. 1995. Acid sulphate soil-landscape relationships in an undrained, tide-dominated estuarine floodplain, Eastern Australia. Catena 24, 177–194.

Magdon, P., González-Ferreiro, E., Pérez-Cruzado, C., Purnama, E., Sarodja, D. & Kleinn, C. 2018. Evaluating the potential of ALS Data to increase the efficiency of aboveground biomass estimates in tropical peat–swamp forests. *Remote Sensing* 10(9): 1344.

Millard, K., Thompson, D., Parisien, M. A. & Richardson, M. 2018. Soil Moisture monitoring in a temperate peatland using Multi-Sensor Remote Sensing and Linear Mixed Effects. *Remote Sensing* 10(6): 903.

Miranda, N. & Meadows, P.J. 2015. *Radiometric Calibration of S-1 Level-1 Products Generated by the S-1 IPF, ESA-EOPG-CSCOP-TN-0002*. Paris, France: European Space Agency.

Mouche, A. & Chapron, B. 2015. Global C-Band Envisat, RADARSAT-2 and Sentinel-1 Measurements in Co-Polarization and Cross-Polarization. *Journal of Geophysical Research: Oceans* 120(11): 7195–7207.

Nordin, S.A., Latif, Z.A. & Omar, H. 2018. Individual tree crown segmentation in tropical peat swamp forest using Airborne Hyperspectral Data. *Geocarto International*, (just-accepted): 1–40.

Nunes, M.H., Ewers, R.M., Turner, E.C. & Coomes, D.A. 2017. Mapping aboveground carbon in oil palm plantations using LiDAR: A comparison of tree-centric versus area-based approaches. *Remote Sensing* 9(8): 816.

Osaki, M. & Tsuji, N (eds). 2016. *Tropical peatland ecosystems*. Tokyo, Japan: Springer.

Sammut, J., White, I. & Melville, M.D. 1996. Acidification of an estuarine tributary in Eastern Australia due to drainage of acid sulfate soils. *Mar. Freshw. Res* 47: 669–684.

Subagjo. 2006. Lahan rawa pasang surut. Dalam *Karakteristik dan Pengelolaan Lahan Rawa*: 23–98. Bogor: Balai Besar Penelitian dan Pengembangan Sumberdaya Lahan Pertanian.

Schlund, M., von Poncet, F., Kuntz, S., Schmullius, C. & Hoekman, D.H. 2015. TanDEM-X data for aboveground biomass retrieval in a tropical peat swamp forest. *Remote Sensing of Environment* 158: 255–266.

Soil Survey Staff. 2014. United States Department of Agriculture. Natural Resources Conservation Service. Twelfth Edition.

Widjaja-Adhi, I.P.G., Suriadikarta, D.A., Sutriadi, M.T., Subiksa, I.G.M. & Suastika, I.W. 2000. Pengelolaan, pemanfaatan, dan pengembangan lahan rawa. Pp. 127–164. In: A. Adimihardja, L.I. Amien, F. Agus, dan D. Djaenudin (eds.). Bogor: Sumber Daya Lahan Indonesia dan Pengelolaannya. Pusat Penelitian Tanah dan Agroklimat.

Widyatmanti, W. & Sammut, J. 2017. Hydrogeomorphic controls on the development and distribution of acid sulfate soils in Central Java, Indonesia. *Geoderma* 308: 321–332.

Application of ALOS PALSAR for mapping swampland in South Kalimantan

D. Cahyana & Y. Sulaeman
Indonesian Center for Agricultural Land Resources Research and Development (ICALRRD), Bogor Indonesia

R. Tateishi
Chiba University, Chiba, Japan

ABSTRACT: Developing countries, like Indonesia, encounter difficulties in agriculture development due to the decrease in available dry lands. They are left with no choice other than to make use of problematic soils for wetland agricultural activities. South Kalimantan is one of such areas in Indonesia with a large region of swamplands and has a history of utilizing wetlands for agricultural purposes. There is an urgent need for the evaluation and quantification of the base map for wetland development. Remote sensing data have been applied to the study of wetland in many countries because it is economical compared to the conventional mapping technique. Currently many satellite datasets, both optic and microwave ranges, are available for public use. The objective of this study is to map the wetland and swampland that are suitable for agriculture activities in South Kalimantan using remote sensing data. In particular we used the SRTM and ALOS PALSAR images which are radar data that can detect elevation and wet areas. This research first used DEM SRTM 30 m to obtain the baseline of wetlands or alluvium areas. Afterwards, ALOS PALSAR Mosaic data was used in combination with the random forest method to estimate areas suitable for agriculture. Results showed that the baseline area of wetlands in South Kalimantan is 999,768 ha. Swamplands suitable for agriculture are about 669,185 ha or 67% of the wetland. This method can be readily applied to other areas in the tropical regions as a cost-efficient method for wetland mapping.

1   INTRODUCTION

Wetland is a broad term used to define terrestrial areas affected by water, including plant and animal life. As a broad ecosystem, it has an important value for hydrological, biogeochemical, and ecological functions (Tana et al., 2013). It has important hydrological functions, including storing water during rainy seasons and gradually releasing it in dry seasons (Bullock and Acreman, 2003). With regards to biogeochemical functions, it plays a significant role in sequestering carbon and emitting methane (Matthews and Fung, 1987). Its ecological functions include supporting rich biodiversity of flora and fauna in terrain and aquatic species that differs from dryland species (Margono et al., 2014).

However, in the past, wetlands were called 'wastelands' without values unless drained. Wetlands are categorized as vulnerable regions that should be conserved to keep up their value and function. A good number of countries such as Indonesia, Thailand, and Vietnam have converted these regions into shrimp ponds, agriculture area, agro-industrial plantation, and settlement (Adger and Luttrell, 2000). The vulnerability of wetlands also increases with the effect of climate change effect such as long periods of drought which raises the risk of forest fire in wetlands area (Winter, 2000).

Developing countries with huge population tend to encounter difficulty in choosing agricultural regions because available arable dryland area is decreasing. Lands were converted and used for non-agricultural purposes during economic and population growth. This is an essential topic of discussion in China and other countries in the world (Rong Tan, 2009). In Indonesia, agricultural regions associated with drylands decreased by 90,417 ha/year from 1981-1999, and 187,720 ha/year from 1999-2002 (Agus, 2006).

Many countries have no option other than to utilize wetlands. Recently the Ramsar Convention allowed 'wise use of wetlands' as long as the integrity and health of its ecosystem remained intact. Many regions in the world from ancient time to current, such as the Euphrates and the Tigris, the Rhine, the Mississippi, the Danube, the Po, the Yangtze, and the Ganges are examples of floodplains that are used successfully for agriculture (Verhoeven and Setter, 2010).

There is an undeniable fact that some types of wetlands such us deep peatland or mangrove are unsuitable for agriculture purpose as they have high conservation value. The base map and evaluation quantification of the wetlands feasibility and suitability are urgently needed. Remote sensing data have been applied in many countries such as Vietnam, Brazil, Amerika, India, and Malaysia to study their wetland distribution (Evans et al., 2010; Hestir et al., 2008;

Ibrahim and Jusoff, 2009; Luong et al., 2015; Prasad et al., 2002).

The Indonesian Government through The Ministry of Agriculture, Ministry of Forestry, and Ministry of Public Works had identified and mapped wetland area from 1952 until 2000. All institutional research from these ministries applied conventional and manual method such as field survey and manual digitization. Their research produced various results that are difficult to validate and update. Wetland International, a Non-Profit Organization (NPO), and the Ministry of Agriculture collaborated from 2000-2005 to map a wetland known as peatland area (ICALRRD, 2006).

Ultimately, in 2013 the Ministry of Forestry applied remote sensing data such as Landsat and Palsar to map out wetland regions in Indonesia (Margono et al., 2014). Furthermore, the more detailed regional and local map is required to support specific purposes, such us agriculture planning that would be produced by policymakers.

For agriculture purposes, the term 'wetland' is broad because the definition includes natural and artificial lakes, mangrove swamps, and wet forest unsuitable for crop and plant cultivation. The most potential wetlands for agriculture region are found only in the transition area between upland and water body (river, lake, or ocean) and are known as swamplands.

South Kalimantan is one of the provinces in Indonesia with wide swampland areas and a long history of utilizing these regions. In this province, the Banjar ethnic had applied traditional agriculture for more than 200 years while combining their farming with other related activities such as extensive fishing, livestock, craft produce and trading of specific commodities (Hamdani et al., 2014; Sulaeman et al., 2019). The presence of swamplands map in South Kalimantan is necessary for policymakers to support agricultural activities.

Mapping swamplands with adequate accuracy is still difficult in tropical regions using current methods. This is because optic remote sensing data such as MODIS and Landsat are unable to penetrate cloud cover in this locality, especially during rainy seasons. The extent of swamplands also differs in rainy and dry seasons. On the other hand, there are active remote sensing data that can penetrate cloud cover to produce images such as ALOS PALSAR Mosaic 25 cm. The main objective of this study is to map suitable swamplands for food agriculture activities in South Kalimantan using ALOS PALSAR Mosaic 25 m, DEM SRTM, and ancillary datasets.

1.1 Study area

The study was located in South Kalimantan Provinces within Kalimantan at Geographic coordinates of 114° 19' 13" - 116° 33' 28" E and 1° 21' 49" – 4° 10' 14" S. The eastern part of the study site is near the Strait of Makassar that separate Borneo and Sulawesi Islands. In the southern part lies the Java Sea, while Central Kalimantan and East Kalimantan are located in the western and northern regions. The site was approximately 37,530.52 km² in area, which is about 6.98% of the Borneo Island and 1.96% of the total Indonesia area (Statistics of Kalimantan Selatan Province, 2011). The study area comprises 40.22 % of South Kalimantan. The alluvium is the main mineral soils with 1,186,913 ha or 31.8 % of the total terrestrial area. There are two types of swampland in South Kalimantan; the first is located near estuarine and affected by sea water movement, while the second is situated far from estuarine and not affected by sea tide movement. Both are located in Barito River (BPS-Statistics of Kalimantan Selatan Province, 2011).

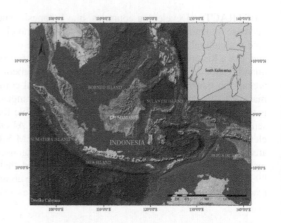

Figure 1. Study area was located in South Kalimantan. In this area there are two types of swampland: tidal swampland and non-tidal swampland.

1.2 Datasets

Since 2015, Japan Aerospace Exploration Agency (JAXA) has produced a global PALSAR mosaic product. This image series were collected by ALOS PALSAR from 2007-2010 using accurate SAR processing strategies. The images at 25 m resolution are available freely. This research made use of the 2010 data series, which happened to be the most current data image. Other datasets were obtained from the Indonesia Ministry of Agriculture through the Center for Agricultural Land Resources Research and Development (ICALRRD), which include: soil data compilation, topography maps, and supporting data in Indonesia is collected. This research also used Swampland Map of South Kalimantan in 2013.

2 METHOD FOR CLASSIFICATION OF WETLAND

2.1 Image classification

In South Kalimantan, the climate factor is the precipitation rate, which is relatively homogenous meaning that all area are potential wetlands. Therefore, the most important factors used to determine wetlands are topography and position. Based on the above reason, this research focuses on lowland regions that are less than 10 m asl, obtained using a digital elevation model (DEM) from the SRTM in regions around the coastal, estuary and big rivers. This research used two types of DEM SRTM: 90 m and 30 m. Both of them were classified below 10 m asl (swamp landscape) and above 10 m asl (non-swamp landscape). ALOS Palsar 25 m mosaic datasets were processed using the latest version of Land Use Land Cover (LULC), an open source software launched by JAXA launched in 2014.

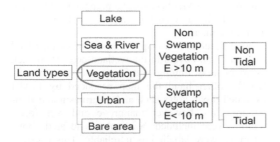

Figure 2. A framework to define wetland area.

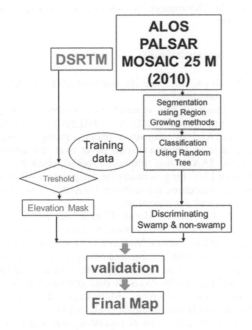

Figure 3. Flowchart of the research.

Afterward, SRTM 30 m was applied to masked ALOS PALSAR Mosaic 25 m. The detail processing steps are as follow.

2.2 ALOS PALSAR 25 m GLOBAL MOSAIC DATA Classification

JAXA launched ALOS PALSAR 25 m global mosaic data that uses Land Use Land Cover (LULC) software. This research made use of HH and HV polarization data obtained from the JAXA website. Afterward the data was processed following the following steps:

1) Mosaicked HH and HV polarization scene using ENVI 4.8.;
2) Resize the mosaics dataset to study the area using ENVI 4.8.;
3) Collected training data (307 point) from Google Earth;
4) Segmented HH and HV dataset using various weights combinations by region growing algorithm using LULC;
5) Created truth image using training data (point) at segmentation map (polygon) making it the training data using LULC;
6) Classified segmentation image (derived from 50% training data) using the random forest algorithm;
7) Evaluated the results of 3 combinations of HH and HV by visual assessment;
8) Evaluated the accuracy of map using 50% training data. Reevaluated accuracy of map using 500 random sampling points;
9) Refined the map using DEM SRTM 30 (area higher than 10 m asl).

2.3 Training data

This research used the supervised classification method with point form training data as references to classify the ALOS PALSAR data. A total of 307 observation points were collected using Google Earth based on the Swampland Map of South Kalimantan 2013. The data were identified as: urban (27), mining (15), water (33), dry vegetation (43), swampland (120), and wet vegetation (69). All training data were combined using HH and HV segmentation technique. This processing changed the training data point to a polygon. Afterward the data points were separated into two: calibration and validation. The model was also further validated using 500 random samples.

2.4 Classification

Random Forest and Region Growth Segmentation algorithms were used to classify ALOS PALSAR 25 m. Random Forest is an algorithm developed by Leo Breiman in 2001 capable of constructing many decision trees. The trees were then combined to give the best classification (Bradski and Kaehler, 2011; Breiman, 2001). Random Forest were found to have the best classification accuracy and processing speed compared to other methods such as Naïve Bayes, AdaBoost, multilayer perceptron, and support vector machine. JAXA recommended it for classifying ALOS PALSAR 25-m mosaic data using Land Use Land Cover (LULC) software that was released in 2014 (Shiraishi et al., 2014).

Region Growing method is one of the simplest approaches used to segment images. This algorithm clustered pixels starts at a limited number of single seed point. Afterward, the region grows in adjacent points depending on homogeneous criteria such us texture, pixel intensity, or color. The main goal of this algorithm is to classify the similarity amongst image into regions. Its merits lie in its ability to correctly separate regions with the same properties and to clearly define the edges of borders (Kamdi and Krishna, 2011).

3 DISCUSSION

3.1 *Basin landscape*

The most important factor in predicting swampland is topography. In this research, the basin landscape, also known as concave area, has the highest probability value. It is located in a five landscape system: marine, estuarine, riverine, lacustrine, and palustrine bounded by upland regions. The landscape system in lowland regions is also known as alluvial landform. The approach used to define the boundary between low and upland regions is using a digital elevation model (DEM) map. The slope position of alluvial landform is shown in Figure 4.

In Indonesia, lowlands are classified as regions with dry ecosystem located lower than 200 m asl. It is not compatible to delineate the edge of the basin area and in defining a wet ecosystem. Its numerous inconsistencies were overcame using a rapid assessment by tracing the edge of swampland in the contour map, using the Google Earth imagery elevation, and a landform map. Manual tracing was carried out

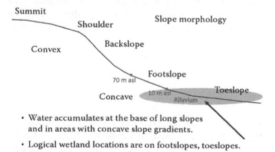

Figure 4. Hillslope profile position. Wetland as a landform is favored at the position where water volumes are maximized and slope gradient are low (modified from Schoeneberger, P.J. et al. 1988. field book for describing and sampling soils. national soil survey center, natural resource conservation service, USDA, Lincoln, NE.).

in the northern and western area of South Kalimantan Province because both are upstream adjacent to the upland. The probability of alluvial landform is located in an area lower than 10 m asl as described in Figure 5. This value is used as a threshold for wet and non-wetlands. On the other hand, this research also defined area below 71 m asl as lowland ecosystem and those above it as upland ecosystem.

This approach produced a wetland map of South Kalimantan Province of about 999,768.78 ha (Figure 6.a). The result is similar to the swampland map, with a scale 1:50,000 produced by the Indonesian Swampland Agriculture Research Institute (ISARI), Indonesia in 2014 (Figure 6.b). In 2013, ISARI mapped swampland in South Kalimantan as 930,362 ha. The result of this research is larger than the area obtained by ISARI because the purpose of this stage is to obtain a wetland map as a template or a baseline map.

As a baseline this map comprises of various land uses and land covers in the wetland ecosystem such as urban areas, mining, wet forest, and water bodies. On the other hand the map produced by ISARI excluded urban areas such as Banjarmasin and Banjarbaru City, however mangrove and wet forest regions were included. Swamplands located above 10 m asl were labelled as unreliable. This research also illustrated that the alluvial area in South Kalimantan is 27.46% of the total mainland which was similar to the alluvial landform area published by the Statistic Department, BPS-Statistics of Kalimantan Selatan.

3.2 *Swampland map*

3.2.1 *Swampland map derived from ALOS PALSAR mosaic 25 m*

The result of this research is a swampland map derived from ALOS PALSAR Mosaic 25 m. PALSAR is an active remote sensing, L band, and able to penetrate cloud cover over the atmosphere. JAXA also provided special software as known LUC for processing the image. According to some researchers, the software is capable of separating wet and dry vegetation.

ALOS PALSAR Mosaic 25 m did not only separate swampland and non-swampland, but it also attempted to separate lands use in swamp areas such as urban regions, mining, and others. JAXA reported Random Forest as the best classification method for classifying ALOS PALSAR data compared to Support Vector Machine, Multi-Layer Perceptron, Bayes, and Boost (Shiraishi et al., 2014).

This research aims at separating wet vegetation and swamplands to subtract wet vegetative regions such as mangrove, orchard, plantation, or vegetation. Mining areas were also separated, because for the past 10 years it has expanded in upland and swampland areas in this province. This research also tried

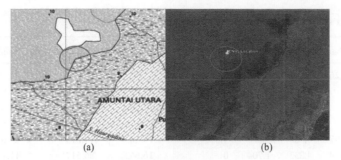

(a)   (b)

Figure 5. a) Topographic map of Sungai Pandan Hulu (Bakosurtanal, 1991); b) Google Earth (Imagery Date: 5/30/2014). This research defined the threshold of swampland and non swampland as 10 m asl.

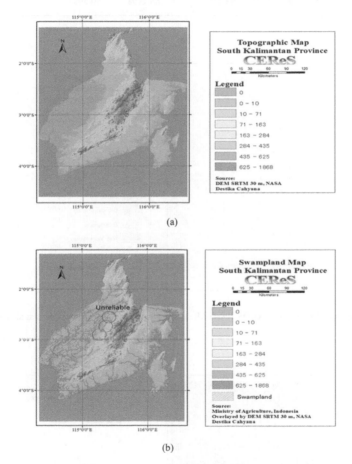

Figure 6. a) Topographic map showed alluvial area as area located below 10 m asl. This map also described the baseline of wetlands; b) Swampland map was created by the agriculture ministry of Indonesia in 2014 (unpublished). Some areas that are located above 10 m asl are unreliable.

various types of combination of HH and HV images to get the best segmentation and classification.

The 9 combination of HH and HV are 10:90; 20:80; 30:70; 40:60; 50:50; 60:40; 70:30; 80:20; and 90:10. Three types of combination, 30:70; 40:60; and 50:50, showed good and almost similar results. The visual comparison showed that the best combination of HH:HV is 50:50 and the overall accuracy is 85.25% as shown in Table 1. Using that combination the area of swampland suitable for food agriculture activities is 669,185.37 ha or around 66.93% of the total alluvium area. Figure 7 depicted the swampland and other land use.

This research also produced swampland and non-swampland maps. The result showed an

41

Table 1.   Error matrix swampland map derived from ALOS PALSAR 25 m.

Class	Urban	Mining	Water	Dry Vegetation	Swampland	Wet Vegetation	Total	User's Accuracy
Urban	35,263	0	0	0	1,447	0	36,710	96.06%
Mining	0	17,046	0	0	4,783	0	21,829	78.09%
Water	53	0	46,922	0	1,336	0	48,311	97.12%
Dry Vegetation	0	0	0	79,115	876	1,690	81,681	96.86%
Swampland	3,597	3,780	11,908	4,899	187,031	19,273	23,0488	81.15%
Wet Vegetation	1,954	4,233	1,704	0	16,643	86,517	11,1051	77.91%
Total	40,867	25,059	60,534	84014	212,116	107,480	530,070	
Prod's Accuracy	86.29%	68.02%	77.51%	94.17%	88.17%	80.50%		85.25%

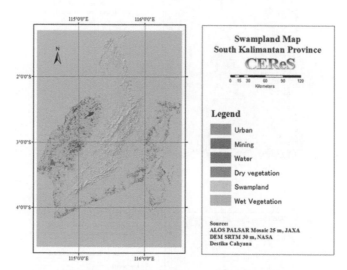

Figure 7. Swampland map derived from ALOS PALSAR 25 m. The area of swampland calculated from this image is 67% of total alluvial area.

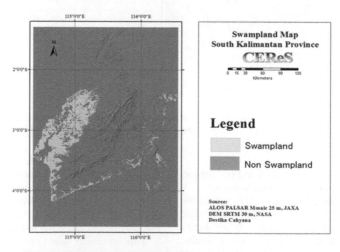

Figure 8. Swampland and non-swampland map derived from ALOS PALSAR 25 m.

Table 2. Error matrix of swampland and non-swampland map derived from ALOS PALSAR 25 m.

	Swampland	Non-Swampland	Total	User Accuracy
Swampland	251	49	300	86.67%
Non-Swampland	25	175	200	87.50%
Total	276	224	500	
Producer Accuracy	90.94%	78.13%		85.20%

Kappa Coefficient = 0.6977

overall accuracy of 85.20% as shown in Table 2. The user and producer accuracies were 90.94% and 83.67% respectively which is in line with a Kappa coefficient of 0.698. Those indicators described that ALOS PALSAR 25 m is useful for mapping swampland in Indonesia. Unfortunately the PALSAR images were only freely available for the year 2007-2010.

4 CONCLUSION

This research attempts to obtain baseline maps of wetlands and to obtain the swamplands suitable for agriculture activities. This study illustrates that DEM from SRTM 30 m is capable of creating an adequate baseline map of wetlands because the result is almost similar to the previous data from the Agency of Statistics and Agriculture of Ministry, Indonesia with an alluvium area of 999,768 ha. Secondly, ALOS PALSAR Mosaic 25 m is useful for rapid mapping swamplands for agriculture activities. Swamplands suitable for food agriculture in South Kalimantan is about 669,185 ha or around 67% of the wetlands area.

REFERENCES

Adger, W.N. & Luttrell, C. 2000. Property rights and the utilisation of wetlands. *Ecol. Econ* 35: 75–89. https://doi.org/10.1016/S0921-8009(00)00169-5.

Agus, F.I. 2006. Agricultural land conversion as a threat to food security and environment quality. *J. Penelit. Dan Pengemb. Pertan* 25: 90–98.

Bradski, G.R. & Kaehler, A. 2011. Learning OpenCV: [computer vision with the OpenCV library], 1. ed., [Nachdr.]. ed, Software that sees. Beijing: O'Reilly.

Breiman, L. 2001. Random forests. *Mach. Learn* 45: 5–32.

Bullock, A. & Acreman, M. 2003. The role of wetlands in the hydrological cycle. *Hydrol. Earth Syst. Sci. Discuss* 7: 358–389.

Evans, T.L., Costa, M., Telmer, K. & Silva, T.S.F. 2010. Using ALOS/PALSAR and RADARSAT-2 to map land cover and seasonal inundation in the Brazilian Pantanal. *IEEE J. Sel. Top. Appl. Earth Obs. Remote Sens* 3: 560–575. https://doi.org/10.1109/JSTARS.2010.2089042.

Hamdani, H., Hanafi, I., Fitrianto, A., Arsyad, L.F. & Setiawan, B. 2014. Economic-ecological values of non-tidal swamp ecosystem: case study in Tapin District, Kalimantan, Indonesia. *Mod. Appl. Sci* 8. https://doi.org/10.5539/mas.v8n1p97.

Hestir, E.L., Khanna, S., Andrew, M.E., Santos, M.J., Viers, J.H., Greenberg, J.A., Rajapakse, S.S. & Ustin, S. L. 2008. Identification of invasive vegetation using hyperspectral remote sensing in the California Delta ecosystem. *Remote Sens. Environ* 112: 4034–4047. https://doi.org/10.1016/j.rse.2008.01.022.

Ibrahim, K. & Jusoff, K. 2009. Assessment of wetlands in Kuala Terengganu District using Landsat TM. *J. Geogr. Geol* 1: 33.

ICALRRD, 2006. Karakteristik dan pengelolaan lahan rawa, Ed.1. ed. Balai Besar Penelitian dan Pengembangan Sumberdaya Lahan Pertanian, Badan Penelitian dan Pengembangan Pertanian, Departemen Pertanian, Bogor.

Kamdi, S. & Krishna, R.K. 2011. Image segmentation and region growing algorithm. *Int. J. Comput. Technol. Electron. Eng. IJCTEE* 2.

Luong, N.V., Tateishi, R. & Hoan, N.T. 2015. Analysis of an impact of succession in mangrove forest association using remote sensing and GIS technology. *J. Geogr. Geol* 7. https://doi.org/10.5539/jgg.v7n1p106.

Margono, B.A., Bwangoy, J.-R.B., Potapov, P.V. & Hansen, M.C. 2014. Mapping wetlands in Indonesia using Landsat and PALSAR data-sets and derived topographical indices. Geo-Spat. *Inf. Sci* 17: 60–71. https://doi.org/10.1080/10095020.2014.898560.

Matthews, E. & Fung, I., 1987. Methane emission from natural wetlands: Global distribution, area, and environmental characteristics of sources. *Glob. Biogeochem. Cycles* 1: 61–86. https://doi.org/10.1029/GB001i001p00061.

Prasad, S.N., Ramachandra, T.V., Ahalya, N., Sengupta, T., Kumar, A., Tiwari, A.K., Vijayan, V.S. & Vijayan, L., 2002. Conservation of wetlands of India-a review. *Trop. Ecol* 43: 173–186.

Ramsar, 2013. The Ramsar Convention Manual: a Guide to the Convention on Wetlands (Ramsar, Iran, 1971), 6th ed.pdf. Ramsar Convention Secretariat, Switzerland.

Rong Tan, V.B. 2009. Governing farmland conversion: comparing China with the Netherlands and Germany. *Land Use Policy* 26: 961–974.

Shiraishi, T., Motohka, T., Thapa, R.B., Watanabe, M & Shimada, M. 2014. Comparative assessment of supervised classifiers for land use #x2013;land cover classification in a tropical region using time-series PALSAR Mosaic Data. *IEEE J. Sel. Top. Appl. Earth Obs. Remote Sens* 7: 1186–1199. https://doi.org/10.1109/JSTARS.2014.2313572.

Sulaiman, A.A., Sulaeman, Y. & Minasny, B., 2019. A Framework for the Development of Wetland for Agricultural Use in Indonesia. *Resources* 8: 34.

Tana, G., Letu, H., Cheng, Z. & Tateishi, R. 2013. Wetlands mapping in North America by decision rule classification using MODIS and Ancillary Data. *IEEE J. Sel. Top. Appl. Earth Obs. Remote Sens* 6: 2391–2401. https://doi.org/10.1109/JSTARS.2013.2249499.

Verhoeven, J.T.A. & Setter, T.L. 2010. Agricultural use of wetlands: opportunities and limitations. *Ann. Bot* 105: 155–163. https://doi.org/10.1093/aob/mcp172.

Winter, T.C. 2000. The vulnerability of wetlands to climate change: a hydrologic landscape perspective1. *JAWRA J. Am. Water Resour. Assoc* 36: 305–311. https://doi.org/10.1111/j.1752-1688.2000.tb04269.x.

Part B. Wetlands use and management: Global perspectives

Wetland development for agriculture in Indonesia 1935 to 2013: Historical perspectives and lessons learned

Y. Sulaeman
Indonesian Center for Agricultural Land Resources Research and Development, Bogor, Indonesia

ABSTRACT: This paper discusses 78 years of Indonesian swampland development to support food production and regional economic growth. We analysed each stage of development and discussed lessons learned. We divided the development into five periods, i.e., pre-independence, 1956-1958, 1969-1995, 1995-1999, and 2000-2013 period. For each period, we see that the government's policy determined the success of the swampland development. The development of innovative technology is one major keys for success. We elaborate the lessons learned into six aspects: water management, soil understanding, local participation, sectoral vs. integrative development, management, and indigenous knowledge. For future development, three aspects must be considered: land-soil-water characterisation, landscape and landuse design, and development approach.

1 INTRODUCTION

Indonesia has one of the highest density population in the world which is currently at 264 million (BPS,2018). However, with an annual growth rate of 1.3%, the arable land is highly relied on to achieve food security. The country also faces a shrinking land area per capita while rice consumption continually increases. Moreover, irrigated rice field shows leveling productivity while some arable lands have been converted to non-agricultural uses making agricultural areas becoming fragmented.

The limitation of arable lands has led to the development of several options, one of which is the use of wetlands as a way to boost the local economy. Originally, the wetland is not an ideal land resource for agriculture due to its environmental and ecosystem services, low fertility and ability to create lots of soil and crop issues. However, due to the scarcity of arable lands, its profitable and sustainable management is unavoidable. There are many types of wetlands, and in Indonesia, the main focus is on swampland.

Swampland has been discovered to have many problems such as a high-water table making it prone to flooding for an extended period, saltwater intrusion for coastal tidal swamps, low pH of 3 to 4.5 and could be lower for the pyrite-affected acid sulfate soils, high concentrations of toxic elements, and deficiencies in nutrients. However, there are also advantages in developing swamplands as they are relatively available, abundant water availability, resilient to seasonal drought, and provides a longer cropping period (Haryono et al. 2013).

Furthermore, both tidal and non-tidal swamplands have great potential and a big opportunity for an integrated farming system for food crops, estate crops, and animal husbandry by considering the land condition and making use of available environmental-friendly technology (Suryana, 2016).

The indigenous people of Indonesia, notably the Banjar in Kalimantan, has settled in the swamp area since the early period of their lives. They built small canals called handil (Dutch: anndeel) with the ability to protrude into the stretch of a river for 2-3 km at a width of 2-3 meters and a depth of 0.5-1.0 meters. The Handil can also be extended up to 5 km in order to reclaim around 20-60 hectares of land. These indigenous community practices have succeeded in developing swamp areas for the production of rice, secondary crops, vegetables, and other horticulture crops in Kalimantan and Sumatra. This, however, inspired the government to reclaim more swamp areas for agriculture.

The aim of this paper therefore was to review the development of swamp reclamation in Indonesia for agricultural use and draw some lessons for future sustainable agricultural development. First, the swampland was outlined and the history of the development was reviewed through a five-period division including before independence (pre-1950), 1956-1958, 1969-1995, 1995-1999, and 2000-2013. Finally, lessons learned from these periods were discussed and suggestions were made on its sustainable application in recent times.

2 SWAMPLANDS IN INDONESIA

The swampland is mainly divided into two types and they include tidal (Indonesia: rawa pasang surut) and non-tidal swamp or inland swamp land (Indonesia: rawa lebak). It is, however, possible to find peats on these lands but for the purpose of mapping, they are separated from peatland despite the presence of organic matters in both of them.

There are about 34 million hectares of swampland (BBSDLP, 2015) covering 18% of Indonesia's land area with the largest observed in Sumatra with 13 million hectares followed by Kalimantan with 10 million hectares. Furthermore, based on a land evaluation conducted by the ministry of agriculture, it was estimated that 7.5 million hectares of swampland are available for cropping, with 5 million hectares for paddy rice, 1.5 million hectares for horticultural crops and about 1 million hectares for perennial crops (BBSDLP, 2015).

3 SWAMPLAND DEVELOPMENT HISTORY

3.1 *Pre-independence (before 1945)*

Indigenous people, notably the Banjar in Kalimantan and Bugis in Sumatra, built small canals called Handil as well as ditches which were made constructed to be perpendicular to the Handil and used for transportation. They were both made relatively shallow at about 1 m to avoid excessive drainage. Moreover, the water surface level in the plot of rice fields is controlled with a gate called tabat placed along the primary channel and the intersection with the secondary channel. This, however, gave the Banjar community the opportunity to maintain land productivity for up to 20 years (Suwardi et al., 2005).

Furthermore, the Dutch colonial government initialized the development of swamp lands in South Kalimantan by digging large canals, known locally as anjir (Houterman and Haag, 2015). Anjir Serapat and Anjir Kelampang were hand-dug around 1896 and, in 1935, the Dutch completed Anjir Serapat, which is 28.5 km long connecting the Kapuas Murung River to the Barito River. Initially, the canal was built for transportation, but the locals settled around the banks and the farmers made ditches (handil) extending up to 5 km with the distance between them being about 200-500 meters and this made the structure resemble a herringbone. Moreover, during the period between 1920-1962, around 65,000 hectares of swamps were independently cultivated by farmers (Idak, 1982).

In 1941, the Dutch colonial government continued the development by building Anjir Tamban which ran perpendicular to the Barito river. The area was a Dutch transmigration sponsored program with settlers from East Java. However, the work was halted because of world war and resumed in 1952, with 25 km long canal extending to the Kapuas river. Moreover, the farmers started to build secondary canals to drain areas far from the anjir. Up to 1940, there were 216 families or 989 inhabitants in these areas including Purwosari Village, Tamban Subdistrict, Kapuas Regency, Central Kalimantan which is well-known as a center for coconut and rice production.

In the 1950s, H.J. Schophyus, a Dutch agricultural expert with H. Idak, from Banjarmasin continued the development of swamplands in Kalimantan. Schophuys proposed the swampland be divided into separate water management units called polders. They both designed the construction of polders for the development of swamp area in the Alabio, North Hulu Sungai Regency in South Kalimantan, which is a non-tidal zone along Negara river covering an area of 6,500-7,000 hectares. This polder area is now a center for rice production. However, it is only possible to cultivate around 3,000 hectares once a year and there were some periods when cultivation could not be conducted due to high standing water.

3.2 *1956-1958 period*

The reclamation of swamp lands using the Anjir System was initiated by Ir. Pangeran Mohammad Noor, the Indonesian Minister of Public Works and Labour between 1956-1958 through the Dredge, Drain and Reclamation Project. This project did not only construct but also perfected several large canals connecting two large rivers, and opened access to swamplands between them.

In this period, Anjir Serapat was further developed, and a bridge was built to connect the Barito River (Banjarmasin city) and Kapuas Murung (Kuala Kapuas city). The anjir was also used as a transportation medium between the two cities through the use of large boats and cargo ships. In addition, the locals built secondary canals called handil and the area became the center for rubber and rice production.

The anjir system was replicated in several other areas of South, Central, and West Kalimantan, and South Sumatra. However, due to the lack of infrastructure and funding, these projects could not be realized. Nevertheless, in this era, transport via water became the key to the development of regions in Kalimantan. The Ministry of Agriculture launched a Three-Year Rice Production Plan to achieve food self-sufficiency in 1958. However, due to the unstable political conditions, this was unsuccessful.

3.3 *1969-1995 period*

In the first five-year development program of Indonesia (1969-1974), the government began reclaiming large areas inland from coastal rivers and rice was planted on about 558,000 ha of tidal swamps area comprising 7% of the total rice area in late 1970-1980. This massive expansion of the transmigration program funded by the world bank was aimed to reclaim more than 30,000 coastal areas in Sumatra.

The project initially covered 5.25 million hectares of the swamp in Sumatra and Kalimantan for 15 years. However, up to the end of the project, swamp areas successfully reclaimed reached only 1.2 million hectares with a fork water system network of 29 water management schemes in South and Central Kalimantan. Moreover, a comb system of 22 network schemes spread across Sumatra, West Kalimantan and a small portion was found in Sulawesi and Papua. This period also saw the difficulty in inland development to be potential acid sulfate and peat soils.

The reclamation of tidal swamps was started by the government in Kalimantan and Sumatra from the early 1970s up to the mid-1990s, directed at developing rice fields with the main goal of increasing rice production. Unfortunately, this was not well managed as observed in the botched logging and land clearing, excessive draining of lands via canals and inadequate water management strategies. Therefore, several transmigration areas developed on swamplands failed and were abandoned.

In the 1985-1990 period, the government focused on rehabilitation of canals and building of water control structures to satisfy drainage needs for agriculture. Furthermore, the government intensified agricultural research and development activities, especially on food crops in tidal swamps through several projects such as Swamps II and ISDP Projects. However, the reclamation and development of swamps in this and the previous era have not produced good results due to the inadequate location, partial developmental approach, many problems in recognizing soil problems, inadequate application of technology, and limited support.

Therefore, a staged development program was introduced to overcome these problems. The first stage focused on smallholder farmers' rice production with an open tidal canal system in unripe and organic soils. The second stage involved the introduction of soil maturity, social cohesion, water management, crop diversification, mechanization, and public and socio-economic services. The final stage employed a full water management system.

Reclamation in this era was supported by a transmigration program which placed residents from Java, Bali and Nusa Tenggara in Kalimantan, Sumatra, Sulawesi, and Papua swamps. The swamp area built was expected to use cultivation technology with the rice specifically covering around 400 thousand hectares in Sumatra and 150 thousand hectares in Kalimantan. At present, some of these swamp areas have succeeded in becoming centers of rice, vegetable and plantation production.

Within 25 years (1969-1994), swamps planned and phased out reached almost 1 million hectares and of the 900 thousand hectares opened, around 715 thousand hectares are on tidal and 185 thousand hectares on non-tidal swampland.

The expansion of swamp use for agriculture and transmigration succeeded in placing 1,717,610 households from the target of 2 million households. Up to 1995, around 1.18 million hectares of swampland has been successfully reclaimed in seven major provinces with the potential.

3.4 1995-1999 period

Development of swamp areas in this period was in accordance with the Presidential Instruction (Inpres) No. 6 of 1995, known as the One Million Hectares Peatland Development Project in Central Kalimantan. This project is better known as the Mega Rice Project in Central Kalimantan. The Instruction was issued on June 5, 1995, followed by a Presidential Decrees No. 82, 83, and 84 of 1995.

The project was based on concerns about the leveling of the trend in rice production of the country and increasing land conversion due to the increased population growth. Since the policy to achieve food self-sufficiency was made in 1984, Indonesia became an importer of rice from 1988-1989.

However, the Mega Rice Project failed due to mismanagement. For example, drainage networks were built without considering the land, especially the areas with peat domes were drained excessively and the use of domes as water reservoir diminishes water supply. Furthermore, the excessive drainage also exposed some of the acid sulfate soils causing extreme acidification and consequently led to plant toxicity. Therefore, due to strong pressure both from within the country and internationally, a decision was made to close the Mega Rice Project after an evaluation by the Review Team through the use of Presidential Decree No. 80 of 1999. It was, however, unfortunate that a large area of swamplands cleared was not utilized and became abandoned.

3.5 2000-2013 period: rehabilitation and revitalization

During this period, focused was placed on rehabilitating the area of ex-Mega Rice Project in Central Kalimantan Province since some of them possess the potential for agriculture to provide income for 13,500 households (around 600.000 people) that have lived in Block A of the area (Abdurachman and Suriadikarta, 2000).

The activities were conducted based on Presidential Instruction No. 2 of 2007 with no attention placed on the agricultural sector. The uncertainty of efforts to revitalize the Mega Rice Project area resulted in land degradation due to land fires occurring every dry season, especially on uncultivated land.

The next period also declared the opening of swamp land in Papua through a project known as the Merauke Integrated Food and Energy Estate (MIFEE) in 2006. However, this project was hampered and constrained, therefore, no satisfactory results have been observed.

4 LESSONS LEARNED

Indonesia has learned a lot of lessons and gained much knowledge on swampland management through numerous reclamation projects, research and development. The initial development of swamplands for transmigration settlement was conducted haphazardly with a low budget. This is further challenged by unknown soil conditions in the new areas.

Furthermore, staged development with iterations of review and redevelopment ensure sustainable development. However, the lessons learned were elaborated into five aspects, namely soil understanding, local citizen participation, sectoral vs. integrative development, unified management, and indigenous knowledge.

4.1 Soil understanding

Reclamation focusing on draining creates adverse impact due to shallow groundwater of the dry season. Moreover, pyrite under inundation is stable but upon exposure to the air, it can be easily oxidized leading to severe acidification of soil and water. Furthermore, pyrite, Al and Fe are toxic to plants. In addition, haphazard drainage also has the ability to cause waterlogging in the rainy season (Anda et al., 2009).

There are two options in managing acid sulphate soils and they include leaching out pyrite oxidation products by rainfall or tidal flushing and water management. The first option is hard to achieve due to the required amount of fresh water and its slow process while the second is better due to its ability minimize pyrite oxidation by keeping the groundwater table below the pyrite position and use of irrigation during the dry season.

Therefore, based on the soil knowledge, the use of land can be design in accordance with its type and crop requirements and Surjan System (Nursyamsi et al. 2015) has been observed to be providing appropriate techniques for landscaping and land-use design.

4.2 Local citizen participation

Local residents who have lived in swampland for a long time have a balance between their lives and the nature around them. Their experiences are based on knowledge which is an accumulation of several generations of experience which are useful in the development of swamplands. Furthermore, they have been able to develop a system of behavior to regulate the relationships between them and their environment as a landscape.

The development of swamps in the 1950s involved many local residents in planning and implementation while they were ignored in the 90s due to the excessive trust in knowledge and technology. However, local communities have policies in managing the sustainability of channels and gates.

Therefore, the development of future swamplands should involve local communities at each stage of development and the society should be a subject not an object of development. For example, the handil institution in the Banjar community proves how local culture has the ability to maintain and secure the sustainability of swamps.

4.3 Sectoral vs. integrative development

During the reclamation periods, it appears swampland reclamation was only directed towards rice production and this sectoral objective had the tendency to fail as observed in The Mega Rice Project and Merauke Projects which tend to be sectoral-oriented.

The integrative development aims to improve the welfare of people living in the area, as compared to sectoral development which only aims to obtain a certain output such as rice production. Sectoral development only looks at crop productivity as opposed to the focus of integrative development on land productivity. Anwarhan and Sulaiman (1985) studied farming system in tidal land under Surjan System and they compared rice-based, fish-crops as well as mixed citrus and coconut-based farming systems and found mixed citrus and coconut with rice farming to have produced the highest farmer income.

In addition, integrative development considers drainage channels should be made not only to regulate the water system but also to provide new communication and transportation lines for closed areas and bring agricultural products to a broader market. Integrated approach using the integrated farming system has been reported to be a major factor for success in swampland development (Suryana, 2016).

Therefore, future development of swamplands should be integrative from the beginning and more people should be involved in order to obtain better outcomes and greater opportunity for sustainability from swampland development.

4.4 Unified management

Management in this context means the reclaimed swamplands should be managed for maximum output and increased welfare. Many ministries are involved in swampland management, and each has its own rules and interests. In most cases, the importance of an institution is often incompatible with the interests of others. Therefore, unified management is crucial in order to ensure they all complement each other to reach the same goal.

Centralized management should be established to coordinate the development and all the stakeholders should be involved in accordance with an agreed protocol. This is necessary because past failures are largely due to weak coordination and communication between agencies and communities. Therefore, local communities and young people need to be involved in the management of this development land area.

4.5 Indigenous knowledge

Swamp and water culture are very attached to the indigenous people of Banjar, Bugis, Malay, Dayak, and Papua. They have been in balance with their environment in the form of land, plants, and climate and possess a work culture specific to them. They also have beliefs and rules which are allowed and those not allowed in the development of swamplands and this is a local decision that must be respected. However, documentation on indigenous knowledge is still rare and should be a subject of focus in the future.

5 CONCLUSIONS

Indonesia has over 34 Million ha of swampland with high potential to support Indonesia's food security program. The crop production and productivity can also be increased through optimization of existing reclaimed wetland and abandoned swamplands. There is the possibility of achieving this by improving water management via the construction of new and revitalization of existing infrastructures, especially the drainage and water control system. It is also possible to improve land productivity by using crop-fish-poultry diversification.

Furthermore, the lessons from 78 years of experience showed the importance of involving local communities and wisdom in the development and management of swamplands. Going forward, the focus should be placed on revitalizing existing drainage networks. Moreover, opening new and optimizing existing areas requires precise planning of land-soil-water characterization and mapping, landscape and land-use design, and adaptive development approach. This effort has the ability to increase crop production and productivity as well as community welfare.

REFERENCES

Anwarhan, H. & Sulaiman S. 1985. Pengembangan pola usahatani di lahan pasang surut dalam rangka peningkatan produksi tanaman pangan, *Jurnal litbang* 4(4),91-95.

Abdurachman, A. & Suriadikarta, D.A. 2000. Pemanfaatan lahan eks-PLG Kalimantan Tengah u pengembangan pertanian berwawasan lingkungan. *Jurnal Penelitian dan Pengembangan Pertanian* 19 (3), 77-81.

Anda, M., Siswanto, A. & Subandiono, R. 2009. Properties of organic and acid sulfate soils and water of a 'reclaimed' tidal backswamp in Central Kalimantan, Indonesia. *Geoderma* 149, 54–65.

[BBSDLP] Balai Besar Penelitian dan Pengembangan Sumberdaya Lahan Pertanian. 2015. Sumberdaya lahan pertanian Indonesia: luas, penyebaran dan potensi ketersediaan. Laporan Teknis Bogor (ID): Balai Besar Penelitian dan Pengembangan Sumberdaya Lahan Pertanian. 100p.

Haryono, Noor, M., Syahbuddin, H., & Sarwani, M. 2013. Lahan rawa; penelitian dan pengembangan. Jakarta (ID): IAARD Press. 103p.

Houterman, J. & Haag, A. 2015. Dutch involvement in tidal swamp development (1930–2014). In Irrigation Revisited: An Anthology of Indonesian-Dutch Cooperation; Kop, J.H., Ravesteijn,W., Kop, K.J., Eds.; Eburon: Delft, The Netherlands; pp. 173–218.

Idak, H., 1982. Perkembangan dan sejarah persawahan di Kalimantan Selatan. *Pemda Tingkat I. Kalimantan Selatan. Banajarmasin.*

Nursyamsi, D., Noor, M., & Haryono. 2015. Sistem Surjan: Model pertanian lahan rawa adaptif perubahan iklim. IAARD Press, Jakarata. 135p..

Suryana. 2016. Potensi dan peluang pengembangan usahatani terpadu berbasis kawasan di lahan rawa. *Jurnal litbang*, 35(2),57-68.

Suwardi, Mulyanto, B., Sumawinata, B. & Sandrawati, A., 2005. Sejarah Pengelolaan Lahan Gambut di Indonesia. *Gakuryoku* 11, 120-126.

Managing wetlands in Vietnam: Current practices and future challenges

H. Van Thang

Central Institute for Natural Resources and Environmental Studies, Vietnam National University, Vietnam

ABSTRACT: Vietnam wetlands consist of inland and coastal wetlands, inhabited by many rare and endangered species, as well as a source of income for the local population. It is estimated that there are more than 12 million hectares of wetlands distributed from north to south, from the mountains to the coast and islands. Of those, five areas are the most important wetlands including: Red river delta, central coastal and lagoons, Mekong delta, lakes and ponds and others. The country has put a lot of effort into managing and conserving the wetlands. Vietnam joined the Ramsar Convention in 1989 with a strong commitment by establishing 8 Ramsar sites to date. Vietnam has also developed several legal documents for wetland management and conservation, while building its capacity for implementing wetlands related policies mechanisms. Several wetland management models have also been developed and applied in different types of wetlands and wetland protected areas. Nevertheless, the wetlands are under degradation in both quality and quantity due to a number of threats and challenges from natural and human causes. This paper presents the general status of wetlands in Vietnam and their current management practice at all levels, as well as future challenges.

1 INTRODUCTION

There are many ways to manage wetlands, depending on the purposes of the managers (Mitsch and Gosselink, 1993; Keddy, 2000). Sometimes, there could be conflicts among management purposes, for instance, the prevention of waste input into the wetland and the use of wetlands for liquid waste treatment. Wetland management also depends on mechanisms and policies related to wetland conservation.

Managing wetlands on purposes based on wetland functions and values is often the choice of managers (Isozaki et al., 1992).

In general, wetland management can be understood as a coordinating and guiding process for relevant stakeholders in implementing and assessing the effectiveness of wetland-related conservation, wise utilization, and sustainable development activities.

The wetlands of Vietnam are diverse and range from streams, ponds, and lakes to large deltas and coastal lagoons, islands, and archipelagos. For generations, Vietnamese people have practiced wet rice cultivation, fishing, and aquaculture for their living in and surrounding wetlands. As for that, most of the Vietnamese population is concentrated in the deltas and coastal areas. Consequently, the wetlands have been converted into rice paddies, constructed lands, etc. As a result, many habitats and species have been destroyed and become rare and endangered, and some even extinct.

Acknowledging the functions and ecosystem services that wetlands can bring to the country, as well as pressures and threats to the wetlands, Vietnam has put a lot of effort into managing and conserving its wetlands by different means at different levels from central government to wetland sites and the community. A number of policies, mechanisms, and practicing models have been developed and applied. Research has been carried out together with capacity building and threat identifying.

In this paper, we will present an overview of the status of wetlands of Vietnam. We will then describe the management practices and lessons learned, as well as threats to wetlands in the future.

2 VIETNAM WETLAND AND MANAGEMENT

2.1 Vietnam wetland

According to the inventory in 2016, Vietnam wetland covers an area of more than 12 million ha, including inland wetlands and coastal wetlands. Inland wetlands are distributed in all three regions and eco-regions of the country (north, center, and south). The wetlands are diverse in types, resources, function, and value. Inland wetlands consist of deltas, fresh water streams, and canals, ponds, lakes, peatlands, saltwater lakes, hot springs, aquaculture ponds. Coastal wetland includes estuaries, lagoons, mangroves, sea grasses and coral reefs.

Of those, five areas are the most important: Red river delta, central coastal and lagoons, Mekong Delta, lakes and ponds and others.

2.1.1 Red river delta wetland

The wetland accounts for 229,762 ha (76,01 % total land area of Vietnam). Among them, saltwater wetland is 125,389 ha, which can be found in Nam Triệu, Cấm, Lạch Tray, Văn Úc, Thái Bình, Ba Lạt,

Lạch Giang, cửa Đáy river mouths (estuaries). These wetlands are mainly used for agriculture and aquaculture. Freshwater wetland is at 103,373 ha, which is mostly used for agricultural practices.

2.1.2 *Mekong delta wetland*
The Mekong Delta of Vietnam covers an area of about 3,9 million ha (about 12% of the total acreage of the country) administrated by 13 provinces and cities: Long An, Tiền Giang, Đồng Tháp, An Giang, Kiên Giang, Hậu Giang, Cà Mau, Bạc Liêu, Sóc Trăng, Trà Vinh, Vĩnh Long, Bến Tre and Cần Thơ City.

The Mekong delta wetland, which covers about 95.88 % of the total area (4,939,684 ha), is one of the richest eco-regions in biological products of the country. The ecosystems are also important spawning and nursing habitats for many aquatic species.

2.1.3 *Central coastal lagoons*
Central coastal wetlands are distributed mainly from Thua Thie Hue to Ninh Thuan with 12 lagoons (Tam Giang - Cầu Hai, Lăng Cô, Trường Giang, An Khê, Nước Mặn, Trà Ô, Nước Ngọt, Thị Nại, Cù Mông, Ô Loan, Thủy Triều). The total acreage of the lagoons is about 447.7 km^2. Of those, Tam Giang - Cầu Hai with a total length of 67 km (216 km^2) is the largest one, and the smallest is the Nuoc Man lagoon in Quang Ngai province (2.8 km^2).

2.1.4 *Other wetlands of Vietnam*
Lacustrine, ponds, and rivers
Vietnam has many lakes and ponds (naturally) and humanmade reservoirs. Also, with a rather dense system of rivers and streams, the wetlands are habitat to many species including 243 species of river fishes in the north, 134 species in the central, and 255 species in the south.

Mangroves: There are about 155,290 ha of mangroves in Vietnam. Of those, 32,402 ha (21%) are natural and 122,892 ha (79%) are planted mangroves. Vietnam mangroves are distributed unequally along the coastline. In the northeast, there are 22,969 ha (14,8%), 20,842 ha (13,4%) in the Red River delta, 20,092 ha (16,8%) in the northcentral and Ho Chi Minh City, and 82,387 ha (53%) in Mekong Delta.

Coral reef: distributed in many places along the coastal zone of Vietnam with a total area of about 1,122 km^2. There are about 350 species of coral identified with about 300 coral related biological species.

Sea grass: distributed from north to south, ranging from 0-20 m deep and salinity of 5-32%. Vietnam has identified 15 species of 16 sea grass species of Southeast Asia.

2.2 *Wetland management*

In Vietnam, the government manages wetlands systematically and through a hierarchy. The Ministry of Natural Resources and Environment (MONRE) is delegated to managing the wetlands (Biodiversity Law, Decree 109). According to Decree 109 of the government (2003), wetland management for sustainable development are as follows:

1. Inventory and study on wetlands.
2. Development of mechanism, policies and laws on conservation and sustainable development of wetlands.
3. Zoning and planning in wetland use for conservation and socio-economic development.
4. Managing wetland areas designated for conservation.
5. Managing wetland relevant activities, including exploitation of resources and services of wetlands from agriculture, fishery, tourism, transportation, energy to others.
6. Inspecting and solving illegal activities.
7. Encouraging communities, especially those people who are living within and around wetland sites, to take part in conservation of biodiversity and protection of the environment.
8. Strengthening international cooperation in conservation and sustainable development of the wetlands.

2.2.1 *Management structure and responsibility*
Central/National level: Ministry of Natural Resources and Environment (MONRE).

Under Decree 109 and Biodiversity Law of Vietnam, MONRE is responsible for natural wetland habitat management, Ramsar Convention, Biodiversity Convention, etc.

Provincial level: Department of Natural Resources and Environment (DONRE), as a line department of MONRE, will be responsible for the same task at the provincial level.

Nevertheless, at all levels, other ministries and departments are also involved in the management of wetlands, such as Ministry of Agriculture and Rural Development (MARD), which manages the protected area system (both inland and marine ones). Of those protected areas, many national parks, nature reserves are wetlands. As a hierarchy system, the provincial Department of Agriculture and Rural Development (DARD) are responsible for the same task at its level.

Site level: It depends on the category of the wetland site whether it is a national park, a nature reserve, or landscape area as to how the management body is organized. It is very often that the management board of a national park is directly under the ministry (MARD or MONRE) or provincial authority. A nature reserve or landscape area is under DARD or DONRE. The management boards are responsible for all tasks that they are assigned but mainly focus on protection, conservation of the site, and assist local sustainable development.

53

2.2.2 Mechanism and policies

Since 1989, when it became a member country of the Ramsar Convention, Vietnam has developed a number of legal documents, policies, and mechanisms related to wetlands and their management.

Of those, Decree 109 (2003), Biodiversity Law, Law of Fishery are the most important for wetland management.

Together with policies and legislation development, other technical documents have been developed, such as Decision No. 1093/QĐ-BTNMT dated 22/8/2016 of Ministry of Natural Resources and Environment, guiding the identification of wetland types and wetland classification of Vietnam.

2.2.3 Establishing wetland protected areas

The important wetland areas have been integrated into the protected area plan by Decision No. 45/QĐ-TTg in 2014, which ratified the master plan to conserve the biodiversity of Vietnam by 2020 visioning to 2030.

46 wetland areas in total have been designated as protected areas. Some are within the existing protected area system (national park, nature reserve, species and habitat protection, and landscape and environment protection).

At present, a legal document guiding the establishing a wetland protected area is being developed by MONRE.

2.2.4 Capacity building and awareness

Together with training courses offered by universities and academies, MONRE has also developed training materials for on-job training of officials.

Posters, visual media programs are also developed and aired.

Annually, World Wetland Day is organized with the participation of many people from different levels and stakeholders.

2.2.5 Wetland management model and lessons learned

Besides the management of protected areas, which follows Biodiversity Law, Law on Forest protection and development, and Law of Fishery, some of the models on wetland management have been developed and applied in Vietnam.

1. Community-based wetland conservation management: The model has been developed and applied in the Ba Che estuarine area (Tien Yen District, northern Quang Ninh province of Vietnam) where the mangroves and mudflat are restored and managed by the local authority and a social organization (Women Union) with technical support from the university. This model has been also applied for nipa palm conservation and management in Cam Thanh commune of Hoi An ancient city.
2. Co-management and benefit sharing: The model has been developed and applied in regeneration

and core zones of the parks of Xuan Thuy in the Red river estuarine (first Ramsar site of Vietnam) and Tram Chim (Ramsar site) in the Mekong delta. Local people are actively involved in managing the wetlands while they can benefit from rational exploitation of certain products (fish, bivalves, crabs) of the wetland without causing harm to the wetland health.

3 THREAT AND CHALLENGE

3.1 From natural process and extreme event

- Vietnam is one of the countries most affected by climate change. Climate change and extreme events have caused serious damage to the wetlands and their biological diversity. The Linda typhoon damaged thousands of hectares of Melaleuca forest in Ca Mau's U Minh Thuong National Park in 1995. Consequently, many species, including rare and endangered storks, lost their habitat. Hailstones damaged more than 20 ha of mangroves in Bai Tu Long National Park of northern Quang Ninh province in 2010.
- Sea level rise would cause great impact on the Mekong delta, where a vast area of wetlands could be inundated, changed in hydrology regime, as well as water quality. In 2010, it was estimated that 168 km^2 of wetlands, especially aquaculture areas, were flooded due to sea level rise. It is also predicted that climate change and sea level rise would be even more serious in the Mekong delta, especially in water supply during the dry season.
- Forest fires in the melaleuca forest (U Minh, Dong Thap) in the dry season is one of the greatest threats to the melaleuca forest and peat land. In 2002, more than 8000 ha core zone of U Minh Thuong national park was burned due to forest fires.

3.2 From human activities

- Destructing and fragmenting the habitats of wetlands: due to the rapid development of the socio-economy, many wetland areas have been destructed or converted into other land use purposes, especially infrastructure development including road, harbor, houses, agricultural cultivation and aquaculture. For instance, from 1943 to 2006, it was estimated that Vietnam had lost at least 200,000 ha of mangroves.
- Many coastal wetland areas, especially lagoons, have been degraded due to the construction of sea dikes. In Thua Thien Hue alone, the Thao Long dam in Phong Dien district has caused the change of hydrological regime, siltation process, and blocked migration of aquatic species, such as *Sewellia albisuera*, which has led to the fragmentation and degradation of the habitats. In addition, in the

south-western Ca Mau, almost 20 benthos species became extinct just after one year of converting the mangrove area into shrimp farming. Some of the water bird species had migrated out of Bac Lieu and Dam Doi bird colonies. Other IUCN red list species, like *Cairina scutulata* and *Lutra sumatrana*, are under great threat of extinction due to the conversion of wetlands to other use purposes.

- Road and reservoir construction is also another cause in wetland fragmentation and degradation. Some examples are the construction of the road through Dong Thap Muoi (Plain of Reed) in the south and the reservoir in Na Hang Nature Reserve of the north. To date, Vietnam has built about 2000 dams all over the country.
- Deforestation and other development activities upstream have caused a significant impact on the wetlands downstream.
- Over exploitation of biological resources: many benthos and crustacean species, as well as fish and reptiles, birds, and other aminals have been overexploited or killed bu destructive means, including explosive and even toxic chemical. As a consequence, many species, such as Asia soft-shell turtle (*Amyda cartilaginea*), *Protunio messageri*, and *Phyllopteryx taeniolatus* are becoming extinct. Overexploitation also leads to the decrease in production of many scientific and economic species, like *Panulirus sp.*, *Halioles sp.*, *Chalamys spp.*, and *Loligo spp.* Even Pinctada margaritifera has become extinct in the north of Vietnam due to overexploitation.

3.2.1 *Overexploitation of water resource*
It is estimated that Vietnam total surface water resource is about 840 billion cm^3. Underground water that can be exploited is nearly 20 million m^3. Both surface and underground water resources are decreasing due to overuse by different users, including agriculture, aquaculture, industry, and households. In the north, underground water is mainly exploited for household uses, while in the south and highland plateau, underground water is mainly used for agricultural cultivation. In the central provinces and Tay Nguyen high plateau, more than 50% of the flow water was being used. In addition, surface water and sediment transport were impacted from upstream development activities, especially from dam construction (MONRE, 2010).

3.2.2 *Pollution*
Agricultural chemicals, industrial and urban waste are the two main sources of pollution to wetlands. World Bank's survey (2007) on the rivers of Vietnam showed that liquid waste from industry, mining, urban and handcraft villages caused pollution in the rivers and coastal zone. Heavily polluted wetlands cause negative

impact on human health, fishery, and agricultural products. It also leads to eutrophication of the wetlands. For example, in 2012, "red tide" occurred in central province of Binh Thuan and Khanh Hoa, killed a lot of fishery products and damaged the economy of the area. Red river delta is under great threat of agricultural chemicals. Many rare and endemic species of fish are under threat of extinction, such as *Schistura spiloptera* and *Cristaria truncate*.

3.2.3 *Introduction of invasive species*
The introduction of invasive and exotic species causes serious threats to the wetlands. To date, 42 of 94 invasive species are found in Vietnam. Some of them include golden snail (*Pomacea canaliculata*), red ear turtle (*Trachemys scripta*), and *Mimosa pigra*. *Mimosa pigra* has become the greatest threat to Tram Chim national park in the Mekong delta (Ramsar site), habitats to Eastern Sarus Crane (*Grus antigone sharpii*) and Bengal Florican (*Houbaropsis bengalensis*), two endangered species of birds. It is predicted that, if this invasive species is not controlled, about 4.000 ha, or 50% of Tram Chim area will be covered by *Mimosa pigra*.

3.2.4 *Shortage of resources*
Although the Government and its ministries, together with support from international and national organizations and universities, technically and financially, have gained many results in managing the wetlands, there is still a need for manpower and finance.

At the national level, the Biodiversity Conservation Agency of MONRE has adequate staffs for wetland management. However, at the provincial and district levels, there is not enough manpower or expertise in wetland management. While at wetland sites, only few staff have good management knowledge on wetlands.

Several GEF and other donor-funded projects have been carried out on wetlands. Nevertheless, much work needs to be financed to help conserve valuable resources.

REFERENCES

Isozaki, H., Ando, M. & Natori, Y (eds). 1992. Towards Wise Use of Asian Wetlands, Asian Wetland Symposium. International Lake Environmental Committee Foundation, pp 285.

Keddy, A.P. 2000. Wetland Ecology: Principles and Conservation, Cambridge University Press.614p.

Mitsch, J.W. & Gosselink, J.G. 1993. Wetlands, Second edition. Van Nostrand Reinhold Company Inc. Pp 543.

MONRE 2010. National annual environmental report in 2010:Vietnam environmental overview.Hà Nội.

The Government of Vietnam. 2003. Decree of the Government No 109/2003/NĐ-CP dated 23Sep. 2003 on Conservation and Sustainable Development of wetlands. Hanoi. 11p.

Managing acid sulfate soil in Thailand: Current practices and future needs

P. Vijarnsorn
Ministry of Agriculture and Cooperatives, Bangkok, Thailand

ABSTRACT: There are two types of acid sulfate soils in Thailand: potential and actual acid sulfate soils. Both were formed from marine and brackish water deposit rich in pyrite along the coastal zone. The potential acid sulfate soils are found to occur on the active tidal flat where mangroves occur and comprise about 459,000 ha. They have low agricultural potential and their uses should be restricted to mangrove for conservation. The actual acid sulfate soils generally occur in the former tidal flat or in the middle part of the delta. The soils are more developed and frequently contain jarosite mottles in the lower subsoil. They are extremely acidic, contain high amounts of Fe and Al but are low in available phosphorus. At present, the management of the actual acid sulfate soil in Thailand is successful for growing paddy rice and other crops like fruit trees or fast-growing trees. Also, water management using proper irrigation systems is the key factor for watering, leaching and maintaining a high groundwater table. Lime application is recommended for the correction of acidity as well as using acid tolerant crops to reduce the cost of production. At present, the management of acid sulfate soil in Thailand is successful for growing various crops, especially paddy rice, at a commercial scale. However, crop diversification system needs to be implemented for better market buffer in the future. At the same time, more studies on the dynamics of soil chemical properties are required for precise management. Additionally, proper water management at low cost and economical use is needed for future planning.

1 INTRODUCTION

In Thailand, acid sulfate soils are one of the major problematic soils for agricultural uses. They commonly occur in the coastal areas along the Gulf of Thailand and the submergence shoreline of the western peninsula. The acid sulfate soil can be broadly defined as the soil resulted from the process of soil formation, where sulfuric acid is either going to be produced, has already been produced, or has been produced in a quantity that has a lasting effect on the main soil characteristics and potential land uses.

There are about 1.2 million hectares of acid sulfate soils in Thailand. Without proper management, they are generally unproductive. Their low productivity is caused by the number of the soil limitations, namely extreme acidity, aluminum, and iron toxicity, decreased availability of phosphorus and other nutrients, and low content of basic plant nutrients. Therefore, several amelioration measures for agricultural uses, especially for paddy rice, have been conducted by various government agencies like the Land Development Department, Kasetsart University, as well as the Royal Initiative Centres. At present, the outcomes of these researches have been transferred to farmers' field for several years and have been proved successful to some extent. This paper aims to highlight some practical amelioration measures for managing acid sulfate soil for paddy rice and other crops in Thailand currently used and future plans.

2 CHARACTERISTICS, OCCURRENCE, AND UTILIZATION IN THAILAND

Acid sulfate soils in Thailand can be grouped into two types, namely potential acid sulfate soil, and actual acid sulfate soil. Brief descriptions of each type are presented below.

2.1 Potential acid sulfate soil

These soils mostly occur on the active tidal flat where sediments are still under a reduced condition, have a high content of pyrite and low content of lime. Upon drainage or oxidation, they will turn into acid sulfate soils.

2.1.1 Physical characteristics

– Formed in reduced, fine-grained sediments.
– Very poorly drained (permanently water-logging).
– A-C profiles, mostly high in clay content.
– Muddy, n-value> 0.7 $1^>$ (unripe or nearly unripe).
– Many macropores due to root penetration.

$1^>$ n-value = A-0.2 R/L+3H where A = field moisture; R = % silt + sand; L = % clay; H = % O.M

2.1.2 Chemical characteristics

- High salinity (EC values>8 mS/cm)
- Low in Ca, K, Mg but high in Na.
- High in total Sulphur (0.7 to 5%), mostly pyrite (FeS_2).
- High in O.C (2 to 12%) and high C/N ratio.
- High CEC value (20 to 40 cmol kg^{-1}).
- High in total Fe and Al
- Under natural conditions, the soil pH is higher than 7.0, but upon drying, the soil pH is lower than 4.0
  $2FeS_2 + 7O_2 + 2H_2O \Rightarrow 2Fe_{2^+} + 4SO_4^{-2} + 4H^+$
- Smell of H_2S ($SO_4^{-2} + 2H^+$ $2CH_2O \Rightarrow H_2S +$ $2H_2O + 2CO_2$).

2.1.3 Agricultural potentials

- High salinity.
- Adverse soil condition.
- High monetary input for development.
- Fragile ecosystem, so it should remain as a natural resource like a mangrove.
- Possible to utilize for charcoal, timber and aquaculture but with proper management.

2.1.4 Occurrence and land utilization
Potential acid sulfate soils in Thailand mostly occur on the tidal flat where the natural vegetation is mangrove. The total area is approximately 459,000 ha with 24,000 ha, 75,000 ha, and 360,000 ha being in the Central Plain, South East Coast, and Peninsula respectively.

Due to the high price of the shrimp, not less than 30% of the mangrove have been converted to shrimp farming. However, an appreciable amount of these shrimp farms have faced several problems that have brought about declining of the production due to adverse soil and water conditions and diseases, therefore, many farms have stopped the business and left the land abandoned.

Apart from shrimp farming, other land utilizations are wood and charcoal production, and salt farms.

2.2 Actual acid sulfate soil

This soil is mostly found to occur in delta areas such as the Chao Phraya delta and areas of former tidal flats. It has developed as a sequence of drainage and oxidation on the potential acid sulfate soil, either by natural phenomena or human-made. The brief characteristics of the actual acid sulfate soil are presented below.

2.2.1 Physical characteristics

- Mostly formed under brackish water environments.
- A-B-C profile with clayey texture in most case.
- B- horizons often show the straw yellow mottles (jarosite mottles: $KFe_3 (SO_4/2(OH)_6)$).
- Unless artificially drained, they are poorly drained soils.
- The underlying material at about 1 m depth consists of pyritic mud clay.

2.2.2 Chemical characteristics

- Low pH values (4.5).
- Low base saturation (<35%).
- High in total Sulphur (>1%).
- Medium to high organic carbon (2 to 6%).
- Medium to high in CEC values (20-30 cmol kg^{-1}).
- Low in available P but high in available K.
- High in total Fe (>400 ppm) and Al (>4 cmol kg^{-1}).

2.2.3 Agricultural potentials

- Considerably low under natural conditions.
- Range of crops to be grown is limited and yields are low due to severe acidity (pH < 4.5), toxicity of Fe, Al and H_2S, decreasing of P availability and N deficiency.
- Due to extreme acidity, engineering problems commonly occur namely corrosion of metal and concrete, low bearing capacity and uneven subsidence, and blockage of field drains and decrease irrigation rate due to iron oxide deposits.

2.2.4 Occurrence and land utilization
The actual acid sulfate soils mostly occur in the area of deltaic plains such as the Chao Phraya delta in the Central Plain and on the coastal zone of the former tidal flat along the southeast coast and peninsular Thailand. Their aerial extents are about 600,000 ha in the Central Plain, 76,000 ha in the South East Coast and 65,000 ha in the Peninsula, respectively.

Apparently, most parts of the actual acid sulfate soil in Thailand have been utilized for agricultural production from quite a long time ago or at least for more than 4 decades. The basic land use of this soil is cultivation of paddy rice because of its favorable landscape. However, rice yields are relatively variable depending upon soil management and adequate irrigation systems. Other upland crops are also grown on this soil under poldering and raised bedding system for correction of impeded drainage and flooding damage. The types of upland crops that can be grown depend on climatic factors. For example, in the peninsula with a high rainfall, oil palms are generally grown. On the other hand, in the central plain, various kind of fruit trees like mango, jack fruit, santol, guava, and banana are commonly grown. Fast-growing trees like Casuarina, Eucalyptus and Acacia are also generally grown because of their low monetary input.

3 CURRENT AMELIORATION PRACTICES ON ACID SULFATE SOIL FOR CROP PRODUCTION

Most current amelioration measures of acid sulfate soils for crop production in Thailand are focusing on actual acid sulfate soils. Amelioration methods are variable depending upon the types of crops to be grown. Among various crops, paddy rice seems to be most appropriate because it needs low monetary input in comparison to other crops and the topographic condition of the soils are preferable to rice cultivation. To grow other upland crops, the landscape should be modified to protect against flooding and to correct for the impeded drainage. Thus, it needs more investment. Of course, it would be much more sustainable if the areas are irrigated and the groundwater table is controlled in order to protect against the oxidation of pyrite in the mud clay layer.

The followings are some current amelioration measures implemented based on types of crops in Thailand.

3.1 *Amelioration measurement for wetland rice under irrigated area and rainfed area*

3.1.1 *Irrigated area*
From a number of the experiments, it has been confirmed that establishing irrigation systems is required for the sustainable use of the actual acid sulfate soil for rice growing. This is because the irrigation system plays an important role in leaching, control of groundwater table, and water consumption of paddy rice. In addition, the following practices are the overall amelioration measurements of the acid sulfate soil for growing paddy rice in Thailand.

a. Leaching
    Leaching of soils induces an increase of the pH, lowering the content of the sulfur compound, soluble iron, and aluminum. In the Central Plain, leaching is introduced in accordance with the agronomic method, as so-called germinated sowing rice method. During one rice-growing circle, leaching is done about 2-4 times. However, it should be noted here that leaching should be made together with groundwater table control.

b. Control of groundwater table
    In order to protect against the oxidation of pyrite in the deeper subsoil, groundwater table should be kept just above the pyritic mud clay layer. In other words, the groundwater table should be kept at a depth of about 1 m below the soil surface. This can be done by several open ditches within each irrigation system. The advantage of the system is to protect against the oxidation of pyrite in the soil. In turn, sulfuric acid will be

less produced, and the residual effect of liming materials can last longer.

c. Liming.
    Liming is the simplest way of reducing the acidity of the acid sulfate soil. It should be more effective if used together with leaching and control of the groundwater level. In Thailand, liming materials commonly used are marl, lime dust, and ground dolomite. Marl is used in the Central Plain where there is a large amount of marl underneath the surface soil in the area of karst topography in the Central Highland, around 150 km from Bangkok. The cost of ground marl including transportation is about 30 – 40 US dollar per ton. The rate of application is about 6-20 ton/ha. However, to be certain, the lime requirement analysis should be made. Under the system of proper water control and leaching, the residual effect of liming is around 6-12 years for each lime application.In the south or in the peninsula, limestone dust is commonly used because it is easy to find in the limestone quarries scattering everywhere. Lime dust is the waste product from the quarries and it is sold at a low price of approximately 20 – 25 US dollars per ton. Some factories also produce ground dolomite sold as liming material for crop production and for aqua-culture ponds. However, the price of this is about 2-3 times higher than that of marl or lime dust.

The equation of $CaCO_3$ to neutralize the acidity can be expressed as follows:

$$CaCO_3 + H_2SO_4 \Rightarrow CaSO_4 + CO_2 + H_2O$$

d. Application of NPK fertilizer, especially phosphate fertilizer
    Using chemical fertilizers to maintain the soil productivity for rice production is a common sense thing to do. In general, the rate of fertilizer applications is as follows:
    – In the Central Plain: basal application with 16-20-0, 150-180 kg/ha and top dressing with urea 30-60 kg/ha at the PI stage.
    – In the Peninsula: basal application with 16-16-8, 180-200 kg/ha and top dressing with urea 30-60 kg/ha at the PI stage.

It is to be noted here that the soil in the deltaic plain of the Central commonly contains high amounts of available potassium (generally more than 80 ppm). Therefore, potassium application is not needed for paddy rice growing.

e. Use of acid tolerant varieties
In Thailand, a number of the experiments have indicated that some rice varieties are acid-tolerant, especially for the local varieties. The rice varieties can reduce the cost of production and thus, make more profit. Luckily, one of the popular rice variety that is acid-tolerant is called

jasmine rice (Khao Hom Dawk Mali 105) but the yield is commonly around 2-3 ton/ha. Others include the hybrid ones that are not sensitive to light, namely the RD41, RD47, and RD49.

3.1.2 *Rainfed area*

For the areas of acid sulfate soil that has no irrigation systems, the general measurements include liming, application of NPK fertilizer, and use of the acid tolerant rice varieties. The detail description of each measurement is similar to what has been mentioned aforesaid. However, it is interesting to note that the residual effect of lime is only around 2 – 4 years. The farmers will apply it again if they find that the rice production decreases or if the rice leaves turn yellow or yellowish brown in most tillers. The amount to be applied depends on their budget. However, they commonly receive liming material from the Land Development Department free of charge, otherwise, they have to buy it themselves.

3.2 *Amelioration measurement for upland crops*

Since the acid sulfate soils in Thailand occur in the low-lying terrain flooding, water-logging, and high level of the groundwater table, they cause serious problems for growing various upland crops. Therefore, it is imperative that water control measurements need to be constructed in order to avoid those problems. In Thailand, the most simple way and low monetary input is to reshape the landscape into raised bedding under the poldering system. Raised bedding consists of dike and ditch. The width of the dike and ditch depends upon kind of crops or fruit trees that are to be grown. Commonly, the dike (raised bedding) is 6 to 8 m wide, and the height of the bedding is 1 – 1.5 m above the original ground surface. Apart from the bedding, the ditch is dug as deep as 1-1.5 m, and the width of each ditch is around 1.5 – 2 m. The soils of the ditch will be used for making raised bedding or the dike. However, it is important that the subsoil that is extremely acidic to not lay on the top of the original dark surface soil where it is less acidic and high in organic matter. To do this, the dark surface soil should be removed and piled up in the middle line of the bedding. Afterward, the subsoil from the ditch will be laid along the side of the bedding (see Figure 1). Machines like back hoe can be used to make this kind of raised bedding easily, but the cost of construction will be 30-50% higher than the one that does not remove the surface soil. However, the cost of lime application, especially for the fruit trees, can be reduced because it needs to be applied only on the middle row where the trees are planted. In the planting hole with the size of 50x50x50 cm, mix the soil before planting with compost, lime and chemical fertilizer (grade 16-16-16) at the rate of 1 kg, 2 kg,

Conventional method   Mango

Watering by water pump on the boat

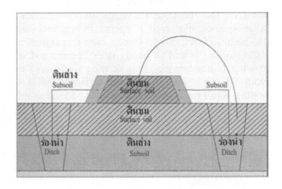

Figure 1. How to make the raised bedding for acid sulfate soil.

and 30 gm, respectively. Afterward, plant the tree together with the mixing soil material as aforesaid.

In addition, in order to protect flooding the dike around the farm, a poldering system is needed together with the installment of a water pump. It is also necessary that the water in each ditch be below the original soil surface (about 40-60 cm) to preventing oxidation of the pyrite in the soil and to enable watering the trees. In Thailand, a method of irrigation is using a boat with a pumping engine floating on the ditch. The pumping engine is installed with the pipes to spray water to the trees on both sides. However, instead of using the boat for irrigation, sprinkling systems can be also installed for each tree. When this system is available, fertilizer application can be also made by means of fertigation.

In the project area of Chaipattana Foundation under the Royal Initiatives, a number of fruit trees such as mango, guava, jack fruit, rose apple, santol, sapodilla, marian plum, plumage, dragon fruit, banana, and rough giant bamboo are successfully grown.

Various kinds of vegetables can also be grown very well on these raised bedding soils such as long bean, sweet corn, chilies, eggplant, kangkong, cabbage, kale, Jerusalem artichoke, lemongrass, basil, squash, and aloe vera.

3.3 Growing of fast-growing trees and certain plants that are acid tolerant

Another alternative of utilization of lands with acid sulfate soils that needs a low monetary input is planting several fast-growing trees and some plants. For instance, the fast-growing trees include Eucalyptus, Casuarina spp, Acacia spp, Eugenia spp, neem tree, and Melaleuca spp. However, the area should have some drainage system like the open ditch, raise bedding and dike around the planting area. The system of water control does not need to be very precise and intensive for reducing the costs of construction. Other species like reed (Cyperus imbricatus) can also be introduced. To grow reed, land preparation is made similar to that of paddy rice. The method of growing is transplanting and keeping water only at about 30 cm above the ground surface. Within 6 months, those reeds can be harvested and used as the raw material for making mats and other kinds of handicraft such as handbags, tissue boxes, sandals, letter boxes, etc.

3.4 Amelioration measurement for aqua-culture production

Since the topographic position and high water table of the acid sulfate soils are favorable for aqua-culture like fish raising, many parts of them have been exclusively used for such purpose. However, the concerning problem is the high acidity in which fish cannot survive. To reduce the problem, the excavation of fishing ponds is made as deep as 1-1.5 m below the original surface. The excavated soil will be used for making dikes around the ponds. Generally, the size of a pond is around 1-2 hectares for reducing the cost of construction. Afterward, lime needs to be applied as soon as possible at about 12-20 ton/ha at the bottom floor of the pond, on the bank of the pond, and dikes around the pond. To protect against soil acidity building up, ponds should be filled with water from the irrigation system as quick as possible. After one or two weeks, if the pH testing of the water is around 6.5-8.0, the ponds should be suitable for fish raising. The types of fish generally raised are the common catfish, tilapia, Thai barb, common carp, and striped catfish.

4 FUTURE PLAN FOR MANAGING ACID SULFATE SOIL IN THAILAND

4.1 Provision of water supply for leaching, water control, and irrigation

Luckily, in the southern part of the Chaopraya deltaic plain in Central Thailand where most the actual acid sulfate soils are found, major irrigation systems have been developed. The systems include open irrigation canals, water regulators, and dams. Yet, more measurements are still needed to distribute the irrigation water for the whole region where there are acid sulfate soils. If the systems work out completely, those acid sulfate soils will become one of the most important rice bowls of the country. Moreover, crop diversification can be made as an alternative agricultural production of the region other than rice production.

4.2 An alternative cropping system should be established for the small land holding farmers

Sometimes, when mono-cropping systems such as paddy rice production may not be appropriate for small holding farmers, the multiple cropping system can be recommended. The practical model of this multiple cropping is to follow His Majesty the Late King Rama the Ninth's concept of Self Sufficient Economy. This royal initiative model is called "the New Theory" of farming.

In the overall practice, The New Theory is to manage the farmland of each farmer to achieve self-sufficiency. In short, it will start with the subdivision of each farmland into 4 parts. The first part, about 30 percent of the farm, is for digging a pond to store water for watering crops or for fish raising. The second part, also about 30 percent of the land, is used for rice cultivation. Hopefully, the farmer can obtain enough paddy rice for his own consumption and the surplus of rice can be sold. The third part, about 30 percent, is utilized for growing fruit trees and annual crops on the raised bedding system. The production can be used for the farmer's own needs and for selling. The fourth part, about 10 percent is for the living quarter and also for gardening as well as livestock raising (see Figure 2). However, those proportional figures are not always fixed at 30, 30, 30, and 10 percent. They can be modified in accordance with the existing water supply and the needs of the farmer. However, the area for rice growing is a must and should provide enough supply for the consumption of the whole family. The introduction of other crops is the farmer's decision. Once the production is beyond need, it can be undertaken for commercial purposes.

As a final remark, one should keep in mind that in order to introduce the "New Theory" farming system

30% of the farm area used for paddy rice

30% of the farm area used for farm pond

30% of the farm area used for orchard farming on raised-bed

10% of the farm for home, gardening, and livestock

Figure 2. Farm layout according to the concept of "new theory farming".

to the acid sulfate soil areas, the soil needs to be improved before making any cultivation.

5 RESEARCH NEEDED

Although the amelioration measurements of the acid sulfate soils in Thailand have been implemented successfully, for the time being, additional works are still needed as the following topics:

1. The physicochemical and biological characterizations of pyrite oxidation and toxic substances like Fe, Al and H_2S need to be studied more.
2. The physicochemical aspects of acidic tolerant plants (why they are tolerant to extremely acidified condition) and how to use them for reducing soil acidity.
3. Soil and water management at a low cost and practical for common farmers. Special attention should be given to:
   – Liming and leaching method
   – Optimizing the use of fertilizer and appropriate techniques for application

   – Use of biological techniques to improve crop production
   – Cropping pattern for best profit
   – How to make economical water control and discarding of the acid water

4. Varietal screening focusing on:
   – Paddy rice
   – Dry land crops after rice harvesting
   – Fruit trees

5. Development of computer crop models for proper use of the acid sulfate soil

6 CONCLUSIONS

– Acid sulfate soils in Thailand have been developed as a result of the drainage or oxidation of soils that are rich in pyrite.
– Soils in the tidal flat or estuary in the coastal zone of Thailand which are high in pyrite are recognized as potential acid sulfate soil. Their natural

state is under mangroves. They should not be reclaimed for agriculture, and their use should be restricted to mangrove conservation.

- On landward (middle part of the deltas), actual acid sulfate soils have developed. Agronomic problems include extreme acidity, Al, Fe and H_2S toxicity, nutrient deficiency, poor drainage, and subject to prolonged flooding.
- Water management is the key factor in the management of acid sulfate soil for agriculture (irrigation, leaching, water control, and drainage).
- The cultivation of paddy rice is the most appropriate option. Growing upland crops and fruit trees needs raised-bedding, polder system, flood-protection, and irrigation.
- Essential soil ameliorations include liming, leaching of acid water, groundwater control, fertilizer application, and appropriate plant varieties.
- Basic and applied researches in various aspects of acid sulfate soil are still needed for the future practical management.

REFERENCES

Dent, D. 1986. Acid sulphate soils. A baseline for research and development. ILRI Publication 39. International Institue or Land Reclamation and Improvement, Wageningen, The Netherlands.

Dost, H and N. Van Breemen 1982. Proceedings of Bangkok symposium an acid sulphate soils. Second International symposium an acid sulphate soils, Bangkok, Thailand. Jan 18 – 24, 198.

Kyuma K, P. Vijarnsorn and A. Zakaria 1992. Coastal lowland ecosystems in Southern Thailand and Malasia. Showado-printing Co., Sakyoku, Kyoto.

Office of the Royal Development Projects Boards 2008. From Royal Initiatives, 60 years of the people happiness under the Royal Aegis. Aroonkarnprim Ltd. Part. Bangkok.

Pons L, J. 1969. Acid sulfate soil in Thailand, studies on the morphology, genesis and agricultural potential of soils with catclay. Soil Survey Division, Bangkok.

Vijarnsorn P and H. Eswaran 2002 The Soil Resource of Thailand. First press Ltd. Part. Bangkok.

Vijarnsorn P, et al. 2010 Handbook of acid sulfate soil improvement for agriculture (In Thai only), Pikurn Thong Royal Development Study Centre Narathiwat, Thailand.

Tropical Wetlands — Innovation in Mapping and Management — Sulaeman et al. (Eds)
© 2020 Taylor & Francis Group, London, ISBN 978-0-367-20964-3

Management package of *Tri-Kelola* plus for increasing production and productivity in wetland development for agriculture

D. Nursyamsi & Y. Sulaeman
Indonesian Central for Agriculture Land Resources Research and Development (ICARRD), Jakarta, Indonesia

M. Noor
Indonesian Swampland Agriculture Research Institute (ISARI), Indonesia

ABSTRACT: Approximately 90% of wetland has a cropping index of 100 (IP 100), while in others crops are no longer planted anymore because their productivity has dropped due to changes in the properties of the soil, its acidity level, Fe poisoning, as well as macro and micro nutrients deficiency. However, it is expected to be a food supplier or substitute for arable land. In the next ten years Indonesia will become one of the biggest importers of rice, if there is no breakthrough in production. Therefore, an increase in production is essential because with the right strategy, nawacita Indonesia will become a world food barn by 2045. At a population growth rate of around 3 million per year and a paddy fields conversion rate of 50-100 thousand hectares per year. On the other hand, the increasing and widespread occurrence of floods, droughts, landslides, tsunamis and others natural disasters as a result of climate change have added to the complexity rice production. Therefore, as a result of these, the government has initiated the Serasi Program, which is the development of 1 million hectares of swamp land in six provinces namely South Sumatra, South Kalimantan, Lampung, Jambi, Central Kalimantan and South Sulawesi in order to increase Indonesia's future food production. The key to developing wetland in the future is appropriate technological innovation. The Indonesian Agency for Agricultural Research and Development (*Badan Litbang Pertanian*) through the Central for Agriculture Land Resources Research and Development (ICARRD) initiated "Trikelola Plus" to support the Serasi Program. The implementation of water management is determined by the type of overflow, while land arrangement depends on the type of commodity, swamplands, tides, and the implementation of balanced fertilization. Furthermore, it is necessary to pay attention to the properties of the soil and its level of fertility along with the obstacles faced. The application of Trikelola Plus innovation, which comprises of water and land management, long with balanced fertilization and the use of various bio pesticides to support the *Serasi* Program, increases cropping index and productivity.

1 INTRODUCTION

Indonesia was the first country to become self-food sufficient in production through the application of 5 (five) new technologies called "Five Enterprises" in 1984. This process is also known as "Panca Usaha", and it comprises of (1) adequate and timely water supply, (2) the use of high yielding variety (HYV) superior seeds, (3) balanced fertilization, (4) integrated pest control (IPC), and (5) improvement in cultivation technology. It is also equipped with an institutional building called *Bimas* (Mass Guidance). The success of self-sufficiency should be utilized in the development of swamps (Noor, 1996: Subagio et al., 2016).

For 25 years (1970-1995) the opening wetland also known as swampland reached almost 1 million hectares. Out of the 900 thousand hectares opened, 715 thousand were on tidal swamp land while the remaining 185 thousand were on swamp land. The development of swamp land expansion for agriculture and transmigration succeeded in placing the population of 1,717,610 households (KK) to a target of 2 million households. However, before now almost 90% of the above areas still have cropping intakes once a year (IP 100), while in some others crops are no longer planted due to a drop in their productivity, changes in the physical, chemical, and soil fertility of cultivated land, including acid, Fe poisoning, and low nutrient status (BALITTRA, 2011; Noor, 2014).

Meanwhile, Indonesia is still at a population growth rate of around 3 million per year, while the rate of conversion of paddy fields is 50-100 thousand hectares per year. On the other hand, the inadequate increase in paddy production has added to the increase and widespread expansion of floods, droughts, landslides, tsunamis and others as a result of climate change which has disrupted rice production nationally. In the next ten years, the country is feared to be the biggest importer of rice, if no significant intensification is conducted. The government views food, especially rice as a strategic commodity to note the availability, distribution and affordability of the community. It plans to build its rice production so that by 2045 Indonesia will become the world's food barn (Ministry of Agriculture, 2017; 2018).

Wetland (read: as swampland) is a potential, strategic, and perspective land resource with an area of 34.70 million hectares, of which 19.99 million hectares are suitable for agriculture across 17 provinces and 300 districts/cities (BBSDLP, 2015). The term wetland in the context of this study is limited to only swamp land which is divided between tidal and fresh water back swamps. This paper addresses increase in agricultural production in wet land using the "Trikelola Plus" technology. The characteristics, potential and availability of swamp lands, requires this technique in order to increase its land productivity and rice to support the Serasi Program.

2 CHARACTERISTICS, POTENCY, AND AVAILABILITY OF WETLAND

2.1 *Characteristics of swamp land*

Swamp land, also known as wetlands, is strongly influenced by the water regime due to overflowing tides from river/seas and floods from upstream areas. It comprises of tidal swamp land (rawa pasang surut) and fresh water back swamp (rawa lebak). Wetland (read: as tidal swamp land) is a swamp area affected by the presence of tidal overflows from the river/sea. When it occurs within two weeks, it is called a single tide, and known as a maximum tide when a full moon occurs which called is *pasang purnama* in Indonesian, at the 15th qomariah and a dead moon 1st qomariah which is called a single tide (*spring tide*). It also occurs twice on a daily basis within 24 hours in the form of small pairs called double or minimum tide also known as pasang ganda (*neap tide*) (Noor, 2014; BALITTRA, 2014a). Based on hydro-topographic overflows, the high-low tide swamps is divided into four types, namely types A, B, C and D (Figure 1).

Type A, is an area that is absorbed by single (spring tide) or double tide (neap tide). This region also experiences daily drainage covering coastal areas along the banks of large rivers.

Type B, is an area covered by single tide (spring tide), and not affected by the neap tide. This area also experiences daily drainage and cover the area at a distance of 50-100 km from the large river

Type C, which is located in an area not completely flooded either by spring or small/neap tide, influences tides through seepage, and has a ground water level of <50 cm. It covers the widest and relatively higher topography.

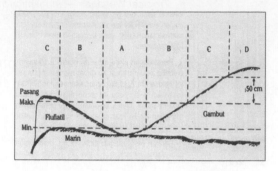

Figure 1. Distribution of tidal swamp land based on its hydrotopography into tidal types A, B, C and D.

Type D, is an area not completely occupied by spring, small/neap tides, but has a groundwater level of >50 cm. Peat domes belong to this type of group.

The criteria or limitation of the distribution of tidal types A, B, C, and D, only apply during the rainy season, in the dry season type B partially turns into C. A part of the fork system for the South Kalimantan has been set up in some tidal swamps such as the Central Kalimantan and the comb system for Sumatra and West Kalimantan. In type B, the areas that experience acute drainage due to the construction of reclamation networks in the form of dense channels changes to type C. Furthermore, the type C area changes to type B with the construction of irrigation channels (handil/ray) to facilitate the flow of water into the rice fields (Noor, 1996; Noor, 2014; BALITTRA, 2014a).

The fresh water back swamp also known as *lahan rawa lebak* is a region which is affected by a high inundation of at least 50 cm and a long inundation time of at least 3 months (BALITTRA, 2014b). In this area there is no more influence of the tide movement, though it is still large. This area covers a portion of the affected region by river overflowing, though the tidal strength has been reduced. Based on the shape or altitude of the inundation hydrotopography and its duration, the swamp area is divided into 4 types of fresh water (lebak), which are shallow, middle, deep and very deep (Figure 2).

Lebak dangkal also known as shallow of fresh water back swamp, is a swamp area with a height of <50 cm and inundation length of <3 months a year. This shallow of fresh water back swamps is often called *lebak pematang*.

Lebak tengahan which called is the middle of fresh water back swamps, is an area with a height of 50-100 cm and inundation length of 3-6 months yearly.

Lebak dalam also known as deep fresh water back swamps is a swamp area with a height of 100-200 cm and inundation length of >6 months a year

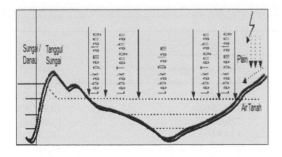

Figure 2. Distribution of fresh water back swamps (rawa lebak) based on the height and duration of inundation to be shallow, middle, deep and very deep.

Lebak sangat dalam also known as very deep of fresh water back swamp is a swollen area that is high >200 cm with an inundation length of one year. This very deep of fresh water back swamps is also called *lebung*.

Some of the fresh water back swamps (rawa lebak) already have reclamation networks and built in polder systems in Hulu Sungai Utara Regency, with an Alabio Polder System covering 6,000 hectares built (Idak, 1982). Mini polders (100-200 shaking) have also been built in Ogan Ilir, South Sumatra in Pamulutan Village (200 ha), Seri Banding (177 ha), Kalang Pangaren (213 ha), Olak Kembang Satu (66 ha), and Arisan Jaya (85 ha). In water management axial pumps (50 HP) is used with the ability to pump water around 2500 m3/hour or 700 liters/second. Rice yields obtained over a long area 3-4 years (opened in year 2015) reached 6 t milled dry grain (MDG)/ha, while new openings areas in year 2016 produced 4 t MDG/ha (Noor, 2018). Mini polders are also being built in an effort to optimize tidal swamp land such as in Jejangkit Muara Village, Barito Kuala Regency, South Kalimantan, and Telang Rejo Village, Telang Makmur, and Telang Raya Village in Banyuasin Regency, South Sumatra; swamp in Tajau Landung village and villages around Banteng, Banjar Regency, South Kalimantan.

From the characteristics of the land above, the problems faced in developing agriculture in tidal swamp land are (1) inability to optimally irrigate rice fields, especially during dry season; (2) acute acidity (pH 2-3), generally during the dry season; (3) poisoning of toxic elements in plants such as Fe^{3+} (iron), Al (aluminum), H_2S (sulfide), and Na (sodium) in the event of salt water intrusion; (4) low nutrient values of K, P, Ca, Mg, and some deficient micro nutrients such as Cu and Zn, especially in peat land. Problems faced in swamp land include (1) waterlogging which is high during rainy season, thereby making the arrival of water difficult to predict; (2) slow drainage channel making it difficult to accommodate rising shipment water so that the inundation is relatively long during the rainy season; (3) sluice gates are not provided on the channels produced (BALITTRA, 2014a,2014b, 2014c).

2.2 Area and distribution of swamp land

The total wetland (read: as swampland) area in Indonesia is 34.10 million hectares, and consists of 8.919 million hectares of tidal swamp land and 25.21 million hectares of fresh water back swamp area (Table 1). The largest distribution is in three major islands, namely Sumatra (12.93 million ha), Kalimantan (10.02 million ha), and Papua (9.87 million ha) and a small portion is in Sulawesi (1.048 million ha), Maluku (162 thousand ha) and Java (94 thousand ha).

On the island of Borneo, it is located along the southern and western coasts of Kalimantan including the provinces of South, Central, East, and North Kalimantan. On the island of Sumatra, it occupies the eastern coast in the provinces of Aceh Darussalam, North, West and South Sumatra, Riau, Jambi, Bengkulu and Lampung. On the island of Papua, they are located along the south and north coast occupying Papua. A small portion of Sulawesi Island covers the south and west coasts occupying the provinces of South and West Sulawesi (Figure 3). It occupies an area of > 0.4 million hectares in 14 provinces or around 300 regencies and cities. Most widely in seven provinces, namely South Sumatra, Central Kalimantan, West Kalimantan, South Kalimantan, Riau, Lampung, Jambi, Bengkulu Bengkulu and West Papua (Table 2).

2.3 Suitability and availability of swamp land

The survey results and land suitability analysis show that swamps have good potential and suitable for the development of rice, horticulture, and annual crops/plantations. Approximately 19.19 million hectares (56%) and 34.01 million hectares are suitable (Table 3). The potential of swamps is suitable for the most extensive rice plants, which is equal to 14.18 million hectares, double the size of the existing national rice fields, followed by horticulture which covers about 3.14 million hectares, and finally for annual crops covering 1.86 million hectares. However, of the 19.19 million hectares of potential swamps for agriculture mentioned above, only 7.5 million are available, with 5.12 million used for padding, 1.47 million for horticulture, and annual crops 0.93 million hectares (Table 4).

From the 1970s to 1995, the development of swamps focused more on opening or reclamation with various forms or reclamation on network systems in South, Central and West Kalimantan and comb systems in most of Sumatra and West Kalimantan. After 1995, the development of swamps turned to its rehabilitation and optimization already available (Noor, 2014). In 2016 the government targeted a program to optimize swamp land of 1 million hectares and in 2019 through the improvement and development of water management and rice field network infrastructure, six provinces were covered, namely South Kalimantan, South Sumatra, Lampung, Jambi, Central Kalimantan and South

Table 1. Extent and distribution of swamps in Indonesia.

Island	Swamp Land (hectare)		Total (ha)
	Tidal swamp	Fresh water back swamp	
Sumatra	3,019,354	9,907,481	12,926,835
Kalimantan	2,986,438	7,037,062	10,023,500
Sulawesi	318,030	730,064	1,048,094
Maluku	74,395	88,159	162,554
Java	94,756	0	94,756
Papua	2,426,376	7,443,070	9,869,446
Indonesia	8.919.349	25,205,836	34,125,185

Source: BBSDLP (2015)

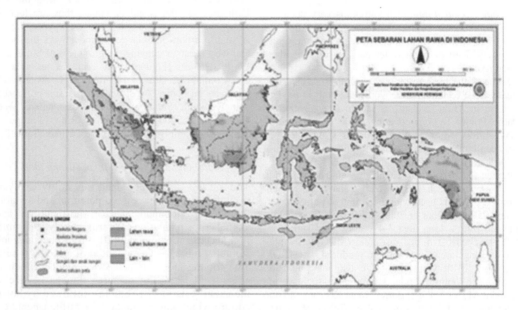

Figure 3. Distribution of swamps in Indonesia (BBSDLP, 2011).

Sulawesi. That same years, the government launched the Serasi Program for 1 million hectares of swamps in the above six provinces with demonstration plots (demfarm) in Tajau Landung Village and around Banteng, Banjar Regency, South Kalimantan and Telang Rejo Village, Makmur, and Telang Raya, Banyuasin Regency, Sumatra The South is implementing mini polder and pumping systems.

3 STRATEGY FOR AGRICULTURAL DEVELOPMENT IN WETLAND

The paradigm of swamp land development has changed since 1995, after cases of large fires, especially the 2015 outbreak that burned 2 million hectares of land, including peat lands with the increasing climate change and global warming (Noor, 2010). Furthermore, Therefore, Indonesia is required to implement its intervention in an effort to reduce GHG emissions as agreed upon or planned at a meeting of climate change expert organizations (COP). The old paradigm of swamp land development is focused on "how to drain? " swamp land for it to become a cultivation area for agricultural crops. Therefore, most reclaimed tidal swamps then experience over-drainage (Noor, 1996; 2016).

In preventing excessive drainage, sluice gates or the dam overflow (tabat) are needed to hold water remain on the channel and land. Under inadequate water management, swamp land is only planted in the rainy season once a year (IP 100). Without any form of blockage, the incoming water will come out and become runoff. In other places, some tidal swamps do not have a complete water supply network and are not equipped with sluice gates

Table 2. Area and distribution of swamps based on provinces in Indonesia.

Provincis	Tidal Swamps		Fresh Water Back Swamps		Total (Mllion Hectare)
	Mineral	Peat	Mineral	Peat	
1. D.I. Aceh	0.128	-	0.468	0.215	**0.812**
2. Sumut	0.309	0.084	0.553	0.180	**1.127**
3. Sumbar	0.077	0.002	0.300	0.098	**0.477**
4. Riau	0.768	-	0.556	3.850	**4.894**
5. Kepri	0.057	0.000	0.012	0.008	0.077
6. Jambi	0.267	0.415	0.331	0.225	**1.239**
7. Bengkulu	0.029	0.000	0.084	0.008	0.121
8. Sumsel	0.855	-	1.246	1.257	**3.358**
9. Babel	0.138	0.012	0.119	0.046	0.300
10. Lampung	0.153	0.003	0.317	0.046	**0.519**
Sumatera	2.501	0.517	3.988	5.919	12.927
11. Kalbar	0.664	0.450	0.755	1.215	**3.084**
12. Kalteng	0.333	0.139	1.117	2.537	**4.126**
13. Kalsel	0.258	-	0.550	0.107	**1.896**
14. Kalitim	1.045	0.095	0.522	0.233	**1.896**
Kalimantan	2.301	2.986	2.944	7.037	10.023
15. Sulut	0.019	-	0.014	-	0.033
16. Gorontalo	0.016	-	0.011	-	0.028
17. Sulteng	0.047	-	0.013	0.010	0.187
18. Sulbar	0.020	-	0.095	0.013	0.128
19. Sulsel	0.118	-	0.296	-	**0.414**
20. Sultra	0.098	-	0.159	-	0.257
Sulawesi	0.318	-	0.706	0.730	1.048
21.Papua Barat	0.910	0.164	0.112	0.021	**1.207**
22. Papua	1.355	0.007	3.803	3.505	**8.662**
Papua	2.265	0.164	3.916	3.527	9.869
23. Maluku Utara	0.009	-	-	-	0.009
24. Maluku	0.066	-	0.088	-	0.154
Maluku	0.074	-	0.881	-	0.162
25. Jabar & DKI	0.017	-			0.017
26. Jateng & DIY	0.018	-			0.018
27. Jatim	0.060	-			0.060
Jawa	0.947	-			0.947
Indoensia	**7.552**	**1.366**	**11.643**	**13.563**	**34.125**

Source: BBBSDLP (2015)

(*flapgates*). Especially in basin areas during the rainy season, water is confined thereby creating an opportunity for waterlogging to occur because it cannot be thrown out. Some farmers use pumps to dispose water in their farms, however, this is an ineffective and inefficient technique because water management is partially conducted, requiring cross-farms in a management area.

The influence of excessive drainage in the farming field leads to acidification, increase in the acidic (pH 3-4) content of soil and water, iron poisoning, and drop in ground water level (subsidence). In areas filled with water, organic matter is accumulated and composted, resulting in iron fouling and poisoning in stagnant conditions. Soil cultivation is usually followed by a decrease in land productivity and a decrease in the yield of cultivated rice and even crop failure due to acute poisoning. Over time various results of research, experience strategic issues both outside and within the country. However, Indonesia has developed suitable ways of managing swamps so that it is sustainable or in other words

Table 3. Potential area of swamps suitable for agricultural development.

| The Island | The Potential of Swampland for Agricultural (hectare) | | | |
	Paddy Rice	Horticulture crops	Annual crops	Total (Million Hectare)
Sumatra	6,851,446	1,488,656	834,163	9,174,265
Java	94,756	0	0	94,756
Kalimantan	3,268,706	900,639	800,497	4,969,842
Sulawesi	681,911	0	23,429	705,340
Maluku	100,336	0	0	100,336
Papua	3,188,403	753,720	204,194	4,146,317
Indonesia	14,185,558	3,143,015	1,862,283	19,190,856

Source: BBSDLP (2015)

Table 4. Area of swamps available based on commodities.

| The Island | Availability of Swampland (hectare) | | | |
	Paddy Rice	Horticulture crops	Annual crops	Total (Million Hectare)
Sumatra	1,655,079	342,800	266,083	2,263,962
Java	0	0	0	0
Kalimantan	848,668	529,758	475,483	1,853,909
Sulawesi	62,370	0	0	62,370
Maluku	83,006	0	0	83,006
Papua	2,468,991	599,378	187,170	3,255,539
Indonesia	5,118,114	1,471,936	928,736	7,518,786

Source: BBSDLP (2015)

"how to irrigate water conservation" (Noor, 2014; Suwanda and Noor, 2014).

Indonesia targets a national GHG emission reduction of 26% or 0.008 Giga (Gt) t CO_2 or 40% equivalent to 0.011 Gt t CO_2 eq. with bilateral cooperation, between 9.5-13.0% comes from peat land which incidentally is in swamp until 2020 (Noor, 2016). President Jokowi during the COP meeting in Bali a number of years ago suspended the technique until 2030 with a decline of 30%. The national action plan for reducing GHG emissions (RAN GRK) above has been prepared in Presidential Regulation No. 61 of 2011 concerning the reduction of National GHG emissions by 355 million t CO_2 (Noor, 2016).

In the above plan, several strategies will be taken, including optimization of swamp land, application of low emissions cultivation technology, development of abandoned plantation area, degraded (APL), utilization of municipal and agricultural waste for biogas, peat land management for sustainable agriculture,, livestock and horticulture sectors (Noor, 2016).

The following description will reveal the "Trillion-Plus" Technology in the management of swamps to increase food production which comprises of water management technology, land and balanced fertilization based on the nature and constraints faced.

4 TECHNOLOGY OF TRIKELOLA PLUS FOR RICE PRODUCTION IN WETLAND

4.1 Pengelolaan air

4.1.1 Handil system
Swamp farming was originally community agriculture with a family farming system approach in the form of extensive agriculture. In groups farmers open swamp land by making simple channels with 0.5-1.0 m and 2-3 meters wide along 2-3 km from the edge of large rivers called handil or ditches (Idak, 1982). The *handil* or ditch canal functions as an irrigation canal to push water into the business area as well as a drainage channel to remove it in the event of inundation (Noor, 2001; 2004; 2012).

Hundreds of handil were built by the community both in Kalimantan, Sumatra and several other places. Along with the development of society and a farming business system in swamps from polyculture which later turned into monoculture with food crops or rice as the main commodity. This is in accordance with the nature and condition of the land and swamp environment which is strongly influenced by the water regime. Most (> 90%) swamps are planted with local and superior rice (HYV) (Noo, 1996).

The government was impressed with the techniques adopted by the local community in clearing swamps which became the inspiration for wider swamp land clearing by building water management networks in Kalimantan, Sumatra and other places using the fork and comb systems with a target of 5.25 million hectare. This effort was considered quite successful in opening isolated swampy regions to agricultural land even though only 1.24 million hectares of the target was opened (Coller, 1982). In the same year, other swamps were also independently opened by the community around 2.36 million hectares (Manwan, 1992).

4.1.2 *One-way flow system*
One-way flow system is a water management technique designed in such a way that water moves in and out of paddy fields through different channels, especially in tidal swamp areas which are directly installed, in types Λ and B. This is however, the opposite of handil where water enters and exits through the same channel. The purpose of this one-way system is to create water conditions in the rice fields in accordance with its requirements for plants. It also creates different water level from the two tertiary or secondary channels to enable the continuous flow of groundwater through the subsurface to the drainage channel to dispose the toxic substances for plants.

Water management of one-way flow system is made possible by the installation of a swing gate ("flag-gate") on each tertiary or secondary channel estuary which is for proper irrigation flow and removal of water from the drainage canal. The flap-gate at the mouth of the irrigation canal is made open for fresh water thereby making it easy for the irrigation canal to close automatically when the water recedes in order to have it retained. The flap-gate at the mouth of the drainage channel opens to remove dirty water containing toxic substances especially at low tide and closes inside to hold the water entering into the drainage canal. Furthermore, water from irrigation channels through quarter channels is included in the plot of land to be circulated in one direction as well and released into the drainage channel, to enable the installation of several functions as height regulators and low water levels in the plot of lands (Widjaja Adhi et al., 1992; Darmanto, 2000).

The application of one-way flow system significantly increases land productivity compared to two-way flow systems. The results showed a one-way system increases rice yield by 50% compared to two directions. Furthermore, rice yields on land managed with a one-way water system combined with other technologies such as lubrication, drainage, liming, fertilization, and restriction of flow from acidic areas with interceptor channels reaching 4.0-4.6 t MPD/ha compared to the system two-way water system which reaches 2.5-3.0 t MPD/ha (Noor and Saragih, 1992; Noor, 2014). Other research shows what one-way water management in tidal swamplands in South Sumatra produces (Harsono, 2001).

4.1.3 *Cascade dam overflow system*
Cascade Dam Overflow System also known as Sistem *Tabat Bertingkat* is a water management technique designed to increase and maintain water levels in drains and groundwater in field-farms with installation and runoff or multilevel tabs on handil channels. Unlike the tidal swamp land with the tide type of A and B on land and type of C and D in rain water during the rainy season. With the installation of multilevel sacks, water comes from tides with rain accommodated in the canal. In this blocking system, tertiary or secondary channels are used as collector and conservation channels, which are equipped with over-flow tabs whose height are adjusted according to the water requirements of the plants. This blocking system reduces the risk of flooding during the rainy season and the risk of drought in the season (Noor, 2007).

4.1.4 *Pump system (mini polder)*
Swampland has strong characteristics and dynamics with the water regime, if the rainy season is flooded, and vice versa in the dry season it becomes drought. The river water overflows in such a way that the swamp area is flooded thereby, making it necessary for water to be removed from the farmland. On the other hand during the dry season there is chaos, but water is in the canal to take or raise it from the river/canal to the field farm which requires power from the pump. However, the pump alone is not enough, because the paddy fields need to be given a dike/shipyard to prevent water from returning to farmland. A polder system is a method of handling or managing water in a farming area (Noor, 2007; BALITTRA, 2014b).

In practice a stretch of rice farm covering an area of between 100-500 hectares is protected with a perimeter dike, inlet and out channels, and dividers. It is equipped with doors and water pumps that work to isolate and remove water from farms into primary/secondary canal/river. This water management model is called the Mini Polder System. Mini polders (100-200 shaking) have been built in Ogan Ilir, South Sumatra in Pamulutan Village (200 ha), Seri Banding (177 ha), Kalang Pangaren 213 ha, Olak Kembang Satu (66 ha), and Arisan Jaya (85 Ha). They are also being built in an effort to optimize tidal swamp land such as in Jejangkit Muara Village, Barito Kuala Regency, South Kalimantan; and Telang Rejo Village, Telang Makmur, and Telang Raya Village in Banyuasin District, South Sumatra; and lebak swamp in Tajau Landung Village and Village around Banteng, Banjar Regency, South Kalimantan (Figure 4).

Figure 4. Pump on the Jejajngkit village in the mini polder system, Barito Kuala Regency, South Kalimantan.

Figure 5. Tidal swamp rice fields (up) and fresh water back swamp land (bottom).

4.2 Land arrangement

One of the advantages of swamp land in becoming a food barn compared to other agro ecosystems is the availability of vast expanses of land and flat topographical conditions, making it easier to open, prepare, manage and develop land in an integrated manner as expected. Other supportive conditions are the social, economic and cultural states of the swamp community.

In addition, residents of swamp areas are built to occupy one location or colony thereby, making it easy to gather, organize and communicate. Farmers who are diligent and tenacious also plant crops or vegetables asides rice. Crops or vegetables are planted in the upper part of the sunken bed, while rice is planted in the bottom bed (Noor, 2004; Nursyamsi et al., 2014). In general, there are two forms of land management systems in swamp land, namely (1) paddy field and (2) surjan systems.

4.2.1 Rice field system

The paddy system is recommended for tidal land types A and B overflow and mid-range. In principle, the plot of rice fields is laid out and equipped with a mobile channel, consisting of an irrigation and drainage quarter, stop log doors for the application of a one-way water system (Figure 5). Adequate perimeter and quarter channels are 1-1.5 m wide and within 1-2 meters. In addition, in a stretch of 1 to 2 hectares, farms and bridges are made to facilitate mobility of tractor and transport of crops.

4.2.2 Surjan system

The surjan system is recommended for tidal land types B and C with overflows, and shallow and middle swamp. In the surjan system, the tabukan/ paddy fields are planted with rice and with crops, vegetables and fruits in the ridge (Figure 6). Its size is 2-3, 1-1.5, and the 14 meter high respectively therefore a 1 hectare farm area comprises of 6 surjan lines with a distance of 14 meters surjan. A total of 500 men is needed to produce it and its installment is supported with 1 x 1 meter lava and 1-1.5 m high, depending on the maximum water level (Noor, 1996; Nursyamsi et al., 2014). Support is gradually expanded as the plants grow. However, it is connected to a surge after the fruit plants/annual plants are 4-5 years old (Rina and Subagio, 2017).

Figure 6. Surjan system in tidal swamp land (up) and fresh water back swamp land (bottom).

4.3 *Fertilizer and fertilization*

Swamp land needs to be mixed because generally the availability of macro nutrients and some micro or low nutrient status is low or less available such as P, K, Cu and Zn. Fertilization is based on nutrient availability or soil nutrient status to avoid waste or overdose. In addition to fertilizers, ameliorants are also needed, by agricultural lime or dolomite, especially in swamps with a soil pH <4.5. Determination of the amount of fertilizer and ameliorant materials uses the DSS compiled by Balittra and accessible on its website or the Balitbangtan (IAARD). The Swamp Soil Test Tool (PUTR) made by the Indonesian Soil Research Institute (ISRI) can also be utilized.

In addition, the three management components above which are the domains of the Central for Agriculture Land Resources Research and Development (CALRD/BBSDLP). Other components that requires to be considered, are the selection of rice varieties cultivated and integrated management of plant pest and disease which are the domain of the hall commodity port, namely the Center for teh Central for Rice Research Institute (CRRI/BB Padi) and the Assessment Institute for Agricultural Technology, Indonesia (AIAT/BPTP) in each province.

The development of agriculture on swamps in a sustainable manner needs to be carried out in a regional approach, through holistic, selective and integrated and participatory. The approach is based on a unit of development area, a holistic approach implies that technology and infrastructure related to the development of their agriculture are carried out comprehensively. While the selective and integrative approach means that the development of regions, commodities, technology and institutions should be selectively and synergistically integrated. In terms of management, planning and implementation of agricultural business development in swamps should involve active community participation, especially in the area of development (Rina dan Subagio, 2017).

From various development experiences and results of agricultural research, the development of agricultural models in swamps is specific, dynamic and based on local resources. But in general, the development of an agricultural pattern developed sustains an integrated agricultural model that integrates various agricultural commodities harmoniously according to the characteristics and arrangement of land as well as people's preferences and the results of marketing prospects (Alihamsyah, 2004).

5 CLOSING

Nearly 90% of swamp land still has a crop index (IP) of 100, and some others are not planted anymore because their productivity has dropped due to changes in its properties, including being sour, Fe poisoning, and the presence of macro and micro nutrients. It is planned that in the Serasi Program, the development of 1 million hectares of swamp land in 6 (six) provinces (South Sumatra, South Kalimantan, Lampung, Jambi, Central Kalimantan and South Sulawesi) efforts to optimize land and clear new land.

It is hoped that through a harmonious program an increase in productivity and cropping index (IP) from IP 100 to 200 or 300. The IAARD through the Central for Agriculture Land Resources Research and Development (ICARRD) initiates "Trikelola Plus" as support in the Serasi Program. Technological innovations called "Trikelola Plus" include water management, land arrangement and fertilization plus variety selection and control of plant pests and diseases by utilizing other natural and biological control enemies. Its implementation of requires integration and infrastructure support. Agricultural development in swamps must be integrated, comprehensive and synergistic to ensure the success and sustainability of its development, so that coordination and commitment as well as the participation of the community and stakeholders are needed.

REFERENCES

BALITTRA, 2011. State of the Art & Grand Design of Swamp Land Development. Editor Muhammad Noor et al. Center for Research and Development of Agricultural Land Resources (BBSDLP), IAARD. Bogor. 44 p.

BALITTRA, 2014a. General Guidelines for Sustainable Management of Acid Sulfate Land for Agriculture. Editor Dedi Nursyamsi et al. IAARD Press. Jakarta. 58 p.

BALITTRA, 2014b. General Guidelines for Sustainable Swampyland Management for Agriculture. Editor Dedi Nursyamsi et al. IAARD Press. Jakarta. 72 p.

BALITTRA, 2014c. General Guidelines for Sustainable Management of Peatlands for Agriculture. Editor Dedi Nursyamsi et al. IAARD Press. Jakarta. 68 p.

BBSDLP (Agricultural Research and Development Center for Agricultural Land Resources). 2015. Indonesian Agricultural Land Resources. IAARD-Press. Bogor. 100 p.

Darmanto, 2000. Flashback of Swamp Land Development in Indonesia: History of Swamp Land Reclamation. Speech of Inauguration of Head of Civil Engineering at Faculty of Engineering. Univ. Gadjah Mada. Yogyakarta. 40 p.

Dent, D. 1986. Acid Sulphate Soils: A Baseline for Research and Development. ILRI Publ. No. 39. Wageningen. 204 p.

Idak, H. Development and History of South Kalimantan Rice Fields. South Kalimantan Regional Government. Banjarmasin. 40 p.

Ministry of Agriculture, 2017. Success Self-Sufficiency: Indonesia Becomes World Food Barn 2045. Ministry of Agriculture. Rep. Indonesia. Jakarta. 272 p.

Ministry of Agriculture, 2018. Swamp Land: The Future of Indonesian Food Barn "Generating Sleeping Giant". Ministry of Agriculture. Rep. Indonesia. Jakarta.

Noor, M. 1996. Rice Margin. Penebar Swadaya. Jakarta. 213 p.

Noor, M. 2001. Peatland Agriculture, Potential and Constraints. Kanisius. Yogyakarta. 210. p.

Noor, M. 2004. Swamp Land: Nature and Problematic Management of Acid Sulfate. Raja Grafindo Persada. Rajawali Press. Jakarta. 241 p.

Noor, M. 2007. Lebak Swamp: Ecology, Utilization, and Development. RajaGrafindo Persada. Rajawali Press. Jakarta. 213 p.

Noor, M. 2010. Peatlands: Development, Conservation and Climate Change. Gadjah Mada University Press. Yogyakarta. 212 p.

Noor, M. 2012. History of the Opening of Peatlands for Agriculture in Indonesia. In Edi Husien, et al (eds.) Pros. Sem Nas. Peatland Management. Balitbangtan, Ministry of Agriculture. Jakarta. P. 399-412.

Noor, M. 2014. Water management technology supports land optimization and intensification of agriculture in lowland swamps. Magazine of Agricultural Innovation Development: Vol. 7 (2), June 2014. p. 95–104. Pertannian Research and Development Agency, Ministry of Agriculture. Jakarta.

Noor, M. 2017. Peat Debate: Economy, Ecology, Politics and Policy. Gadjah Mada University Press. Yogyakarta. 219 p.

Sinar Tani, 2018. Water Harvests remain the main concern. Sinar Tani edition 28 February-6 March 2018. pp. 18.

Subagio, H, M. Noor, Wahida .A Yusuf, and Izhar Khairullah. 2016. Swampland Agriculture Perspective: Supporting Food Sovereignty. IAARD Press. Jakarta/Bogor. 108 p.

Subagio, H, and M. Noor, 2017. Swamp Land Perspectives in Supporting the World Food Barn. In Masganti et al. (eds) Swamp Agroecology. IAARD Press - Press Release. Jakarta/Bogor. Pp. 653–677.

Susanto, R. H. 2010. Swamp Land Management Strategy for Sustainable Agricultural Development. Faculty of Agriculture. Sriwijaya University. Palembang. 172 p.

Suwanda, M, H. and M. Noor. 2014. The policy of utilizing tidal swamps to support national food sovereignty. Journal of Land Resources. December 2014 Special Edition: 31–40 page. BBSDLP. IAARD. Ministry of Agriculture. Bogor.

Rina, Y. D and H. Subagio. 2017. Swamp Farming: Analysis and Implication. IAARD Press-GMU Press. Bogor/Yogyakarta. 238 p.

Widjaya Adhi, IPG. K. Nugroho, D.A. Suriadikarta and A. Syarifuddin. 1992. Swamp Land Resources: Potential, Limitations, and Utilization. In the pros. Sem Nas. Integrated Development of Agricultural Tidal Swamp Land and Lebak. SWAMPS II. Bogor.

Peatland management for oil palm in Indonesia

B. Sahari, Y.A. Adhi, R. Rolland, P.P. Utama & I. Ismail
Indonesian Palm Oil Association, Jakarta, Indonesia

ABSTRACT: Responsible peatland management for sustainable oil palm production is the main concern for growers in Indonesia and it is regulated through the Government Decree No 57/2016 junto No. 71/2014 regarding the Protection and Management of Peat Ecosystems. This applies to all companies and farmers running businesses on peatlands. Staged approaches are required to build better knowledge and appropriate technology to bridge agronomic needs and conservation. One of the major peat management techniques is the maintenance of high moisture level to avoid fires. The management of peat requires an understanding of its types and characteristics, water behavior both on the surface and subsurface, soil moisture, and hydro-topography. Management approaches based on the understanding of tropical soil characteristics can be used to formulate appropriate practices to manage peatland for oil palm. An example included drainage systems that were built based on hydro-topography information and zonation systems with controlled water gates can maintain moist water content yet still achieve good yields. Some examples of successful management can be seen in many areas of Sumatera including Jambi, Riau, and North Sumatra. An oil palm plantation on peatland in Labuhan Batu, North Sumatra which was established since 1910, produced high yield for many years with no record of land fire. As this plantation was built based on a good understanding of peat soil characteristics, this model promises a glimmer of hope in supporting responsible peatland management scheme.

1 INTRODUCTION

In recent years, tropical peatland has become a subject of major concern in many climate change-related discussions, especially for the development of oil palm plantations in Indonesia. The total global peatland area was estimated to be around 400 million Ha (Lappalainen, 1996; Minasny et al., 2019), and are mostly located in North America and Europe. Peatland in Indonesia is approximately 14.9 million ha with Sumatra is the largest area with 6,436,649 ha or 43% of the national peatland, followed by Kalimantan with around 4,778,004 ha (32%) and Papua with 3,690,921 ha (25%) (Ritung et al., 2011; Wahyunto et al., 2013).

The use of peatland varies greatly across the globe. For example, in Finland and Ireland, it is utilized for burnings to produce energy (Byrne, 2016) and it is also imported from countries in Europe, USA, Russia, and China to be used as a planting medium, fertilizer, medicine, and energy (Xianmin, 2016). In tropical countries, specifically in Indonesia and Malaysia, peatlands are utilized for forest protection and agriculture. Since the 1970s, large lowland areas of tropical peatland in Southeast Asia have been converted for agricultural use after forest clearance and drainage. In Indonesia, for example, the coastal swamps on shallow peat have been cultivated successfully for many decades using traditional techniques that involves utilizing daily tidal movements to flush toxic organic substances out of rice fields (Notohadiprawiro, 1997;

Sulaiman et al., 2019). This is particularly true for Sumatra and Kalimantan where many agricultural activities are established on peatlands. Moreover, in the Riau Province, peatlands comprise about 70% of the total land, and the economic development of the area is supported by agricultural activities on this land such as the pineapple, paddy, coconut, sago, and oil palm. However, oil palm is a miracle plant species which has significantly contributed to the alleviation of poverty of the local people and economic development of the country.

The use of peatland in Indonesia for agricultural cultivation and peat conservation is regulated through Government Decree No. 57/2016 juncto No. 71/2014 concerning the protection and management of peat ecosystems, and Article 23 subsection 3 states a minimum water level of 40 cm from the surface of the peat must be maintained. This regulation applies to all companies and farmers operating on the land and directed towards maintaining moisture to avoid fire hazards. Furthermore, peatlands with extensive and long-term drainage (Van Lanen et al., 2004) have been found to cause low groundwater level, peat subsidence, rapid peat compaction (Rieley et al., 2007) and drought, which consequently increases the vulnerability of peat to fire (Szajdak and Szatylowicz 2010). It was further reported that an increase in the vulnerability of fire is observed when the groundwater level falls below 40 cm from the surface of the peatland (Wösten et al., 2008).

Overdrained peatland for oil palm cultivation is frequently mentioned in many discussions as the main factor triggering water loss, thereby, making it more susceptible to fire during the dry season. This presumption reflects the lack of information on the current development of peatland management for oil palm plantations. In fact, knowledge, science, and technology developed rapidly and helped the industry to manage peatland from marginal land to be one of the most important arable land generating economic benefits for the nation and poverty alleviation without losing its ecological functions.

From a plant physiology point of view, oil palm is expected to grow well in peatlands maintained with groundwater level ranges between 40 cm and 60 cm. In contrast, the plant suffers when peat soil gets flooded or over-drained leading to a significant decrease in yield, therefore, growers ensure any of these does not happen. However, science and technology have provided an appropriate strategy to prevent oil palm from falling into such situations through the application of a water gate system. This is a complex process which includes not only topography information and climate dynamics, but also subsidence parameters and managing these dynamics among those interacting factors is a challenge for growers to maintain high yield and, at the same time, the ecological functions of the cultivated peat soil. Therefore, the purpose of this paper was to provide an overview of peat management for oil palm cultivation by discussing the key factors to reduce the risk of drought and also to prevent the decline of peatland quality.

2 METHODS

This study was conducted through the use of policy analysis, literature studies, and field observations. Moreover, two oil palm plantation locations in Jambi and Riau provinces with peat thickness varying from 1 m to 6 m were used as a case study. The water management system at the study site involved the use of permanent water building consisting of overflow, sluice gates, and canals.

The observations were made on oil palm plantations on approximately 6,697 hectares of mature peatland with groundwater level (GWL) checked manually every week. In the Jambi area, 70 blocks (30 ha/blocks) were selected with 5 observation points for each block (350 piezometers). In Riau plantation area, sampling was conducted on 32 selected blocks (160 piezometers). Furthermore, GWL was automatically observed through the use of water level data loggers, with 7 blocks representing each zone based on topographic information, such that areas with similar heights were grouped into one zone. Water was managed by operating canal systems, water gates, and overflows. The canal system was built along 917 km to include the main canal and canal block while water gate and overflow were

built in each zone with a total of 25 water gate units and 657 overflow units. Information on rainfall was recorded every day and collected as a one-month observation while peat subsidence was observed through the use of subsidence pole method on 20 blocks and recorded once a month. The peat moisture was monitored using an EC5 soil moisture sensor, values were recorded every 1 hour with a data logger. Correlation analysis was conducted to understand the relationship between each parameter for 4 years (2012 – 2015).

3 RESULTS AND DISCUSSION

3.1 Economic importance of oil palm for Indonesia

Based on the data obtained from the Director General of Agro-Industry, the potential peatland area for oil palm cultivation is 1.02 million ha and the largest estimated area distribution was found in Riau Province with 187.1 thousand ha, followed by South Sumatra Province with 125.8 thousand ha and the rest spread across Jambi Province (23.14 thousand ha), West Kalimantan (21.5 thousand ha), Central Kalimantan (45.51 thousand ha), South Kalimantan (9.15 thousand ha) and Papua (7.95 thousand ha). The remaining 599.69 thousand ha are scattered across other provinces with peatlands designated for oil palm plantations.

A study conducted by the University of Indonesia Economic and Community Research Institute (LPEM) in 2017 estimated the impact of the growing national economy due to oil palm plantations on peat after the consideration of the national gross domestic product (GDP) of around 37.65 trillion rupiah, the incomes of those dependent on peat-situated oil palm being around 10.3 trillion rupiah and the absorption of a 500 thousand strong workforce.

3.2 Indonesian regulation

Indonesia has a strong commitment to implement sustainable production of palm oil through several government regulations, particularly those related to peatland such as Government Decree No 57/2016 junto No. 71/2014 regarding the Protection and Management of Peat Ecosystem and this was followed by the establishment of Peat Restoration Agency through Presidential Decree No 1/2016. These regulations are aimed at conserving peatland through the implementation of responsible management. Furthermore, the government has also issued several regulations including the Decision Letter from The Ministry of Environment No 129/2017 (regarding Map of Peatland Hydrological Unit) and No 130/2017 (regarding Map of Peatland Ecosystem Functions—protection function and cultivation function), Minister Decree No. 14/2017 (regarding peat inventory), No. 15/2017 (regarding water level monitoring) and No.16/2017 (peatland rehabilitation) to implement these policies on the ground. They, however, apply to all companies and farmers operating on peatlands, and any of them found to have breached the regulations on environmental protection could be prohibited from operating.

3.3 Environment impact of palm oil on peat

Oil palm development on peatland is frequently blamed as a high contributing factor to GHG emissions. However, the results of scientific studies comparing CO_2 emissions between oil palm plantations, forests, and sago showed different evidence with the forest ecosystem found to have a higher CO_2 flux than oil palm with a production rate of 2.1 kg C m^{-2} year^{-1} and 1.5 Kg C m^{-2} year^{-1} respectively (Melling et al., 2005). This study shows the CO_2 emissions from the oil palm developed on the peat depends on the management.

Peat subsidence is another issue that has gained a lot of attention from environmentalists and oil palm development on peatland has been suspected to be its causative factor. A plantation in Labuhan Batu-North Sumatera, which was first opened in 1919 to have only experienced a decline in peat (subsidence) of around 70-80 cm up to the present time. Assuming a constant subsidence rate of 5 cm/ year (Hooijer, 2012), after 100 years, peat of 500 cm would disappear. Therefore, it would be wrong to assume a linear decrease in peat.

3.4 Understanding peat for management

The increasing understanding of the tropical peatland characteristics has triggered the development of appropriate technologies to support agricultural activities on peat and, at the same time, maintain the ecosystem's supporting capacity.

Peat characteristics in Indonesia are very diverse depending on the distribution area and peat materials, thus, it is managed by the type of peat. Moreover, the key to proper management is not dependent on the groundwater level but on the maintenance of high moisture content to avoid fire. In principle, it requires an understanding of the type and characteristics of the peat, water behavior both on the surface and subsurface, soil moisture, and hydro-topography.

In Indonesian context, majority of the peat are clearly visible in original form, especially those derived from wood and leaves, except for only a small portion of the plant component. However, through careful observation, wood has been discovered to be the material dominating peat in the country. Moreover, part of the accumulation process of peat material is influenced by the "erosion-transport-deposition" process of fine-textured weathering products originating from Tertiary Facies (Furukawa and Sabiham, 1985). Therefore, the peat is often in a mixed state with minerals whether it is deposited at the bottom layer or dispersed around the river. This mixture is, therefore, referred to as peaty

clay (Sabiham, 1988) and found to be approximately 30-65% organic material content.

Peatlands have the ability to absorb and store water at a much higher capacity than mineral soils due to the dominant component of organic matter. Verry et al. (2011) states that the water contained in peat soil has the ability to reach 300 - 3000% of its dry weight, much higher than for mineral soils whose ability to absorb water is only around 20-35% of its dry weight. Moreover, the bulk density, is the physical property of the soil which shows the weight of the solid mass in a certain volume and it is generally expressed in units of g cm^{-3} or Mg m^{-3} or t m^{-3}. It is most often analyzed due to its usefulness as an initial description of other soil physical properties such as porosity, bearing capacity, and the potential to store water. Dariah et al. (2012) reported the level of peat maturity has an effect on its BD, such that more mature peat is expected to give the higher value of BD.

Capillary water in peatlands during dry conditions (dry season) play a very important role in providing water for the root zones of plants because the actual groundwater levels are very dependent on the water levels in areas with shallow water levels. Therefore, the increase in capillary water in the soil is a phenomenon visualizing the movement of water through soil pores from low to higher elevations. Qiang et al. (2014) also explained the concept to be a common phenomenon in nature happening due to the presence of surface tensions which causes water to be higher than the groundwater levels. Moreover, capillary water also has the ability to replace water lost by evapotranspiration in the upper layers (Yazaki et al., 2006; Chesworth, 2008; McCarter, 2012). Lu and Likos (2004) reported three basic physical characteristics related to capillarity water increase to include maximum height, storage capacity, and rate.

From the graph in Figure 1, the capillary water occurred even though the groundwater level (GWL) reached the lowest level of 68 cm and soil moisture at a depth of 30 cm is still at 320 - 400%. Moreover, no significant effect of decreasing soil moisture was observed with the varied height of groundwater level between 25 - 70 cm, as shown in Figure 2. Montemayor et al. (2015) reported capillary water rises to 40-50 cm while Schindler et al. (2003) and Schwärzel et al. (2006) explained it can reach the root zone even with a drainage depth of up to 70 cm. The results of these studies were obtained on peat soils in subtropical regions with different characteristics from those in the tropics.

Furthermore, the irreversible drying process becomes accelerated in a situation where the peat water content is below the critical moisture content which is 73% (hemic) and 55% (sapric) in Central Kalimantan (Masganti et al., 2001). However, the critical soil water content for the emergence of hydrophobicity of peat soils at the research location (Meranti Paham) was in the range of 118-126% and 184-213% respectively for hemic and sapric peat soils. Frandsen

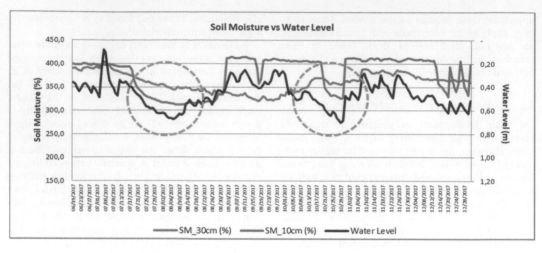

Figure 1. Relationship between water level and soil moisture from June to December 2017.
Note: SM = Soil Moisture

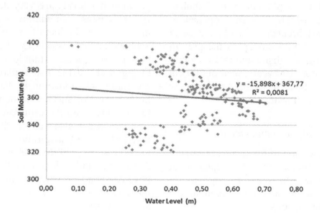

Figure 2. Water level value used for correlation with soil moisture. N for GWL = 196; N for soil moisture = 196. Y=-15,898x+367,77.

(1987) reported peat soils to have the ability to experience ignition in soil moisture content below 110% while Putra (2003) found a water content of less than 117.39% as a starting point of burning and the foundation for a larger scale fire. Rein et al. (2008) also added that peat soil water content below 125% also can trigger ignition.

3.5 *Peatland management model*

Despite it has been in existence for 4 generations, the production of palm oil on peatlands remains high, reaching an average yield of around 22-23 tons/ha/year.

23 tons/ha/year. Rainfall pattern seems to be similar Rainfall pattern seems to be similar from year to year with lower found between July and October, and rise between November to January, as shown in Figure 3.

However, the result indicated GWL is positively related to rainfall since the increasing value of rainfall is followed by increasing GWL, as shown in Figure 4. In the rainy season, GWL was observed to reach 14 cm under the soil surface, thereby making the plantation to get flooded while, in contrast, it was 157 cm in the dry season.

However, the use of water gate and overflow can effectively maintain GWL around 40 cm and 60 cm in order to keep the soil wet and not flooded or over-drained. This can be seen from Figure 5 where a large number of GWL values were found in the range of 40 cm to 60 cm, followed by 60 cm to 80 cm, and 20 to 40 cm with only a few values occupying lower than 80 cm. This showed the successful application of a water management system to control the water level.

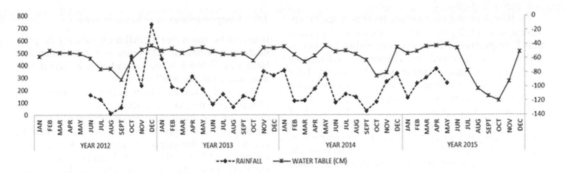

Figure 3. Ground water level and rainfall in the study area from 2012 – 2015. Ground water level correlation with soil moisture. N = 3.5.

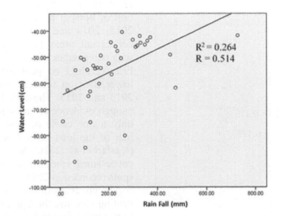

Figure 4. Relationship between ground water level (GWL) and rainfall from 2012 to 2015. Water level value used for correlation with rainfall is from the average of 707 samples in the same month. N for GWL = 36; N for rainfall =36. Y=64,876-0,045X; F=12,19; P=0.001.

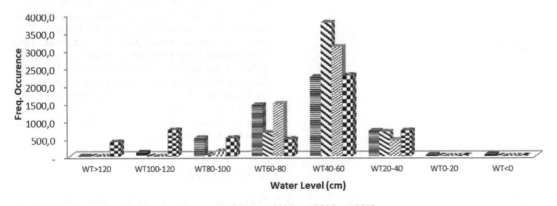

Figure 5. Relative Frequency of occurrence of plots with GWL values in the same GWL range. Number of plots with GWL in the same range relative to all GWL values. Distribution pattern of GWL values from 60 selected locks with ages range from 12 to 15 years old.

Furthermore, the peat surface level was negatively related to rainfall and GWL, as shown in Figures 6 and 7, respectively. Such that increasing amount of rainfall was followed by decreasing subsidence level, and the same situation was also recorded for subsidence and water level. This shows subsidence strongly influenced GWL since lowering ground water level means peat soil loses some amount of water, making the organic material becoming more compact and the surface of peat lower. Moreover, the oil palm yield of the study site was recorded to be 24.2 ton/year on average, and the older plants were observed to have a higher yield. In conclusion, the peat management system applied in the study area ensured high yield and, at the same time, maintained water level as expected.

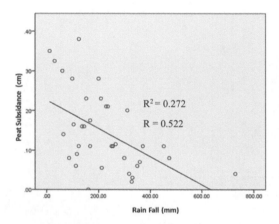

Figure 6. Relationship between rainfall and peat subsidence from 2012 to 2015. Subsidence value used for correlation with rainfall is from the average of 707 samples in the same month. N for subsidence = 36; N for rainfall =36.

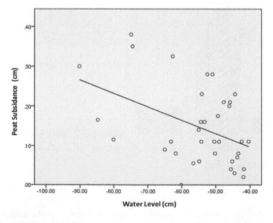

Figure 7. Relationship between water level and peat Subsidence from 2012 to 2015. Subsidence value and GWL are from the average of all samples in the same month. N for subsidence = 36; N for GWL =36.

3.6 *Fire prevention in relation to peat*

It should be noted that the palm oil industry on peatlands started almost 100 years ago in the Labuhan Batu area of North Sumatra, and no fire-related incident was recorded at all times.

In recent years, fires have occurred in many areas of Indonesia due to the susceptibility of peat during the dry season. Moreover, the utilization of peatlands for oil palm development is often associated with the occurrences of fire-related incidents. However, information available in the public domain shows different facts. From the spatial distribution data of land fires reported in the country by several public domains such as Global Forest Watch Fires (GFWF), Sipongi (under the Ministry of Environment and Forestry) and LAPAN (Indonesian National Institute of Aviation and Space), in 33 provinces, the distribution of hotspot patterns was found to be similar in 2015, 2014 and 2013 with fire warnings in non-concession areas higher than in concession areas with percentages of 61%, 59% and 58% respectively. Fire warnings on peatlands were discovered to be lower than on mineral soils at 49%:51% between 2013 and 2015. These findings indicate there is not enough evidence to show peatland cultivation is the only major cause of land fires because more than half of the area burned is located on mineral soils (Kasfari et al., 2016). Moreover, hotspots in oil palm concessions comprise about 10% of the total hotspots recorded in 2015.

The government has a strong commitment to preventing land fire through several policies such as the Presidential Instruction No 11/2015 regarding the strengthening of forest and land fire prevention. This was followed by Ministerial Decree No P32/2016 regarding forest and land fire control, and recently Ministerial Decree No 5/2018 regarding land development without burning. Furthermore, a Zero burning policy has been ratified by ASEAN members since 2003, and it has been applied in oil palm companies to prevent land fire accidents. Learning from the past, many companies in Indonesia have strengthened their prevention systems to ensure all potential fire incidents are well managed. They have also expanded their efforts through the development of Community Cares about Fire (Masyarakat Peduli Api/MPA) with local communities living around their concessions and, in 2017, 664 MPAs were established in the country.

4 CONCLUSIONS

Water management system is a major factor in managing peatland for oil palm plantation due to the influence of water level in preventing fire.

Water management was also observed to prevent the impact of subsidence on peatlands. This is achieved through the use of effective means to minimize the risks of floods and fires due to the subsidence. It was also found to have the ability to increase the production of oil palm plantations on peatlands and to consequently increase the Indonesian economy.

The role of the government is also important in preventing fire disasters, especially on peatlands as indicated by the formulation of several regulations related to fire and its prevention as well as sustainable peat management to prevent fire hazards. In addition, the government and the community were also found to be collaborating to prevent and minimize fire incidents through the establishment of Community Cares about Fires (MPA).

REFERENCES

Byrne, C. 2016. Planning For the End of Energy Peat Production in Bord Na Mona. International Peat Congress. Sarawak - Malaysia.

Chesworth, W. 2008. Encyclopedia of Soil Science. Dordrecht (NL): Springer.

Dariah, A., Susanti, E., Mulyani, A. & Agus, F. 2012. Faktor Penduga Simpanan Karbon Pada Tanah Gambut. In: Husen E, Anda M, Noor M, Suwanda M, Maswar, Fahmi A, Sulaeman Y, editor. 2013. Prosiding Seminar Nasional: Pengelolaan Lahan Gambut Berkelanjutan [Internet]. 2012 Mei 4. Bogor, Indonesia. Pp. 213–222. Bogor (ID): Balai Besar Penelitian dan Pengembangan Sumberdaya Lahan Pertanian. Tersedia pada: http://bbsdlp.litbang.pertanian.go.id/

Frandsen, W.H. 1987. The Influence of Moisture and Mineral Soil on The Combustion Limits of Smouldering Forest Duff. Canadian Journal of Forest Research. 17 (12): 1540–1544. doi: 10.1139/x87-236.

Furukawa, H. & Sabiham, S. 1985. Agriculture landscape in the lower Batang Hari, Sumatera. In: Stratigraphy and geomorphology of coastal swampylands (in Japanese). Southeast Asian Studies. 23(1):3–37.

Hooijer, A., Page, S., Jauhiainen, J., Lee, W.A., Lu, X.X., Idris A. & Anshari, G. 2012. Subsidence and Carbon Loss in Drained Tropical Peatlands. Biogeosciences. 9 (3): 1053–1071. doi: 10.5194/bg-9-1053-2012.

Kasfari, et al. 2016. Is Peatland Utilization The Main Cause Of Land Fire In Indonesia. International Peat Congress. Sarawak - Malaysia.

Lappalainen, E. 1996. Global Peat Resources. Finland: International Peat Society. pp 53–53281.

Lu, N. & Likos, W.J. 2004. Rate of Capillary Rise in Soil. Journal of Geotechnical and Geoenvironmental Engineering. 130(6): 646–650. doi: 10.1061/(ASCE)1090 0241(2004)130:6(646).

Masganti, Notohadikusumo, T., Maas, A. & Radjagukguk, B. 2001. Hydrophobicity and its impact on chemical properties of peat. In: Rieley JO, Page SE, editor. 2002. Jakarta Symposium Proceeding on Peatlands for People: Natural Resources Functions and Sustainable Management. Pp. 109–113.

McCarter, C. 2012. The Hydrology of The Bois Des Bel Bog Peatland Restoration: A Tale of Two Scales [thesis]. Ontario (CA): University of Waterloo.

Melling, L., Hatano, R., Goh, K.J. 2005. Soil CO_2 flux from three ecosystems in tropical peatland of Sarawak, Malaysia. Journal Tellus B (Vol 57). Doi: doi.org/ 10.1111/j.1600-0889.2005.00129.x. Page 1–11.

Minasny, B., Berglund, Ö., Connolly, J., Hedley, C., de Vries, F., Gimona, A., Kempen, B., Kidd, D., Lilja, H., Malone, B. and McBratney, A., 2019. Digital mapping of peatlands–A critical review. Earth-Science Reviews Volume 196, September 2019, 102870.

Montemayor, M.B., Price, J. & Rochefort, L. 2015. The Importance of pH and Sand Substrate in the Revegetation of Saline Non-Waterlogged Peat Fields. Journal of Environmental Management 163: 87–97. doi: 10.1016/j.jenvman. 2015.07.052.

Notohadiprawiro, T. 1997. Twenty-five years experience in peatland development for agriculture in Indonesia. In: Rieley, J.O., Page, S.E. (Eds.), Biodiversity and Sustainability of Tropical Peatlands. Samara Publications, Cardigan, United Kingdom, pp. 301–310.

Putra, N.S.S.U. 2003. Hubungan Kadar Air dengan Konsentrasi Emisi Gas Rumah Kaca pada Kebakaran Gambut [Thesis]. Institut Pertanian Bogor, Bogor.

Qiang, L., Yasufuku, N., Jiali, M. & Jiaguo, R. 2014. An Approach For Quick Estimation Of Maximum Height Of Capillary Rise. Soils and Foundations 54(6): 1241–1245. doi: 10.1016/j.sandf.2014.11.017.

Rein, G., Cleaver, N., Ashton, C., Pironi, P. & Torero, J.L. 2008. The Severity of Smouldering Peat Fires and Damage to The Forest Soil. Catena. 74: 304–309. doi: 10.1016/j.catena.2008.05.008.

Rieley, J.O. 2007. Tropical peatland – the amazing dual ecosystem: co-existence and mutual benefit. In: J. O. Rieley, C. J. Banks, & B. Ragjagukguk (Eds.), Carbon climate-human interactions on tropical peatland: Carbon pools, fires, mitigation, restoration and wise use. Proceedings of the International Symposium and Workshop on Tropical Peatland, Yogyakarta, August 27 – 29, 2007, p.15.< http://www.geog.le.ac.uk/carbopeat/media/pdf/yogyapapers/yogyaproceedings.pdf>

Ritung, S., Wahyunto, Nugroho K., Sukarman, Hikmatullah, Suparto & Tafakresnanto, C. 2011. Peta Lahan Gambut Indonesia Skala 1:250.000. Balai Besar Penelitian dan Pengembangan Sumberdaya Lahan Pertanian. Bogor (ID): Badan Penelitian dan Pengembangan Pertanian.

Sabiham, S. 1988. Studies on peat in the coastal plains of Sumatra and Borneo: I. Physiography and geomorphology of the coastal plains. Southeast Asian Studies 26 (3):308–335.

Schindler, U., Behrendt, A. & Müller, L. 2003. Change of Soil Hydrological Properties of Fens as A Result of Soil Development. J. Plant Nutr. Soil Sci. 166(3): 357–363. doi: 10.1002/jpln.200390055.

Schwärzell, K., Šimůnek, J van., Genuchten, M.T. & Wessoeek G. 2006. Measurement and modeling of soil-water dynamics and evapotranspiration of drained peatland soils. J. Plant Nutr. Soil Sci. 169(6): 762–774. doi:10.1002/jpln.200621992.

Sulaiman, A.A., Sulaeman, Y. and Minasny, B., 2019. A Framework for the Development of Wetland for Agricultural Use in Indonesia. Resources, 8(1), p.34.

Szajdak, L. & Szatylowicz, J. 2010. Impact of Drainage on Hidrophobicity of Fen Peat-Moorsh Soils. Di dalam: Klavins M, editor Mires and Peat. Volume 6. Riga (LV): University of Latvia Press. pp 58–1. http://www.lu.lv/fiload min/user_upload/lu_portal/apgads/PDF/mirespeat.pdf.

Van Lanen, H.A.J., Kašpárek, L., Novický, O., Querner, E. P., Fendeková, M. &and Kupczyk, E. 2004. Human Influences. In Hydrological Drought – Processes and Estimation Methods for Streamflow and Groundwater, ed. L. M. Tallaksen and H. A. J. Van Lanen, 347–410.

Developments in Water Science 48. Amsterdam, the Netherlands: Elsevier.

Verry, E.S., Boelter, D.H., Päivänen, J., Nichols, D.S., Malterer, T. & Gafni, A. 2011. Physical Properties of Organic Soils. Di dalam: Kolka R, Sebestyen S, Verry ES, Brooks K, editor. Peatland Biogeochemistry and Watershed Hydrology at the Marcell Experimental Forest. Florida (US): CRC Press. hlm 135–176; http://www.nrs.fs.fed.us/pubs/jrnl/2011/nrs_2011_verry_003.pdf.

Wahyunto & Dariah, A. 2013. Pengelolaan Lahan Gambut Tergedradasi dan Terlantar untuk Mendukung Ketahanan Pangan. Di dalam: Soeparno H, Pasandaran E, Syarwani M, Dariah A, Pasaribu SM, Saad NS. 2013. Politik Pengembangan Pertanian Menghadapi Perubahan Iklim [Internet]. Jakarta (ID): Badan Penelitian dan Pengembangan Pertanian. hlm 329–348; http://www.litbang.pertanian.go.id/buku/politik-pembangunan/.

Wösten, J.H.M., Clymans, E., Page, S.E., Rieley, J.O. & Limin, S.H. 2008. Peat–water Interrelationships in A Tropical Peatland Ecosystem in Southeast Asia. Catena. 73(2): 212–224. doi: 10.1016/j.catena.2007.07.010.

Xianmin, M. 2016. The Next Huge Peat and Growing Media Market In The World. International Peat Congress. Sarawak - Malaysia.

Yazaki, T., Uranoa, S. & Yabeb, K. 2006. Water Balance and Water Movement in Unsaturated Zones of Sphagnum Hummocks in Fuhrengawa Mire, Hokkaido, Japan. Journal of Hydrology. 319(1–4):312-327.doi:10.1016/j.jhydrol. 2005.06.037.

Part C. Water management

© 2020 Taylor & Francis Group, London, ISBN 978-0-367-20964-3

Water management in tidal swamps farming: From indigenous knowledge to improved technology

M. Noor & H. Sosiawan

Indonesian Swampland Agriculture Research Institute (ISARI), Indonesia

ABSTRACT: Indonesia has developed tidal swamp area for food production, within an era of 25 years - 1969 to 1994, tidal swamp lands were developed, up to a million hectares, to boost food production in the nation, mainly rice. This development and its expansion were supported by the Transmigration Program. However, there is a major concern that Indonesia might be importing rice again, in the next ten years, if a proper system is not put in place. This is due to the fact that the population growth rate is 3 million per year and a high rate of paddy fields conversion to other uses, about 50 - 100 thousand hectares per year. Also, various natural disasters like earthquakes, landslides, floods, tsunamis, droughts and land fires have negatively affected the national rice production. The government considers rice as a strategic and political commodity and therefore targeted to bounce back production in 2019 and hopes that the country will become the world's food barn in 2045. The plan is to optimize one million hectares of swamp land in six provinces, namely South Kalimantan, South Sumatra, Lampung, Jambi, Central Kalimantan and South Sulawesi. Wetlands are strongly influenced by water regimes due to river or sea tidal overflows as well as floods from upstream areas. Therefore, the success of swamp development lies in water and land management. This paper presents hydrological and hydrotopographic conditions, as well as the problems faced in water management. The importance of water and land management, the approaches and rules needed to be well established to achieve good production.

1 INTRODUCTION

The Food and Agriculture Organization (FAO) declared Indonesia as a self-sufficient nation in 1985, in terms of food production. This achievement was partly because the government diversified into food production using swamp as a source of irrigation. The commencement of swamp rice fields through the Tidal Swamps Rice Opening Project (P4S) in 1979 till 1994, which involved about 400 thousand hectares of swampy area in Sumatra and 150 thousand hectares in Kalimantan, helped in this cause. And according to Noor et al. (2011), this was supported by a transmigration program which involved people from Java, Bali, and Nusa Tenggara in Kalimantan, Sumatra, Sulawesi, and Papua swamp areas.

The efforts of the government towards achieving self-sufficiency in rice production yielded the desired result in 1984 based on the fact the production was able to meet the country's need, reaching 25.8 million tons of rice (ICFORD, 1995). The application of five new technologies known as "Panca Usaha," helped in achieving this status, these are (1) adequate and timely water supply; (2) the use of high yielding superior seeds (VUB), (3) balanced fertilization, (4) integrated pest control (IPM), and (5) improvement in cultivation technology. Aside from all of these, institutions like Bimas (Bimbingan Massal) and Inmas (Intensifikasi Massal) changed to Insus (Intensifikasi Khusus) in 1979, and Supra Insus in 1987, and according to Noor (1996) and Arifin (1994), these finally led to Indonesia receiving the FAO award from a rice importer to a self-sufficient nation.

However, there is a major concern now that Indonesia might be importing rice in the next ten years considering the population growth rate of the country, which is around 3 million people per year and the rate at which paddy fields were being converted for other uses which are 50 to 100 thousand hectares per year. The effect worsens with an increase in floods, droughts, landslides, tsunamis, and other effects of climate change affecting rice production generally.

To prevent this, the government plans, in the nearest future, to put in place structures that could make Indonesia the world's food barn in 2045 (Ministry of Agriculture, 2017). The potential swamp area to execute this project has an area of 34.70 million hectares, of which 14.99 million hectares are tidal swamp land in Pasang Surut and 20 million hectares of swampy land in Lebak. And with good management, the level of productivity can reach 7 - 8 tons of Dry Mill Rice (DMR) per hectare. According to the Agriculture Minister, as reported by Subagio et al. (2015), the harvest of August 10, 2015 on the tidal swamps in these four villages - Karang Buah,

Belawang, Barito Kuala, and South Kalimantan, stretching to 14,558 hectares of rice fields, produced an average of 4.5 tons of DMR per hectare (Margsari Variety) and 6.7 tons of DMR per hectare (Inpara No. 3 Variety). Then the harvest of August 11, 2015, on the tidal swamps in Terusan Mulia Village, Bataguh, Kapuas District and Central Kalimantan, with 3,099 hectares of paddy fields produced an average of 7.5 tons of DMR per hectare (Hybrids Variety).

This paper presents the hydrological and hydrotopographic conditions, problems, research, and development of water management systems as an important part of the developing swamps for agriculture in the future.

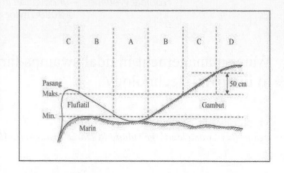

Figure 1. Distribution of tidal swamp land-based on its hydrotopographic overflow (Types A, B, C, and D).

2 HYDROLOGY AND HYDROTOPOGRAPHY

A swamp is a wetland usually caused as a result of water regimes from river or tidal sea overflows. It could also be a waterlogged area caused as a result of floods from upstream areas. These tidal overflows occur when there is a full moon (the 15th qomariah) or dead moon (1st qomariah) usually called a single or spring tide. Also, tidal flooding occurs daily in the form of small or double pairs known as a neap tide. And according to Noor (2004), low tide swamps can be divided into four types - A, B, C, and D, based on hydrotopographic overflows, as shown in Figures 1 and 2.

Type A is a swamp area usually covered by both large and small pairs. The area also experiences daily drainage, and it includes coastal areas and banks of large rivers.

Type B is only covered by large pairs and also experiences daily drainage. It usually covers a distance of 50 to 100 km from the big rivers.

Type C is not completely flooded either by large or small pairs. It has influential pairs through seepage, but its groundwater level is below 50 cm. Also, it has the widest coverage and a relatively higher topography compared with others.

Type D is also an area not completely covered by large or small pairs, but its groundwater level is above 50 cm. Examples of this are the peat domes.

The above distribution of overflow - A, B, C, and D, are only applicable during raining seasons, in the dry seasons, Type B partially changes to C since it experiences acute drainage due to the construction of reclamation networks in the form of channels. However, the type C can be changed back to B by constructing irrigation channels which allow the flow of water into the rice fields.

The first step towards developing swamps in the reclamation process is to regulate high water levels to grow crops. Swamp lands are usually in stagnant conditions for months or throughout the year. However, the development of swamps, according to history, originated from local communities in South

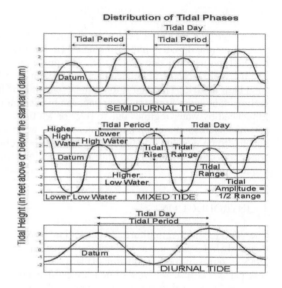

Figure 2. Dynamics of tide or tater level in tidal swamps.

Kalimantan where a small drainage channel was constructed, known as handil (from Dutch anndeel). It is made up of a small channel from a river which is 2 - 3 km far from the field. Also, its width is usually 2 - 3 m while its depth between 0.5 to 1 m. And according to Idak (1982) and Noor (1996), the handil length could be increased or expanded to reach between 20 to 60 hectares.

Various communities have succeeded in applying these swamps in planting rice and other crops including vegetables at a radius of 200 - 500 meters on the left and right side of the handil throughout the year. The handil is commonly used in Kalimantan's general Malay community, the Dayak people regard it as tatah, and in Sumatra, it is called a congress ditch introduced by Bugis (Sulawesi) people who have swamps on the coast of Sumatra and Kalimantan. The success recorded by these communities in the swamp reclamation process motivated the government at that time to develop more swamp lands known as anjir/canal, fork or comb system. In the classification of swamp land infrastructure and development, handil is categorized as tertiary canal or channel, but a wider, deeper and longer version of it is categorized as a secondary channel, while a wider and deeper one is the primary canal.

According to Noor (2016), the development of swamps before 1995 focused more on opening various reclamation network systems but afterward, it changed to rehabilitation and optimization of swamps already available. In 2016, the government launched a program targeted towards optimizing swamp land of 1 million hectares by 2019. This is to be achieved through the improvement and development of water management and rice field network infrastructure covering six provinces, namely South Kalimantan, South Sumatra, Lampung, Jambi, Central Kalimantan, and South Sulawesi.

3 PROBLEMS OF WATERLOGGING AND DRAINAGE

Tidal swamp land is a waterlogged area, but the project of opening swamps through the handil system in1920, the anjir from 1956 - 1958, the fork and comb system from 1969 -1985, and one million hectares reclamation system between 1995 and 1999, all prioritized on accelerating drainage or draining swamps. The process of irrigation or channeling water from the river to the canals or plots only relies on the tidal movements of the rivers or seas regularly, during a full moon, high or spring tide. And when any of these is not happening, most tidal swamps then experience over-drainage. And in preventing excessive drainage, sluice gates or tabs are needed to be able to hold water on the channel. This is what usually results in the tidal swamp land being used for planting once in a year, usually during the rainy season (IP 100). And without blocking the doors of the gates, the incoming water will runoff.

However, there are tidal swamps in other places without a complete water supply network and not equipped with sluice gates. And during rainy seasons, water is confined in the basin areas which leads to waterlogging because there is no channel through which the water can leave. Hence, farmers that have the means always use pumps to get rid of this water, but in the real sense, this is ineffective and inefficient because water management cannot be done partially, it should involve all the farms in the area.

On the other hand, excessive drainage on farmlands makes it very acidic, with pH in a region of 3.0 to 4.0. This could also result to iron poisoning and subsidence as a result of pyrite oxidation. And this could affect areas dense with water as organic matters are accumulated and composted there, which could result in iron fouling and poisoning in the stagnant area, as shown in Figure 3. The whole condition

Figure 3. Conditions of channels with heavy sedimentation, desa Manyahi, Lamunti, Kapuas District, Central Kalimantan (2017) and closed by water-weed, Rasau Jaya, Kubu Raya District, West Kalimantan (2018).

would affect the productivity of the land as well as the general yield of crops.

The following lessons were learned from various experiences in swamp land development for the purpose of agriculture:

(a) The agro physical conditions of swamps are so diverse and fragile, hence, they need to be carefully prepared, and serious attention should be on the nature and dynamics of these changes.
(b) The reclamation and management systems should not be based on budget targets but rather on the appropriate science and technology. And the characteristics of the land should be reviewed from the beginning, which would help in the planning and implementation stages.
(c) Food crops or superior commodity-based agricultural business systems need to be carried out to obtain optimal results and reduce the risk of failure in the agricultural development systems.
(d) Assembling agricultural business systems needs to be in line with the land condition and the socio-economic conditions of the community.
(e) There is a great need to understand the differences between locations so that the technology can be adjusted to solve the problems faced.
(f) There is a need to improve the water management infrastructure, provision of seeds, fertilizers, pesticides, storage facilities.

(g) In developing swamps for agriculture purposes, the community and stakeholders must work together in an integrated, comprehensive, and synergic manner to ensure its success and sustainability.

4 RESEARCH & DEVELOPMENT OF WATER MANAGEMENT

4.1 Swamp farming system

Some of the reasons why the government develop more tidal swamp lands are to discourage importing rice and to increase its production. However, this has not been achieved because planting has not been consistent; it is done once a year (IP 100). But in reality, the potential productivity of these tidal swamps is around 7 - 8 tons of DMR per hectare and can be planted upon twice a year (IP 200). And the government, under the Directorate General of Agricultural Infrastructure and Facilities (PSP), gave high priority to revitalizing agriculture by developing swamps. This was achieved by optimizing an area of 134,700 hectares using 515 units of large excavators - 215 units were used in 2017 and 300 in 2018. There was also various assistance in 2018 covering 112,525 units, in the form of 2 and 4-wheel tractors, planting tools, harvesting equipment, water pumps and so on (Sinar Tani, 14-20 February 2018).

Also, compared with other agroecosystem, swamps stand out in food production due to its flat topography and the availability of large expanse of land. These make it easier to open, prepare, manage, and develop in an integrated manner. The social, economic, and cultural conditions of the swamp community, as well as organization and communication among the residents of the areas, are well handled because the residents are in one location or colony.

The system of swamp farming was originally a small venture usually managed through a family or traditional farming system, but later, through the transmigration program, an intensive farming system was introduced which covers the aspect of producing food crops as well as an agricultural business system. And considering the Serasi Program, a corporate farm was formed consisting of cooperatives initiatives to support the agribusiness systems of the swamp farmers.

More so, farming involving tidal swamp is generally for rice owing to the nature and condition of the land. Also, the swamp environment is usually very influenced by the water regime. More than 90% of the swamps in the Kalimantan region are planted with rice once a year (IP 100), but through counseling and community development from the government, they have planted twice a year (IP 200) in some locations. Those farmers who plant once a year generally use the local varieties of siam rice, pay (long-lived) and/or pandak variety (rather short-lived), while those who plant two times a year use the new high-yielding varieties like Inpara, Margasari, Ceherang, Kapuas, etc. And aside rice, some diligent farmers plant other crops or vegetables. These are planted at the upper part of the sunken bed, while rice is at the bottom part. The distance between surjan is set to 14m so that the tractor does not disturb other activities.

Considering the results of various developmental experiences and agricultural research, it can be deduced that the development of agricultural models in the swamp areas is specific and dynamic and based on local resources. But according to Alihamsyah (2004), Suwanda and Noor (2014), Subagio dan Noor (2017) a sustainable agricultural model should integrate various commodities according to the characteristics and arrangement of land as well as people's preferences and the results of marketing prospects. And with the fact that the swamp farmers' conditions and capabilities are generally limited, the farming system developed needs to be focused more on crops mainly for their food security. Also, a superior commodity-based farming system done on a large scale can be developed. And in line with this, Alihamsyah (2004) and Nursyamsi et al. (2014) reported that exploiting various crop commodities is possible in the swamp land because its structure and the overflow types could support the growth of various crops. For instance, through the surjan system, rice can be planted in the tabukan or paddy field, while other crops, vegetables, and fruits can be planted in the gulud. And according to Widjaja Adhi (1995), Alihamsyah (2004), and Susanto (2010) the arrangement, as well as the management of land and water in the tidal swamp is one of the keys to the success of agricultural development. Therefore, this technology should be applied at the beginning of the land reclamation.

Also, Noorginayuwati and Rina (2003), Rina and Nursyamsi (2013), Rina and Syahbuddin (2013), and Rina and Subagio (2016) reported that the development of various agricultural models through innovative cultivation of both monoculture and polyculture commodities in an integrated manner, is economically feasible with an R/C level > 1. The formation of a farming model is specific to the land and socio-economic conditions of the community, hence, it is important to know these two factors. And to develop a sustainable farming system in swamp land, it needs to be done through an integrated, participatory and regional manner. Also, the planning and implementation of agricultural businesses in swamps should involve the active participation of both the residents and stakeholders.

According to BALITTRA (2001) and BALITTRA (2011), various agricultural commodities like food crops, vegetables, fruits, plantation, and livestock can be developed in swamps. Alihamsyah, (2004) stated that those food crops that can grow and provide good results are rice, corn, soybeans, peanuts, green beans and tubers; vegetables include tomatoes, chili, red onion, cucumber, eggplant, cabbage, beans, spinach, slada, mustard greens, kale, and waluh; while the fruits are oranges, rambutans, papaya, banana, watermelon, and melon. Noor (2001), Alihamsyah (2004) and Noor (2007) also gave the types of plantations that can grow with good results as coconut, oil palm, rubber, pepper, coffee, ginger, and kencur while the livestock includes cows, buffaloes, goats, sheep, chickens, and ducks.

4.2 *Water management system in swamp land*

The technology employed in managing water in tidal swamp lands is adjusted in accordance with the overflow type. For example, the management system for type A or B is a One-Way Flow System (Noor, 2014). This system is designed such that water is flowing in and out of the paddy field area through different channels. The purpose of such water management system is to create water conditions mapped with rice fields by the water requirements of the plants. It also creates differences in water level from two tertiary or secondary channels such that groundwater can continuously flow through the subsurface to the drainage channel to remove toxic substances for plants. This flow system is arranged by installing a swing or flap gate at the estuary of each tertiary or secondary channel, which controls water in and out of the drainage canal. The flap gate is made open so that fresh water from tides can easily enter into the irrigation canal and automatically close inside when the water recedes, thereby retaining the water in the irrigation canal. And then opens out so that dirty water containing toxic substances can get out of the drainage channel, especially at low tides. In general, the one-way flow system in water management can significantly increase land productivity compared to two-way flow systems.

Unlike the water overflow A and B, the type C and D generally get water only from rain, hence, with excessive water during rainy seasons and little or no water during the dry seasons. Such conditions require the conservation of water such as blocking the available ones from flowing off. And in this blocking system, the tertiary or secondary channels function as water collector and conservation channels equipped with an over-flow block, whose height can be adjusted according to the water requirements of the plant. This system helps to reduce the risk of drought during dry seasons.

In general, the water treatment system is vital to farming activities in swamp areas. It has been effective in revitalizing swamp land by increasing its cropping intensity (IP 200) as well as enhancing its productivity in a sustainable manner. And it is necessary to make a mobile embankment and pumping equipment so that the polder/mini polder system can be applied as shown in Figure 4.

HPS Jejangkit Muara, Kab. Barito Kuala, Kalsel
-3°12'17", 114°46'9", 46,0m, 169°
31 Jul 2018 10.46.48 AM

Figure 4. Water management infrastructure and facilities (secondary canal, embankment, pump axial) in Jejangkit Muara Village, Barito Kuala District, South Kalimantan (2018).

5 CONCLUSION

Considering the conditions, research, and development of both water and land management in tidal swamp land, as well as their problems, the authors conclude as follows:

1. The management of water and land is very important in tidal swamp farming and to acheive desired results, re-designing the existing network system is needed. This could be achieved by assessing the topographical, hydrological and suitability of the swamp land.
2. The development of water and land management involving tertiary channels require different long-term approaches based on tidal types.
3. In developing the management system at coastal regions, the focus should be on the protection land as well as fish from the mangrove forests.

REFERENCES

Alihamsyah, T. 2004. Tidal Swamps Lands: Supporting Food Security and Sources of Agribusiness Growth. Indonesian Swampland Agriculture Research Institute (ISARI). Banjarbaru. 53 p.

Arifin G. 1994. Food in the New Orde (Orde Baru). Koperasi Jasa Informasi. Jakarta.

BALITTRA (Balai Penelitian Pertanian Lahan Rawa). 2001. 40 Years BALITTRA (1961-2001). Development and

Research Programe Forward. Indonesian Swampland Agriculture Research Institute (ISARI). Banjarbaru.82 p.

BALITTRA (Balai Penelitian Pertanian Lahan Rawa). 2011. Half A Century of BALITTRA: Swamplans as Food Barm Faced Climated Change. Indonesian Swampland Agriculture Research Institute (ISARI). Banjarbaru. 71 p.

ICFORD (Indonesian Center for Food Crops Research and Development Institute). 1995. Rice Production Technology Supporting Sustainable Rice Self-Sufficiency in Indonesia, Compiled by M. Syam and Hermanto. Bogor.

Idak, H. 1982. Development and History of South Kalimantan Rice Fields. South Kalimantan Regional Government. Banjarmasin. 40 p.

Ministry of Agriculture, 2017. Success Self-Sufficiency: Indonesia Becomes World Food Barn 2045. Ministry of Agriculture. Rep. Indonesia. Jakarta. 272 p.

Noorginayuwati and Y. Rina. 2003. Socio-economic aspect of farmer in acid sulphate land. In isdjanto, et al (Eds) Proc. Seminar Result of Research and Assessment of Inovation Technology in Tidal Swamp Land. Kuala Kapuas, Central Kalimantan, July 31- August 3, 2003. Center Resecarh and Development of Socio-Ecomic Agricultural Institute (PSE). Bogor. 120–136 p.

Noor, M. 1996. Rice of Marginal Land. Penebar Swadaya. Jakarta. 213 p.

Noor, M. 2001. Peatland Agriculture, Potential and Constraints. Kanisius. Yogyakarta. 210 p.

Noor, M. 2004. Swamp Land: Characteristic and Management of Acid Sulfate Soils Problem. Raja Grafindo Persada. Rajawali Press. Jakarta. 241 p.

Noor, M. 2007. Lebak Swamp: Ecology, Utilization, and Development. RajaGrafindo Persada. Rajawali Press. Jakarta. 213 p.

Noor, M. 2014. Water management technology supports land optimization and intensification of agriculture in lowland swamps. Magazine of Agricultural Innovation Development: Vol. 7 (2), June 2014. p. 95–104. Indonesian Agricultural Research and Development Agency (AARD), Ministry of Agriculture. Jakarta.

Noor, M. 2016. Revitalization of lowland swamps for food crops production. Swadaya, Media of Agribusinis Vol 5 (42), February 2016 p. 28–30. PT. Swadaya Agro Utama. Jakarta.

Noor, M, K. Nugroho, IGM Subiksa, Wahyunto, Sukarman, T. Alihamsyah, E. Ananto, R. Shofiyati, D. A. Suriadikarta, I. Ar-Riza I, S. Saragih, dan M. Alwi.

2011. State of the Art & Grand Design of Swamp Land Development. Center for Research and Development of Agricultural Land Resources (BBSDLP), AARD. Bogor. 44 p.

Nursyamsi, D. M. Noor and Haryono. 2014. Sorjan System: Agricultural Swamplands Model Adapter Climate Change. IAARD Press. Bogor. 135 p.

Rina Y and D. Nursyamsi. 2013. Financial analysis of citrus farming on Sorjan System in Tidal Swampsland. In Edy Husen, et al (eds). Proc. Intern. Workshop on Sustainable Management Lowland for Rice Production. Banjarmasin Sept 27–28 2013. IAARD. Min of Agric. Jakarta. pp. 351–368.

Rina, Y and H. Syahbuddin. 2013. Suitability zone of tidal swamp land based comvetitive advantage of commodities. J. SEPA Vol 10 (1) Edition sept 2013: pp. 103–117.

Rina, Y. D and H. Subagio. 2016. Swamp Farming: Analysis and Implication. IAARD Press-GMU Press. Bogor/Yogyakarta. 238 p.

Sinar Tani, 2018. Water Harvests remain the main concern. Sinar Tani edition 28 February-6 March 2018. pp. 18.

Subagio, H, M. Noor, Wahida .A Yusuf, and Izhar Khairullah. 2015. Swampland Agriculture Perspective: Supporting Food Sovereignty. IAARD Press. Jakarta/Bogor. 108 p.

Subagio, H, and M. Noor, 2017. Swamp Land Perspectives in Supporting the World Food Barn. In Masganti et al. (eds) Swamp Agroecology. IAARD Press - Press Release. Jakarta/Bogor. Pp. 653–677.

Susanto, R. H. 2010. Swamp Land Management Strategy for Sustainable Agricultural Development. Faculty of Agriculture. Sriwijaya University (UNSRI). Palembang. 172 p.

Suwanda, M. H., and M. Noor. 2014. The policy of utilizing tidal swamps to support national food sovereignty. Journal of Land Resources. December 2014 Special Edition: 31–40 pp. BBSDLP. IAARD. Ministry of Agriculture. Bogor.

Widjaja Adhi, IPG. 1995. Soil and water management on development resources of swampland for sustainablity farming system and enviromentally sound. Paper presented on Training for Agricultural Development on Tidal Swamp Area, South Sumatra, Palembang, Juny 26-30 1995.

Water management methods affected the growth and yield of several rice varieties in a tidal swampland in Jambi

Salwati & L. Izhar
Jambi Assessment Institute for Agricultural Technology, Jambi, Indonesia

ABSTRACT: An alternative way to increase rice production in Indonesia is through the use of tidal swamp areas. The potential development for agriculture in tidal swamp areas in the Jambi province reaches 39,538 ha. However, the development of this land is constrained by several factors, resulting in rice productivity of tidal swamplands still being low, with an average of around 3 tons/ha. These obstacles can be overcome through the implementation of water management and the use of new high-yielding varieties of rice for tidal swamplands. The study aimed to determine the effect of two water management models on two new rice varieties favored by local farmers/residents and was carried out on farmers' land areas covering more than 3 ha. Through the introduction of tidal rice cultivation technology and by observing the growth and yield parameters of rice production, the rice production will hopefully increase. The results showed that the one-way water system model was generally slightly better than the two-way water system in the rice cultivation activity. Rice growth such as plant height, number of tillers, and yield components showed that the Inpara 2 variety gave high outcomes of 5.9 tons/ha that tended to be better than other varieties.

1 INTRODUCTION

Swamp area is one type of land that is potentially useful for developing agricultural commodities like food crops, especially rice. In an effort to support food security, the use of tidal swamp areas is an alternative solution to overcome the predicament of fertile lands becoming increasingly scarce. However, utilizing swampland must be done well and in accordance with the specific characteristics of the soil type in swampy areas (Alihamsyah, 2002; Notohadiprawiro, 2006; Mariyono, 2015).

Tidal swamplands will provide high yields of rice production when attention is given to cultivation technologies, good water management, and using technology inputs resulting from recommendations of research activities such as VUB of tidal swampland, good seed treatment, balanced fertilization, integrated pest and disease control, proper harvest and postharvest subscriptions (Parish et al., 2008; Jamil et al., 2016). The results from previous research on several superior technological innovation packages in tidal swamplands have been carried out by the Indonesian Swampland Agriculture Research Institute (ISARI/Balitra-Balitbangtan), while the model of rice farming patterns in paddy fields has been able to increase productivity to more than 10 tons/ha (Ar-Riza, 2002; Susanti et al., 2015).

The agricultural technology components that need to be considered well in the effort of using swamps areas to get high rice productivity are the use of VUB and water management technology (Imanudin et al., 2018). Water management for rice cultivation in tidal swamplands such as two-way water system, one-way water system, tabat/block water system, one-way and conservation block water system, water management for surjan, shallow drainage system, and water quality improvement when combined with other technology packages such as biological fertilizers, iron poisoning–resistant varieties, and so on. will be more effective in increasing rice production (Rinnan et al., 2003; Wildayana and Edi, 2018). Thus, water management to improve the productivity of swampy tidal land is a part of the technology package that must be integrated with other technology packages (Alihamsyah, 2003; Sulakhudin and Hatta, 2018).

Tidal swamplands in the Jambi province cover 39,538 ha (BPS, 2015). One of the tidal rice production centers in the Jambi province is located in East Tanjung Jabung Regency with a land area of about 18,322 ha in 2015. Some existing varieties such as Inpara 2 and Inpara 8 were used in this research. Both varieties are tidal swamp rice varieties, which were adaptive and based on the consumer preferences of the surrounding population in the region. Rice from these varieties is medium and fluffier and is favored by the majority of the residents from the Javanese tribe. Both varieties were already used in an adaptation test by Jambi AIAT researches that showed that both varieties were performing well in the Jambi tidal swamplands. However, the rice productivity of the tidal swamplands in the Jambi province is still low, ranging from 3.00 to 3.90 tons/ha (Adri and Yardha, 2014). This paper aims to inform about the effect of water management on several rice varieties in the tidal lands in Jambi.

2 MATERIALS AND METHODS

The research for comprehension in the tidal swamp area to enhance rice productivity in the Jambi province was carried out from March to November 2017. The research used the farm spaces in the tidal swampland areas of East Tanjung Jabung Regency for demonstration of the Jarwo Super (JS) technology, by involving three farmers as research cooperators. The demonstrations were carried out on an area of three hectares, whereas the total rice areas comprised more than fifty hectares.

Research also exercised the latest rice varieties assembled by the Indonesian Swampland Research Institute and the Indonesian Rice Research Center, to find a new rice variety that would have potential for higher production and would be adopted completely by farmers. In addition, water management technology will be studied to get the best technology combination. The JS technological innovation package will be evaluated in a specific location of the Jambi tidal swamp area. Materials and supporting tools for rice cultivation are used in accordance with both existing agricultural technologies and new recommendations.

Rice is cultivated in the tidal swampland area using the adaptive farm research (OFR) method, which was a large-scale assessment on farmers' land areas. The agricultural technology components both assembled and applied were the JS technologies and water management technology. The experiment used a split-plot design with two factors, repeated three times; plot size was 1 ha and replicated with each farmer. The first factor as the main plot was the water management system and the subplot was the varieties. The treatments are water management (J) and rice variety (V), with water management system as J1 = one way and J2 = two ways; and varieties (V), V1 = Inpara 2 and V2 = Inpara 8. Both these rice varieties have a high preference among the Javanese people (Darsani and Koesrini, 2018).

The data to be collected during activities included agronomic data and farming system application based on economic and social data feasibility as supporting activities in the form of farming system-image descriptions.

3 RESULTS AND DISCUSSION

Water management is a very important thing in managing agricultural land in swamp ecosystems (Tanaka, 1976; Yoshida, 1981; Ramirez and Simenstad, 2018). This water regulation is not only to reduce or increase the availability of surface water but also to reduce soil acidity, prevent soil contamination due to oxidation of the pyrite layer, prevent the danger of salinity, danger of flooding, and washing toxic substances accumulated in the plant root zone (Suryadi et al., 2010; Helmi, 2015).

The importance of water management arrangements is reflected in the following data on plant height (Table 1) and the number of tillers (Table 2) affected by water management for two types of tidal swamp rice varieties. There was no significant difference. in vegetative growth and development of rice plants in terms of of plant height and number of

Table 1. Average height of rice for treatment of water management systems and new superior varieties in tidal swamp areas in Jambi.

Water Management	Varieties	Plant height (DAP)									
		14	21	28	35	42	49	56	63	70	77
One Way	**Inpara 2**	45	58	61	78	87	95	102	104	110	114
	Inpara 8	40	53	59	73	84	93	98	101	103	105
Two Way	**Inpara 2**	39	45	55	65	82	94	107	111	114	114
	Inpara 8	39	49	56	67	83	92	98	101	103	105

Table 2. Productive tiller numbers of rice for treatment of water management systems and new superior varieties in tidal swamp areas in Jambi.

Water Management	Varieties	Number of Productive Tillers (DAP)									
		14	21	28	35	42	49	56	63	70	77
One Way	**Inpara 2**	8	16	18	25	27	28	28	29	30	31
	Inpara 8	7	14	16	17	18	19	19	20	20	21
Two Way	**Inpara 2**	8	10	16	16	18	19	20	20	21	21
	Inpara 8	7	9	14	14	15	16	17	17	19	20

Table 3. Average data on components of rice in the treatment of the water system and varieties in the tidal swamp areas in Jambi.

Water Management	Varieties	Yield Components			
		Panicle Long (cm)	Fertile Grain/Panicle (grains)	Empty Grain (%)	Weight of 1,000 Grain (g)
One Way	Inpara 2	28 a	116 a	14 a	31 a
	Inpara 8	25 a	91 b	24 b	23 b
Two Way	Inpara 2	22 a	99 b	27 b	22 b
	Inpara 8	22 a	80 b	34 c	21 b

tillers, However water management affected the rice yield components in terms of panicle length parameters, and other yield components such as fertile grain per panicle (grains), empty grain percentage, and weight of 1,000 grains (Table 3). The highest rice productivity was observed in the one-way water management application (Figure 1) and was significantly different from other parameters. In general, the one-way water management provides better results compared to the two-way system, for rice production systems in tidal swamp areas in Jambi.

The assessment results showed the growth and yield of rice planted with a one-way water system is better than the two-way system. A one-way water system is a water system designed in such a way that water enters and exits through different tertiary channels. In the field, the input channel (irrigation), the water gate is designed semiautomatically, which only opens inward at high tide and closes itself at low tide. In the drainage channel, the water gate is installed opening outward so that it will only release water coming in from the tertiary channel in the event of low tide. This system creates the circulation of water in one direction, both surface and underground water, due to differences in water level from tertiary irrigation channels and drainage (Alihamsyah, 2003).

The one-way water management model or system requires the presence of sluice gates (flapgate and stoplog) at the mouths of the channels. The sluice in the inlet is designed to open inward so that at high tide, the gate is pushed open and the inlet water can enter the tertiary or quarter channel, while the sluice on the outlet is designed to open outward so that at low tide, the gate opens out so the water from the map or upstream part can come out as the movement of the water recedes. This one-way water system has been widely adopted by tidal swamp farmers. In addition to facilitating the washing of toxic elements, the application of a one-way flow system also allows the development of a variety of cropping patterns as long as it is accompanied by a water management system at the tertiary level that matches the type of overflow and micro water management system at the level of land plots (International Rice Research Institute, 2014; Sarwani et al., 2006).

The two-way water system provides the regulation of the incoming water (irrigation) and drainage from and for the farming area through the same channel so that water changes only occur in estuary areas that are close to the river/secondary estuary. Generally, this practice is applied by farmers at the tertiary level and quarters on tidal land type B. This system has limitations including the limited and less effective level of leaching and flushing from incoming tides (Widjaja-Adhi and Alihamsyah, 1998).

4 CONCLUSION

The one-way water system model was generally slightly better than the two-way system in rice cultivation. Rice growth such as plant height, number of tillers, and yield components showed that the Inpara 2 variety gave high outcomes that tended to be better than other varieties with a productivity of 5.9 ton/ha, highest tiller production, and better other yield components x

ACKNOWLEDGMENTS

We would like to thank the Head of the Indonesia Agency for Agricultural Research and Development

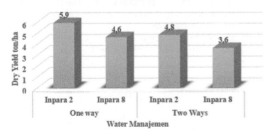

Figure 1. Rice productivity for treatment of water management systems and new superior varieties in swamp area Jambi.

through the SMARTD Program, which has funded this research activity; hence, it could be carried out optimally in 2017. Highly grateful to SMARTD Teams. Thanks to Dr. Ir. Moh. Takdir Mulyadi, MM, who at that time was the Head of AIAT Jambi, for fully supporting the activities of this research. High appreciation to all parties who helped our research activities and in writing this paper.

REFERENCES

Adri & Yardha. 2014. Upaya peningkatan produktivitas padi melalui varietas unggul baru mendukung swasembada berkelanjutan di Provinsi Jambi. *Jurnal Agroekotek* 6(1):1–11.

Alihamsyah, T. 2002. Optimalisasi pendayagunaan lahan rawa pasang surut. Makalah disajikan pada *Seminar Nasional Optimalisasi Pendayagunaan Sumberdaya Lahan di Cisarua tanggal Agustus 6–7, 2002*. Puslitbang Tanah dan Agroklimat.

Alihamsyah, T. 2003. Hasil penelitian pertanian pada lahan pasang surut. Makalah disajikan pada *Seminar Nasional Hasil-Hasil Penelitian dan Pengkajian Teknologi Spesifik Lokasi, Jambi tanggal Desember 18–19, 2003*.

Ar-Riza, I. 2002. Peningkatan produksi padi lebak. *Makalah Seminar Nasional. Perhimpunan Agronomi Indonesia, PERAGI, tanggal Oktober 29–30, 2002 di Bogor*.

Badan Pusat Statistik. 2015. Statistik Indonesia. Jakarta: Badan Pusat Statistik Republik Indonesia.

Darsani, Y.R. & Koesrini. 2018. Preferensi petani terhadap karakteristik varietas padi unggul di lahan rawa pasang surut. *Penelitian Pertanian Tanaman Pangan* 2(2): 85–94.

Helmi. 2015. Peningkatan produktivitas padi lahan rawa lebak melalui penggunaan varietas unggul padi rawa. *Jurnal Pertanian Tropik* 2(2):78–92.

Imanudin, M.S., Bakri, Armanto, E., Indra, B. & Ratmini, S.N.P. 2018. Land and water management option of tidal lowland reclamation area to support rice production (A case study in Delta Sugihan Kanan of South Sumatra Indonesia). *Journal of Wetlands Environmental Management* 6(2): 93–111.

International Rice Research Institute (IRRI). 2014. *Standard Evaluation System for Rice*: 57 p. Manila Philippines: International Rice Research Institute.

Jamil, A.S., Abdulrachman, P., Sasmita, Z., Zaini, Wiratno, R., Rachmat, R., Saraswati, L.R., Widowati, E., Pratiwi, Satoto, Rahmini, D.D., Handoko, L.M., Zarwazi, M.Y., Samaulah, A.M., Yusup, A.D. & Subagio. 2016. *Petunjuk Teknis Budidaya Jajar Legowo Super*. Jakarta: Badan Penelitian dan Pengembangan Pertanian. Kemeterian Pertanian.

Mariyono, J. 2015. Green revolution- and wetland-linked technological change of rice agriculture in Indonesia. *Management of Environmental Quality: An International Journal*. 26 (5): 683–700. doi:10.1108/MEQ-07-2014-0104.

Notohadiprawiro, T. 2006: Mengenali Hakekat Lahan Rawa Sebagai Dasar Pengembangannya Untuk Budidaya Tanaman Pangan. Makalah dalarn Diskusi Panel "Kilas Balik Proyek Pembukaan Persawahan Pasang Surut (P4S) di Kalimantan oleh Tim UGM Th. '68 s.d. '95 Yogyakarta, April 22, 1996, disalin dari makalah yang disajikan dalarn Diskusi Pola Pengembangan Pertanian Tanaman Pangan di Lahan Pasang Surut dan Lebak, Dit. Bina Program Dit.Jen. Pertanian Tanaman Pangan. Palembang, 29 Juli–03 Agustus 1984. Repro: 1 1mu Tanah Universitas Gadjah Made 2006.

Parish, F., Sirin, A., Charman, D., Joosten, H., Minayeva, T., Silvius, M. & Stringer, L. (eds.). 2008. Assessment on peatlands, biodiversity and climate change: Main report. Global Environment Centre, Kuala Lumpur and Wetlands International, Wageningen.

Ramirez M.F. & Simenstad C.A. 2018. Projections of future transitions in tidal wetlands under sea level rise within the port gamble S'Klallam traditional use areas. School of Aquatic and Fishery Sciences, University of Washington. January 2018. p. 35.

Rinnan, R., Silvola, J. & Martikainen P.J. 2003. Carbon dioxide and methane fluxes in boreal peatland microcosms with different vegetation cover-effects of ozone or ultraviolet-B exposure. *Occologia* 137:475–483.

Sarwani, M., Shamshuddin, J., Fauziah, I. & Husni, M.A.H. 2006. Changes in iron-poor acid sulfate soil upon submergence. *Geoderma* 131(1–2):110–122.

Sulakhudin & Hatta, M. 2018. Increasing productivity of newly opened paddy field in tidal swamp areas using a local specific technology. *Indonesian Journal of Agriculture Science* 19(1):9–16.

Suryadi, F.X., Hollanders, P.H.J. & Susanto, R.H. 2010. *Mathematical Modeling on the Operation of Water Control Structures in a Secondary Block Case Study: Delta Saleh, South Sumatra*. Québec City: Canadian Society for Bioengineering (CSBE/SCGAB).

Susanti, M.A., Syaiful, A., Dadang, Irsal, L., Sopiandi, S. 2015. CO_2 and NH_4 emission on different water management and pesticide treatments in rice fields on tidal peat swamp. *J.ISSAAS* 21(2):86–103.

Tanaka, A. 1976. *Comparison of Rice Growth in Different Environment. In Proc. of the Symposium on Climate and Rice*. 429–447. Los Banos, Philippines: International Rice Research Institute.

Widjaja-Adhi, I.P.G. & Alihamsyah, T. 1998: Pengembangan Lahan Pasang Surut; Potensi, Prospek, dan Kendala Serta Teknologi Pengelolaannya Untuk Pertanian. Prosiding Seminar Nasional dan Pertemuan Tahunan Komda HITI, Desember 16–17, 1998.

Wildayana, E.M. & Edi A. 2018. Dynamic of land use change and general perception of farmers on South Sumatera wetlands. *Bulgarian Journal of Agricultural Science* 24(2):180–188.

Yoshida, S. 1981. *Fundamental of rice Crop Science*. Los Banos. Philippines: International Rice Research Institute.

Tropical Wetlands — Innovation in Mapping and Management — Sulaeman et al. (Eds)
© 2020 Taylor & Francis Group, London, ISBN 978-0-367-20964-3

Peat hydrological unit needs to be conserved to support food crop production in Mendol Island, Riau, Indonesia

Suratman & Hikmatullah

Indonesia Centre for Agricultural Land Resources Research and Development, Indonesia Agency for Agricultural Research and Development, Bogor, Indonesia

ABSTRACT: Mendol island (30,995) as a peat hydrological unit, can support many ecosystem functions: conservation (18,592 ha) and biomass production function (12,043 ha). It consists of peat soils (59%) and mineral soils (41%). A total of 4,897 ha of the mineral soils have been cultivated as paddy fields for a long time, in a food granary and rice production center in the Pelalawan Regency, Riau Province. Various attempts have been made to maintain the area as a food crop farming area and the effort is inseparable from its hydrology unit status. Water management and technological inputs have been carried out to balance the conservation and cultivation functions as a mutually supportive system. The biophysical characteristics showed that the mineral soils was generally a clay textured, deep, poorly drained, acid, and some contained sulfidic materials/pyrite. The thickness of the peat soils varied from 2 to >7 m, with a fibric to hemic maturity and very acid. Most of the mineral soils have been used as wetland rice production, while the peat functions as a rainwater storage and an irrigation water supply source for paddy fields. Conservation measures needs to be taken in order to make the forest conservation area a water reservoir. This is in accordance with the government regulation. When considering the Map of The Forest Area of the Kuala Kampar District, Pelalawan Regency, the central part of Mendol Island is included as a Legal Land Use area, which includes a thick peatland ecosystem area. However according to the regulation of the ministry of forestry and environment, the peatland in Mendol island should be conserved to support the hydrology functioning of the island. In summary, the peat and mineral soils in the island must be managed in an integrated manner which mutually support both production and conservation.

1 INTRODUCTION

Mendol Island is a delta formed at the Kampar estuary and located on the eastern part of the Sumatra Island, Indonesia which covers an area of about 31,000 ha. Based on the Reconnaissance Soil Map of Sumatra (Center for Soil Research, 1990) and the Indonesian Peat Land Map (BBSDLP, 2011; BBSDLP, 2017), the Island is dominated by an ombrogenous peat soils which form peat domes with >3 m thickness in the middle and south side of the island. The northern part of the island is occupied by mineral soils, partly containing a sulfidic material (acid sulphate soils). The peat dome has a very important role to play in this ecosystem because it functions as a reservoir and storage for rainwater, used by the local population as a source of irrigation and domestic water use.

Peat soil is formed from an organic matter deposit resulting from the accumulation process of remnants of decayed vegetation under anaerobic conditions. In this condition, the decomposition level of the organic matter is slowed down, and this would result to it forming a thick organic matter accumulation known as peat soil, organosol (Soepraptohardjo, 1961) or Histosol (Soil Survey Staff, 2014). The formation occurred through a geogenic process caused by deposition and transportation which is different from

the mineral soils formation caused by pedogenic processes (Hardjowigeno, 1986). Peat soils in Sumatra and Kalimantan, have formed in the Holocene Period since 11,000 years ago (Polak, 1950). According to Verstappen (1975) and Van Wijk (1951), lowland peat was formed on the coastal plains about 4000-5000 years ago when the sea levels rose, followed by the end of the slowing glacial period.

Peat soil has an organic material thickness of about > 40 cm with a bulk density (BD) of > 0.1 g cm^{-3}, or > 60 cm thick if the BD < 0.1 g cm^{-3}. The maturity level of the peat soil can be divided into: (a) fibric (low-mature) containing fiber >75% volume; (b) hemic (half-mature) containing 17-74% fiber; and (c) sapric (high-mature) containing <17% fiber (Soil Survey Staff, 2014).

According to its formation environment, peat soil is divided into (a) ombrogenous/oligotrophic peat, formed in conditions dependent only on rainwater, forms a dome, has poor nutrients and a low ash content (<5%), (b) topogenous/eutrophic peat, formed in the back swamps of the coastal or river flood plains, receives a mineral enrichment from tidal/flood water runoff, making it more fertile, not too thick, and has a high ash content (> 10%). The peat in between is known as mesotrophic peat, which has about 5-10% ash content (Driessen and Sudjadi, 1984; Widjaja Adhi, 1986).

Lately, various developmental activities, especially in the field of agriculture, have been constrained by the availability of potential lands (Sulaiman et al., 2019). The fertile agricultural land, which supplies 60% of Indonesian's food needs, has changed functions (Masganti et al., 2014). To fulfill the needs for agricultural lands, some parties utilized the sub-optimal lands by applying various methods and innovations (Sulaiman et al., 2019). Peatlands is one of the five sub-optimal land types; the others are acidic, drylands, areas in the dry climate, tidal swamplands, and non-tidal swamplands (Mulyani and Sarwani, 2013). Several factors to be considered in peatland management for agricultural enterprises includes understanding the types and characteristics, its water retention and drainage system, selection of varieties, improvement of the quality, and appropriate cropping patterns.

Among those in Mendol Island, there is a part of the land, which is mineral soils that became a food production center. The existence of peatland cannot be separated from its function of providing a hydrological balance in mineral soil and the management of agricultural lands as a whole. According to the Agriculture Service of Pelalawan Regency (2014) the mineral soils of paddy field with a total area of 4.897 ha, have been cultivated for a long time by local residents, and it has become one of the centers of rice production in the regency. The paddy field is rain fed where the water source comes from rainwater collected by the dome and flows through drainage channels. The economic urge to utilize peatlands for agriculture makes it a challenge to maintain the sustainability of existing agricultural lands. As peatland and agricultural land are one ecosystem, they must be managed in a balanced and a mutually supportive manner.

In this regard, the Agriculture Land Resources Research and Development Center in collaboration with the Pelalawan Regency Agriculture Office, conducted a research on the identification and characterization of land resource potential in Mendol Island, Kuala Kampar District, so as to maintain the balance of the ecosystem in Mendol Island in order to maintain the existence of the agricultural land and also maintain the presence of peatland (Research Team of BBSDLP, 2014). This paper presents the result of the research on the identification and characterization of land in Mendol Island. It provides an analysis of potential linkages in ecosystem unity to support increased food production following the regulation contained in the attachment to the decree of the Ministry of environment and forestry of the Republic of Indonesia. The regulation concerned in the determination of the national peat ecosystem function map and the establishment of a national peat ecosystem function map KHG (Peat Hydrology Unit) of Mendol Island.

2 MATERIAL AND METHODS

Mendol Island (Figure 1) is one of the areas in Kuala Kampar District, Pelalawan Regency of Riau Province. Geographically, it is located in the coordinate of 0°15′00" - 0°20′00" North Latitude and 103° 15′00" – 103°20′00" East Longitude.

It has a wet climate with a rain-fall type (Schmidt and Ferguson, 1951). The average annual rainfall is 2,041 mm, with 260 mm (November) as the highest average monthly, and the lowest is 113 mm (July) (Figure 2). The area has 3-4 consecutive wet months (>200 mm) and without dry months (<100 mm), which is classified as D1 Agroclimatic zone (Oldeman et al., 1978).

The materials used in the study consists of Landsat image data, 30 m resolution of SRTM/DEM (Shuttle Radar Topography Mission/Digital Elevation Model), Indonesian Topographic Map, Geological Map, and location map of the paddy field blocks. The main field equipment consisted of augers for mineral and peat soils, GPS devices, Munsell Soil Color Chart, liquid of Truog pH, Litmus paper pH, and H_2O_2 test.

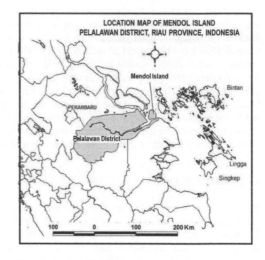

Figure 1. Location of Mendol Island.

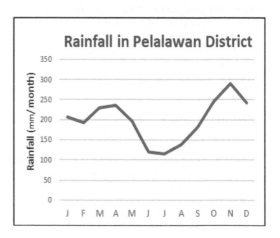

Figure 2. Graph of monthly rainfall of the study area.

Before the fieldwork, the preparation of the Land Unit Interpretation Maps was created using Landsat and DEM images, as well as other supporting data as a basis for observing the soil and the field's physical environment. Soil observations were carried out in a transect system from the coast towards the inland (peat dome) to determine the variation of the soil properties. The observations include the thickness, level of maturity, color, texture of the peats, and the substratum of mineral soil underlying the peat, drainage condition, groundwater table, pH-Truog, sulfidic material or pyrite content, and also pH-H_2O_2 (FAO, 1990). Analysis of the physical and chemical properties of soil samples was carried out in the Soil Research Institute Laboratory according to the procedure outlined by Eviati and Sulaeman (2011). For peat soils, the additional analysis includes fiber content and ash content.

Based on the land biophysical data, the potential of the land is then analyzed for the development of the various prospective food crops, mainly rice. The suitability analysis is carried out by comparing the biophysical characteristics of the land with the requirements needed for growing food crops. The classification is carried out up to the subclass level (class S1= highly suitable, S2 = moderately suitable, S3 = marginally suitable and N = not suitable) coupled with the limiting factors for land suitability subclasses (wa = water availability, rc = media rooting condition, na = nutrient availability, nr = nutrient retention). The land evaluation process was done using computerization software devices SPKL (Land Suit-ability Evaluation System) version 2.0 (Bachri et al., 2016). With this analysis, we could evaluate the type and spread of any food crop commodities that have potential in the region and the extent of the area. Furthermore, it is synergized with the peatland area into a balanced ecosystem.

3 RESULT AND DISCUSSION

3.1 State of the existing land use

Land use observation in the field showed that most of the areas in Mendol Island had been opened for agricultural activities which consist of rice fields, coconut, rubber, mixed garden, and settlement. The rice field is located on mineral soils, while the coconut, rubber and mixed gardens are generally located on peatland. The remaining peat forest is only 14% in the central part of the peat dome (Figure 3). The remaining forest will gradually decline, due to an expansion in agriculture

3.2 Morphological and physical-chemical properties of the soil

The most extensive types of soils found in Mendol Island are Organosol (Histosols), followed by Gleisol (Inceptisols) and Alluvial (Entisols).

3.2.1 Organosol soils

Organosol or peat soils are formed from the deposits of organic material 50 cm thick or more, resulting from the accumulated remnants of decayed vegetation in reductive/stagnant conditions. The results of the study showed that the peat thickness varied between 2 to 7 m and could reach 10 m in the forest area. The peat maturity level varied, generally in the upper layer (0-30/40 cm) was hemic to sapric, while in the lower layer (>40 cm) was fibric. Some peatland cultivated for agricultural use showed experienced subsidence of around 1 m.

Table 1 shows that the soils have very high content of organic matter, low total P and K content, very acidic soil reaction (pH 3,5-4,1), medium to

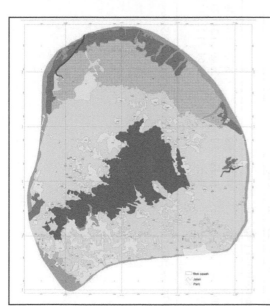

MAP OF LAND USE

MENDOL ISLAND, PELALAWAN DISTRICT,

RIAU PROVINCE, INDONESIA

LEGEND.

SYMBOLS	LAND USE	AREAS	
		Ha	%
Swh	Sawah/wet rice field	4.343	14,12
Ld/Tg	Dry land cultivation	3.768	12,25
Kc	Mix garden	13.993	45,49
Kr	Rubber	47	0,15
Sg	Sago	501	1,63
Smk	Bush	160	0,52
Blkr	Scrub	1.694	5,51
Htl	Peat forest	4.269	13,88
Mgv	Mangrove	1.725	5,61
X1	Sattlement	190	0,62
X2	Water body	72	0,23
Total areas		30.763	100,00

Figure 3. Existing land use map of the Mendol Island.

Table 1. Physical and chemical properties of peat soils from Mendol Island.

Depth	Texture				pH		EC	Org matter		HCl 25%		Cation-exchange (Asetat-ac) pH7							KCl 1N		Content			Fiber 1	Fiber 2	Ash cont
	Sand	Silt	Clay	Class	H₂O	KCl	C	C	N	P₂O₅	K₂O	Ca	Mg	K	Na	CEC	BS	Al³⁺	H⁺	S	Fe	Pyrite			
cm	— % —						dS	– % –		-mg/100g-		— cmol(+)/kg —					%	cmol(+)/kg		– % –			– % –		
Organosol Hemik (Typic Haplohemists)																									
0-40	-	-	-	Hemic	3.8	2.7	-	28.43	1.21	23	44	9.88	3.31	0.88	1.44	86.00	18	0.23	2.96	-	-	-	44	7	2.8
40-150	-	-	-	Hemic	4.0	2.8	-	29.90	0.83	2	5	7.99	3.38	0.17	1.81	86.29	15	0.48	2.61	-	-	-	53	16	2.1
150-200	-	-	-	Hemic	4.0	3.0	-	33.57	0.75	2	2	8.95	3.77	0.06	2.13	115.44	13	0.27	1.73	-	-	-	33	7	1.3
200-260	-	-	-	Hemic	4.3	3.4	-	30.08	1.10	2	2	13.16	2.08	0.05	1.77	108.60	16	0.12	0.97	-	-	-	34	6	2.6
260-300	10	29	61	Clay	3.7	3.3	2.0	8.34	0.31	10	68	10.58	1.48	0.48	1.45	30.40	46	1.58	2.43	1.30	2.47	2.44	-	-	-
Organosol Hemik (Sapric Haplohemists)																									
0-40	-	-	-	Hemic	3.8	3.0	-	38.85	1.64	36	19	13.50	5.66	0.32	1.15	93.72	22	0.17	2.35	-	-	-	38	5	1.4
40-100	-	-	-	Hemic	3.7	2.5	-	36.49	1.31	20	27	3.92	5.32	0.58	7.92	89.50	20	0.29	3.20	-	-	-	26	3	0.8
100-430	-	-	-	Hemic	3.6	2.5	-	28.04	1.22	12	32	3.21	5.31	0.71	6.63	106.11	15	0.36	3.89	-	-	-	22	7	1.2

Note: EC=electrical conductivity; CEC=cation exchange capacity; BS=base saturation; Fiber 1=not destructed; Fiber 2=destructed.

high exchangeable cations (Ca, Mg, K, Na), very high soil CEC, and low base saturation. The substratum of mineral soils underlying the peat soils was generally clayey while some contained sulfidic material or py-rite with 2.44% and showed extremely acid (pH < 2.5) after pouring the H_2O_2 liquid into the soils, but the depth pyrite was >2 m, and therefore had no effect on the plants.

Uncrushed Fiber content varied between 22-53%, according to the criteria of Soil Taxonomy (Soil Survey Staff, 2014) classified as hemic maturity. However, the result of field observations indicated the level of fibric maturity, because the fiber content was still dominant. This condition also occurred on the results of previous research of peat soils in Sumatra and Kalimantan (Hidayat et al., 2011). Ash content was generally low, varied from 0.8 to 2.8% and classified as ombrogenous peat soils with poor Nutrient status. The soils were classified as Organosol Saprik, Organosol Hemik and Organosol Fibrik (Subardja et al., 2016) as equal to Typic Haplosaprists, Typic Haplohemists and Hemic Haplofibrists (Soil Survey Staff, 2014).

3.2.2 *Alluvial Soils*
Alluvial soils were found in fluvio-marine and tidal landforms, formed from fluvio-marine (clay and silt) deposits, stratified due to the differences in particle size, with a structural development in the A-C horizon arrangement. The lower horizons were soft (half-unripe), as the condition might always be in a reductive state. Based on field observation, the soils had gray color, clay to silty clay texture, poor drainage, sticky and plastic consistency, and slightly acidic to alkaline soil reaction, pH-Truog was 6,0 in the upper horizon and 8,0 in the lower horizon.

Table 2 showed that the soils had a clay texture, a very acidic reaction (pH 3,1-5,0), low electric conductivity, moderate to high levels of organic matter, low to high total P, high total K, medium to high base saturation, and low Al saturation. At the bottom layer, the soils contained sulfidic materials/pyrite (1.75-

2.09%) at a depth of 50-100 cm, which is characterized by a high effervescence after been mixed with H_2O_2 liquid and then the soil pH dropped to 1,5-2,5. The soils were classified into Aluvial Sulfik and Aluvial Sulfidik (Subardja et al., 2016) as equal to Sulfic Endoaquents and Typic Sulfaquents (Soil Survey Staff, 2014).

3.2.3 *Gleisol Soils*
Gleisol soils are found on the fluvio-marine landform and classified as having a structural development with A-Bg-C horizon arrangement. The soils were formed from fluvio-marine (clay and silt) deposits, affected by fluctuations of the groundwater level and soil tillage activities (rice field) which then forms mottles, especially at depths <50 cm. The bottom layer (> 50 cm) was generally saturated with water, making the soil have gray colors without mottles.

The soils were generally fine textured, sticky, and has a plastic consistency, very acidic to slightly acidic soil reaction (pH 4.2-6.5) (Table 2), and low electrical conductivity. The organic content was high within 20-40 cm depth and classified as histic epipedon.

The soils showed low to moderate total P content, high total K, high exchangeable bases, and base saturation, and low exchangeable Al. Some subsurface layers contained sulfidic materials with a level of 0.73 to 2.65 %. The soils were classified into Gleisol Sulfik, Gleisol Humik and Gleisol Eutrik (Subardja et al., 2016) as equal to Sulfic Endoaquepts, Typic Humaquepts and Typic Endoaquepts (Soil Survey Staff, 2014). Figure 5 showed the distribution of peat and mineral soils in the Mendol Island.

3.3 *Land potential for food crop development*

Based on the land suitability evaluation, there are several agricultural commodities that have great potentials to be developed in some parts of Mendol Island (Table 3).

Figure 4. Corn and cassava cultivated on peat dome.

Table 2. Physical and chemical properties of mineral soils from Mendol Island.

Depth	Texture				pH		EC	Org matter		ex.HCl 25%		Cation-exchange (NH4oAc) pH 7					BS*	KCl 1N		Content of		
	Sand	Silt	Clay	Class	H$_2$O	KCl		C	N	P$_2$O$_5$	K$_2$O	Ca	Mg	K	Na	CEC		Al^{3+}	H$^+$	S	Fe	Pyrite
cm	—%—						dS	—%—		—me/100g—		—cmol(+)/kg—					%	—cmol(+)/kg—		—%—		
Aluvial Sulfik (Sulfic Endoaquents)																						
0-25	10	33	57	C	4.8	4.1	1.0	2.24	0.21	31	108	7.19	2.50	0.67	3.73	22.77	62	0.34	0.17	-	-	-
25-55	10	37	53	C	3.6	3.1	1.2	3.31	0.31	23	223	3.61	3.06	0.68	4.54	20.96	57	6.53	1.35	-	-	-
55-90	10	37	53	C	5.0	4.6	3.9	2.49	0.23	35	197	9.89	2.16	2.45	8.34	21.18	>100	0.00	0.16	0.93	2.69	1.75
90-120	10	39	51	C	7.9	7.5	2.7	3.18	0.31	47	212	10.54	2.48	2.41	8.45	22.59	>100	0.00	0.09	1.11	2.96	2.09
Gleisol Sulfik (Sulfic Endoaquepts)																						
0-10	12	40	48	SiC	6.0	5.1	3.0	1.41	0.13	20	74	4.97	6.72	0.89	7.37	20.23	99	0.00	0.08	-	-	-
10-35	14	41	45	SiC	5.8	5.0	1.2	3.54	0.33	8	76	5.53	2.89	0.69	4.61	18.10	76	0.00	0.08	0.01	2.00	0.02
35-55	13	43	44	SiC	5.5	4.9	1.6	6.31	0.22	10	100	7.32	2.42	1.08	5.46	24.65	66	0.00	0.08	0.39	2.41	0.73
55-80	11	40	49	C	3.9	3.5	3.0	4.13	0.41	13	135	8.17	2.43	1.40	5.60	20.42	86	0.77	1.42	1.42	3.21	2.65

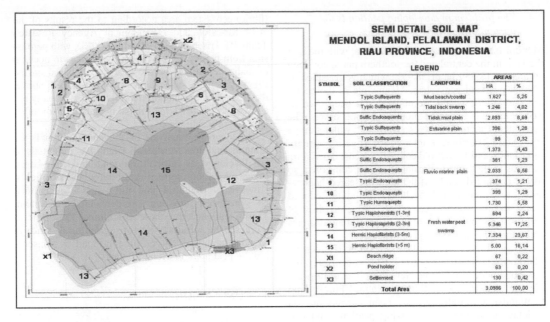

Figure 5. Semi-detailed soil map of the Mendol Island.

Table 3. Land suitability class for several agriculture commodities.

Symbol	Soil Classification (Dominant composition)	Landform	Commodities				Areas	
			Rice	Corn	Vegetable	Sago	Ha	%
1	Typic Sulfaquents	Mud beach/coastal	S3 x	N x	S3 x	N x	1.627	5,25
2	Typic Sulfaquents	Tidal backswamps	S3 x	N x	S3 x	N x	1.246	4,02
3	Sulfic Endoaquents	Tidal mud plain	S2 x	S3 x	S2 x	S3 x	2.693	8,69
4	Typic Sulfaquents	Estuarine plain	S3 x	N x	S3 x	N x	396	1,28
5	Typic Sulfaquents		S3 x	N x	S3 x	N x	99	0,32
6	Sulfic Endoaquepts		S2 x	S3 x	S2 x	S3 x	1.373	4,43
7	Sulfic Endoaquepts		S2 x	S3 x	S2 x	S3 x	381	1,23
8	Sulfic Endoaquepts	Fluvio marine plain	S2 x	S3 x	S2 x	S3 x	2.033	6,56
9	Typic Endoaquepts		S2 n	S2 n	S2 n	S2 nr	374	1,21
10	Typic Endoaquepts		S2 n	S2 n	S2 n	S2 nr	399	1,29
11	Typic Humaquepts		S2 n	S2 n	S2 n	S2 nr	1.730	5,58
12	Typic Haplohemists (1-3m)		S3 r	N r	S3 r	S3 r	694	2,24
13	Typic Haplosaprists (2-3m)	Fresh water	Nr	Nr	S3 r	S3 r	5.346	17,25
14	Hemic Haplofibrists (3-5m)	ombrogenous peat	Nr	Nr	Nr	Nr	7.334	23,67
15	Hemic Haplofibrists (>5 m)		Nr	Nr	Nr	Nr	5.00	16,14
X1	Beach ridge						67	0,22
X2	Pond holder						63	0,20
X3	Settlement						130	0,42
Total Area							**3.0986**	**100,00**

Note: Land suitability class: S2 = moderately suitable; S3 = marginally suitable; N = Not suitable Constraints: x = Toxicity; r = Rooting condition; n = Nutrient availability

3.4 The problem in developing peatlands for agricultural use

Mendol Island has an area of 30,983 ha, with most of the land in the central and the southern parts, covering an area of 18,259 ha or 58.93% which is classified as a thick peatland ecosystem. Meanwhile in the northern part of the island are mineral soil been used as rice fields and for some productive food crops. The ecosystem functions as a reservoir and a water storage source and its resources are used for irrigating agricultural lands and household/domestic needs. Therefore, it needs to be protected/maintained and conserved, so that it doesn't degrade. If the land is cleared haphazardly, the impact will be huge, including reduced water supply, subsidence/decline, prone to drought and fires, the occurrence of the microclimate changes and increase greenhouse gas emissions.

Since Inpres (Instruksi Presiden/Presidential Instruction) No.10, the year 2011 concerning the Postponement of Granting New Permits and Improving the Governance of Primary Natural Forests and Peatlands, a moratorium has been carried out on a delay in opening and utilizing peatland and primary forests for plantations, such as oil palm. The Presidential Instruction was later renewed and strengthened by the issuance of PP (Peraturan Pemerintah/Government Regulation) No.71 of 2014 concerning the Protection and Management of Peatland Ecosystems, to preserve the function in a systematic and integrated manner and also to prevent damage. Following up on the Presidential Instruction and the Government Regulation, the Kepmenhut (Keputusan Menteri Kehutanan/Ministry of Forestry Issued) No. SK.3706/Menhut-VII/IPSDH/2014VII/IPSDH/2014) concerning the Determination of Indicative Maps for Delays in Giving New Permits for Forest Utilization, use of Forest Areas and Changes to Allocation of Forest Areas and Other Use Areas (VI Revision), which states, among others, that there should be a delay in the granting of new plantation business licenses to Other Use Areas if it is in the form of peatland (which is renewed every 6 months).

In the case where the utilization of peatland is permitted for the cultivation of palm oil, if the Plantation Business Permit (IUP) has been obtained and has not expired, then it is regulated by Permentan (Peraturan Menteri Pertanian/Minister of Agriculture Regulation) No. 14/Permentan/PL.110/2/2009 namely; it is in its cultivation area, peat thickness is less than 3 meters, substratum or the mineral layer under the peat is not quartz sand or not acid sulfate soil (containing pyrite>2%), and the maturity of the peat includes saprik (ripe) or hemic (rather mature). Fibric peat (raw) is not permitted for oil palm cultivation.

By paying attention to the Forest Area Map of the Kuala Kampar District in the Pelalawan Regency, it turns out that the middle part of the Mendol Island is included as APL (Area Penggunaan Lain/Other Used Area) which is precisely included in the thick and raw (fibric) ecosystem area according to the results of the 2014 Research and Development Agency Research Team, PT Trisetia Usaha Mandiri (TUM), with permission from the local government, with plans to open an oil palm plantations on the peatland. If based on the PP, Kepmenhut, and Kepmentan mentioned above, then the area of peatland, Mendol island is actually not allowed to be used for oil palm plantation or other land clearing activities.

Based on the analysis of the land potential, the mineral land area in Mendol Island has potentials to be developed as agricultural land and food-based granary for food crops of rainfed lowland rice fields. However, agricultural land and peatland areas must be synergized into one interconnected ecosystem. Taking into account these conditions, the effort to increase the productivity of agricultural land in Mendol Island which can be continued by conserving the peat area on Mendol Island.

4 CONCLUSION

1. Mendol Island is a peat hydrological unity (PHU) which consists of 59% peat soils and 41% mineral soils, has two functions cultivation and conservation.
2. Peat soils of the area were grouped as ombrogenous peat,and classified as Typic Haplohemists and Sapric Haplohemists. The peat thickness varied from 2 to > 7 m, hemic maturity in the upper layer and fibric maturity in the lower layer, low ash content (<5%), very acidic reaction (pH 3.8-4.1) and low nutrient status). The substratum was clay texture and a very acidic reaction (pH 3.7).
3. Mineral soils were classified as Aluvial Sulfidik, Aluvial Sulfik, Gleisol Sulfurik, Gleisol Humik and Gleisol Eutrik as equal to Typic Sulfaquents, Sulfic Endoaquents and Sulfic Endoaquepts, Typic Humaquepts and Typic Endoaquepts. They are generally clay textured, deep cross-sectioned, poor drainage, acid reactions, and some of them contained sulfidic materials.
4. The peat dome of Mendol Island functions as a reservoir and storage for rainwater resources for the purpose of irrigating the rice fields and domestic needs. The availability of water is the key to the success of increasing food production based on rice plants. Therefore, the peatland ecosystem must be conserved (INPRES No.10/2011, PP No 71/2014, Kepmenhut No. 3706/2014, dan Permentan No. 14/2009).

SUGGESTION

The empowerment of the peat hydrological areas unity on Mendol Island must be used as a protected and cultivated area (wet and dry land cultivations) managed in an integrated manner.

ACKNOWLEDGMENT

The authors would like to thank the soil surveyors: Mr. L. Muslihat, Mr. D. Sudrajat, Mr. W. Gunawan and Mr. Soleh for their assistance in collecting raw data and soil samples during the fieldwork.

REFERENCES

Bachri, S., Sulaeman, Y., Ropik, Hidayat, H. & Mulyani, A. 2016. *Sistem Penilaian Kesesuaian Lahan Versi 2.0*. Bogor: Balai Besar Penelitian dan Pengembangan Sumberdaya Lahan Pertanian, Bogor.

BBSDLP (Balai Besar Litbang Sumberdaya Lahan Pertanian). 2011. *Peta Lahan Gambut Indonesia Skala 1:250.000. Edisi Desember 2011*. Bogor: Balai Besar Penelitian dan Pengembangan Sumber Daya Lahan Pertanian, Badan Penelitian dan Pengembangan Pertanian.

BBSDLP (Balai Besar Litbang Sumberdaya Lahan Pertanian). 2017. *Atlas Lahan Gambut Indonesia*. Bogor: Balai Besar Penelitian dan Pengembangan Sumberdaya Lahan Petanian, Badan Penelitian dan Pengembangan Pertanian.

Dinas Pertanian Tanaman Pangan Kabupaten Pelalawan. 2014. Kerangka Acuan Kerja (KAK) Identifikasi dan Karaterisasi Potensi Pening-katan Produksi Padi Lahan Pasang Surut di Kecamatan Kuala Kampar, Kabupaten Pelalawan, Provinsi Riau.

Driessen, P. & Sudjadi, M. 1984. Soils and specific soil problem of tidal swamps. *Workshop on Re-search Priorias in Tidal Swamp Rice*: p143-160. Los Banos, Laguna, Philippines: IRRI.

Eviati & Sulaeman. 2011. *Petunjuk teknis analisa kimia tanah, tanaman, air, dan pupuk. Edisi 2*. Bogor: Balai Penelitian Tanah.

FAO. 1990. *Guidelines for soil profile description. Soil Bulletin No*. Rome, Italy: FAO/UNESCO.

Hardjowigeno S. 1986. *Sumber Daya Fisik Wilayah dan Tata Guna Lahan: Histosol*: 86-94. Bogor: Fakultas Pertanian IPB.

Hidayat, A., Hikmatullah, Sukarman & Wahyunto. 2011. *Survei dan Identifikasi Sumberdaya Lahan Lokasi Dem plot Lahan Gambut di Kalimantan Tengah, Kalimantan Selatan, Riau dan Jambi*. Bogor: Balai Besar Litbang Sumber Daya Lahan Pertanian, Badan Litbang Pertanian.

Instruksi Presiden (Inpres) No. 10 Tahun 2011 tentang Penundaan Pemberian Izin Baru dan Penyempurnaan Tata Kelola Hutan Alam Primer dan Lahan Gambut. Berlaku mulai tanggal 20 Mei 2011.

Keputusan Menteri Kehutanan RI No. SK.3706/Menhut-VII/IPSDH/2014 tentang Penetapan Peta Indikatif Penundaan Pemberian Izin Baru Pemanfaatan Hutan, Penggunaan Kawasan Hutan dan Perubahan Peruntukan Kawasan Hutan dan Areal Penggunaan Lain (Revisi VI). Ketetapan mulai berlaku tanggal 13 Mei 2014.

Masganti, Wahyunto, Dariah, A., Nurhayati, & Yusuf, R. 2014. Karakteristik dan potensi pemanfaatan lahan gambutterdegradasi di Provinsi Riau. *Jurnal Sumberdaya Lahan* 8(1): 59-66.

Mulyani, A. & Sarwani, M. 2013. Karakteristik dan potensi lahan suboptimal untuk pengembangan pertanian di Indonesia. *Jurnal Sumberdaya Lahan* 7(1): 47-55.

Oldeman L.R., Las, I. & Darwis, M. 1978. *Agro-climatic map of Sumatra, scale 1:3.000.000*. Bogor: Contr. Central Res. Inst. of Agric.

Peraturan Pemerintah RI No. 71 Tahun 2014 tentang Perlindungan dan Pengelolaan Ekosistem Gambut. Berlaku mulai tanggal 12 September 2014.

Peraturan Menteri Pertanian RI No.14/Permentan/PL.110/2/2009 tentang Pedoman Pemanfaatan Lahan Gambut untuk Budidaya Kelapa Sawit. Berlaku tgl. 16 Feb. 2009.

Polak B. 1950. Occurrence and fertility of tropical peat soils in Indonesia. 4th Int. Congr.of Soil Sci. Vol 2, 183-185. Amsterdam, The Netherlands.

Pusat Penelitian Tanah dan Agroklimat. 1990. Peta satuan lahan dan tanah tingkat tinjau lembar Rengat skala 1:250.000. Proyek LREP I Sumatera, Puslit Tanah dan Agroklimat, Bogor.

Schmidt, F.H. & Ferguson, J.H.A. 1951. Rainfall types based on wet and dry period ratios for Indonesia with Western New Guinea. Verh.No. 42. Djawatan Meteorologi dan Geofisik, Jakarta.

Soil Survey Staff. 2014. *Keys to Soil Taxonomy. 12nd.ed*: 333p. Washington DC: NRCS-USDA.

Soepraptohardjo M. 1961. *Klasifikasi Tanah di Indonesia*. Bogor: Balai Penyelidikan Tanah Bogor.

Subardja, D., Ritung, S., Anda, M., Sukarman, Suryani, E. & Subandiono, R.E. 2016. *Petunjuk Teknis Klasifikasi Tanah Nasional. Edisi Kedua 2016*: 53 hal. Bogor: Balai Besar Libang Sumberdaya Lahan Pertanian, Badan Litbang Pertanian.

Sulaiman, A.A., Sulaeman, Y. and Minasny, B., 2019. A Framework for the Development of Wetland for Agricultural Use in Indonesia. Resources, 8(1), p.34.

Tim Peneliti BBSDLP. 2014. Laporan Akhir Identifikasi dan Karakterisasi Potensi Peningkatan Produksi Padi Lahan Pasang Surut di Kecamatan Kuala Kampar (P. Mendol), Kabupaten Pelalawan, Provinsi Riau. Kerjasama BBSDLP Badan Litbang Pertanian dan Dinas Pertanian Tanaman Pangan Kabupaten Pelala-wan, Riau.

Van Wijk, C.L. 1951. Soil survey of tidal swamp of South Borneo in connection with the agricultural possibilities. *Contr. Agric. Res. Sta* 123:1-49.

Verstappen H. 1975. On paleo-climates and landform developments in Malesia. In: *Mod. Quat.Res. Inst. S.E. Asia*. The Netherlands: Groningen University.

Widjaja Adhi, I.P.G. 1986. Pengelolaan lahan rawa pasang surut dan lebak. *Jurnal Penelitian dan Pengembangan Pertanian* 5(1):1-9.

Part D. Peatland management

Tropical Wetlands — Innovation in Mapping and Management — Sulaeman et al. (Eds)
© 2020 Taylor & Francis Group, London, ISBN 978-0-367-20964-3

Dissolved organic carbon in tidal and fresh water peatlands

S. Nurzakiah & Nurita
Indonesian Swampland Agriculture Research Institute, Banjarbaru, South Kalimantan, Indonesia

D. Nursyamsi
Indonesian Center for Agricultural Land Resources Research and Development, Bogor, Indonesia

ABSTRACT: Peatlands are ecosystems with high carbon stocks that can be under the influence of both tidal and fresh water swamps. Whereas mineral content is usually low in tidal peat, it can be rather high in the peat of fresh water swampland. The presence of minerals has an effect on the humic acid content and the amount of Dissolved Organic Carbon (DOC) in the system. The purpose of this study was to determine the concentration of DOC in peatlands of tidal and fresh water swampland. The study was conducted by a survey method, followed by field sampling of the soil and water. The study was carried out in Jabiren Village, Central Kalimantan Province (tidal land) and Pulau Damar Village, South Kalimantan Province (fresh water swampland). Water samples were taken from the upper and middle surface level of the water source. Dissolved organic concentrations were measured by thermo-catalytic oxidation on a TOC-VCPN Analyzer (Shimadzu). Humic acid content is determined based on the difference in peat maturity degree with using a gravimetric instrument. The results showed that DOC in fresh water swampland was lower (about 37%) compared to tidal land. The lower end of the DOC concentration can also be caused by the formation process and origin of peat material, peat thickness and peat maturity degree. Besides the loss of carbon as a gas, knowledge of dissolved organic carbon is required as one of the parameters of carbon loss in an ecosystem. Recently, peatlands have been used for many different purposes, which has resulted in being drawn to issues relating to the environment. Peatlands can be productive agricultural lands if with the proper method is utilized. One such method is water management. The application of canal blocking in drainage channels has the effect of reducing the aerobic decomposition of peat and DOC production, thus lowering DOC concentration.

1 INTRODUCTION

Peatlands have unique characteristics and require accurate and precise management since mismanagement can result in soil and environmental damage. The characteristics of peat soils are determined by the composition of the original material (type and variety of plants), as well as the condition of the environment during its formation. Environmental factors that can be considered are climate, hydrology, topography, and time. Peat forming factors of interactivity, so that the characteristics of peatlands are specific and different. Peat contains of a humic substance such as humic acid and fulvic acid that has an important role in the soil properties. This role is influenced by the degree of decomposition of peat. The content of humic acid was higher in sapric than fibric.

Peat soils can generally be found in tidal and fresh water swamplands. In fresh water typology, layers can be observed. These layers are all derived from peat or mineral materials. These conditions will result in differences in the properties of the soil and water. Soil fertility is one of the properties which can be used as a differentiator between the two typologies. Peat soils with mineral substance insertions are relatively more fertile because of the enrichment of nutrients from river deposits. The presence of a mineral substance on peat also affects the production and concentration of dissolved organic carbon (DOC).

DOC is one of the three main components of carbon balance, after CO_2 and CH_4 (Ryder et al., 2012) and represents a significant carbon flux of peatland, which is about 24% of the net C uptake within the ecosystem (Dinsmore et al., 2010). The movement of DOC through the soil is an important process for the transportation of carbon in ecosystems and the formation of soil organic matter. In some cases, the DOC flux may also contribute to the energy balance in the river as well as the transfer of nutrients from terrestrial to aquatic ecosystems. Dissolved organic carbon plays an important role in determining the activity of microorganisms through their carbon supply as well as the distribution of carbon throughout the soil horizon. The dissolved organic carbon that is found in river water or drainage channels mainly comes from soil carbon (Aitkenhead and McDowell, 2000). Due to this effect, if the carbon content in the soil is high, such as can be seen in peat soil, there will be a release of DOC into the river or drainage channel in large quantities. This is an important intermediary for the global carbon cycle. The release of DOC from organic soil stems from the cumulative effect of a few factors: I) Sources of organic matter II) Land use practices III) Hydrology and, IV) Various biogeochemical processes such as the decomposition of organic matter by microbes as well as redox reactions V) The position of the peatlands. The purpose of this study was to determine the concentration of DOC in peatlands of tidal and fresh water swampland.

2 METHODS

The study was conducted using a survey method, followed by a field sampling of the soil and water. The study was carried out in Jabiren Village, Central Kalimantan Province at the coordinates 02°51'48.6" S and 114°17'00.2" E (tidal land); and Pulau Damar Village, South Kalimantan Province at the coordinates 2° 26' 32.93"S, 15° 22' 24.89" E (fresh water swampland) from February-October of 2013. The tidal peatland sites are located between the secondary channels, connecting the primary channel to the Jabiren river, which directly empties into the Kahayan river. The water sample points were determined based on the inlet (1) and outlet (2) of the study area (Figure 1).

In fresh water swampland, there is only one channel, which has the function to regulate water in and out the study area. Water samples were taken from the upper surface and the middle level of the water. Water samples were filtered with a 0.45 μm filter, then DOC measurements were taken with a TOC-VCPN Analyzer (Shimadzu), through the process of burning dissolved organic matter at 680°C. The combustion process is intended to convert CO_2 into DOC by non-dispersive infrared (NDIR) sensors. Soil samples were taken between depths of 0-100 cm using a peat auger (Eijkelkamp model). Humic acid is determined based on the difference in peat maturity degree under the effect of gravity.

Variation data was analyzed with standard error and illustrated with a Sigma Plot program.

3 RESULTS

Peatlands play a role in the sequestering of carbon in the environment. In the natural peatlands, export of DOC can be an important part of the total carbon balance (Dinsmore et al., 2010) and the change in the export of DOC can change the balance of carbon, from carbon sink into carbon source or conversely. The quality and quantity of soil organic matter, as well as the position of peatlands, affect the concentration of DOC. The rivers were formed naturally in the peatlands, and the drainage channels that were created during land clearing will release DOC. Dissolved organic carbon can be lost, either to the surrounding waters or sequestrated into the peat soil. Therefore, DOC plays a role in the distribution of carbon throughout the soil horizon. Carrera et al. (2011) suggested that the concentration and fluctuations of DOC were influenced by biological factors (such as microorganisms and vegetation), environmental factors (temperature, humidity, etc.) and their interactions, while the seasonal dynamics of DOC concentrations are influenced by hydro-meteorological conditions (Clark et al., 2007; Dinsmore et al., 2013; Fasching et al., 2016). It was observed that hydrology had a first-order effect on DOC concentrations for individual samples (Laudon et al., 2011).

The lower concentration of DOC on fresh water swampland is influenced by the characteristics of peat, such as its thickness and maturity degree as well as the formation process and origin of the peat material. Fresh water swampland has a lower concentration of about 37% when compared to tidal peatlands (Figure 2). Thickness and maturity degree of peat affect soil carbon stock (Figure 3). Sapric has a greater carbon stock than hemic and fibric

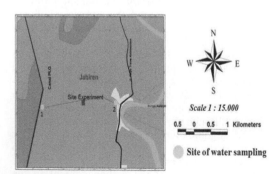

Figure 1. Sites of water sampling on tidal peat.

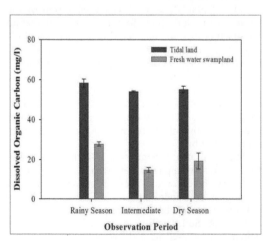

Figure 2. Dissolved organic carbon fluctuations.

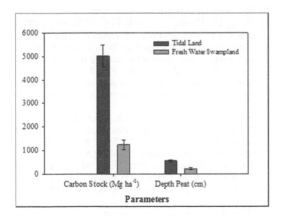

Figure 3. Carbon stock and depth peat at the research location.

From field observations, peat thickness in tidal land ranged from 439 cm to 625 cm, with the dominant peat maturity degree coming from sapric and hemic. The humic acid content ranged between 23.07-26.02%. Peat thickness in fresh water swampland varied between 72 cm to 465 cm, with the dominant peat maturity degree coming from hemic and fibric. The humic acid content ranged between 7.05-7.11%. With the same degree of peat maturity, which is hemic, the humic acid content of both areas of study had different values, up to 30.6%. This indicates that the formation process and the origin of peat substances play a substantial role in influencing the production of dissolved carbon. According to Hope et al. (1994) in Moore et al. (2011). Dissolved organic carbon contains about 50-75% fulvic and humic acid.

Humic acid is the main organic element found widely distributed in peat soil. It contains many functional groups such as carboxylic acids, alcohols, phenols, carbonyl, phosphates, sulfates, amides, and sulfides, and all these groups can interact with metal species (Stevenson, 1994).

The high DOC concentration can be observed from the color of peat water. Black water contains a lot of humic acids while water that is slightly black-brownish has a lower humic acid. This is due to the influence of the overflowing of rivers carrying mineral substances away, resulting in a decrease in dissolved organic carbon concentration. This was seen in the fresh water swampland

The DOC concentration in the rainy season is higher than in the dry season. In the dry season, the carbon is stored in the soil for such a long time, that during the rainy season, there is an increase in precipitation. This "flushing effect" (Hornberger et al., 1994) transfers carbon into aquatic ecosystems. The process of "flushing" increases when the upper layer of the soil (acrotelmic peat), that was previously dry or stagnant, is flooded with a large amount of water as a result of increasing water discharge. This is in line with the study conducted by

Coynell et al. (2005) that there is a positive correlation between water discharge and DOC concentrations on the Congo river with land use types namely pine, savanna and swamp forest. Similarly, in the results, Ryder et al. (2012) reported that hydrological changes, such as increasing water discharge, have the potential to increase DOC exports. In the dry season, the mechanism of loss of DOC occurs through flocculation into particulate organic carbon (POC) or adsorption by existing POC and by mineral particles, resulting in reduced solubility of the DOC with increasing salinity (Battin et al., 2008). Salinity in the tidal peatlands can occur during a long dry season, causing intrusion of seawater that transports dissolved salts in large quantities.

Changes in the season, like the rainy and dry season, that have an impact on fluctuations in the groundwater will also affect the decomposition and mineralization processes. The quantity and quality of DOC (Clark et al., 2005) will also be affected. Thus, any biological or physical factor that increases decomposition will cause an increase in DOC concentration, although Kalbitz et al. (2000) stated that hydrological variation in soil-rich organic horizons might be more important than biotic factors. It should be considered that, along with changes in the hydrology of peatlands, CO_2-C fluxes are the release of C into the air, changing and affecting the amount of residual DOC in peat.

Peat drainage can affect the production and export of DOC, with increased production of DOC. Therefore, the concentration of DOC due to increased peat decomposition in the aerobic zone will become greater. It was observed that drainage on peatlands increases the concentration and loss of DOC (Evans et al., 2016). The application of canal blocking in the drainage channel has the effect of reducing peat aerobic decomposition and DOC production, thereby reducing the concentration of DOC (Turner et al., 2013). Gibson et al. (2009) argued that management intervention techniques do not reduce production but instead changes the results of DOC.

Climate change can affect the balance of DOC from peatland through several mechanisms (Pastor et al., 2003). First, an increase in temperature can increase the production of DOC consumed by microbes, thereby changing the concentration of DOC in drainage water. Secondly, groundwater fluctuations can change the concentration of DOC with regard to redox conditions. Lastly, changes in water balance can control DOC exports independently of any changes in DOC Concentration.

4 CONCLUSIONS

The concentration of dissolved organic carbon in peatlands of tidal and fresh water swampland are influenced by the characteristics of peat, such as its thickness and maturity degree as well as the formation process and origin of the peat material.

REFERENCES

Aitkenhead, J.A. & McDowell, W.H. 2000. Soil C:N ratio as a predictor of annual riverine DOC flux at local and global scales. *Global Biogeochem. Cy* 14:127–138.

Battin, T.J., Kaplan, L.A., Findlay, Hopkinson, C.S., Marti, E., Packman, A.I., Newbold, J.D. & Sabater, F. 2008. Biophysical controls on organic carbon in fluvial networks. *Nat. Geosci* 1:95–100.

Carrera, N., Barreal, M.E., Rodeiro, J. & Briones, M.J.I. 2011. Interactive effects of temperature, soil moisture and enchytraeid activities on C losses from a peatland soil. *Pedobiologia* 54:291–299.

Clark, J.M., Chapman, P.J., Adamson, J.K. & Lane, S.N. 2005. Influence of drought-induced acidification on the mobility of dissolved organic carbon in peat soils. *Glob Change Biol* 11(5):791–809.

Clark, J.M., Lane, S.N., Chapman, P.J. & Adamson, J.K. 2007. Export of dissolved organic carbon from an upland peatland during storm events: Implications for flux estimates. *J Hydrol* 347:438–447.

Coynel, A., Seyler, P., Etcheber, H., Meybeck, M. & Orange, D. 2005. Spatial and seasonal dynamics of total suspended sediment and organic carbon species in the Congo River. *Global Biogeochem. Cy* 19:GB4019.

Dinsmore, K.J., Billett, M.F., Skiba, U.M., Rees, R.M., Drewer, J. & Helfter, C. 2010. Role of the aquatic pathway in the carbon and greenhouse gas budgets of a peatland catchment. *Glob Change Biol* 16(10):2750–2762.

Dinsmore, K.J., Billett, M.F., & Dyson, K.E. 2013. Temperature and precipitation drive temporal variability in aquatic carbon and GHG concentrations and fluxes in a peatland catchment. *Glob Chang Biol* 19:2133–2148.

Evans, C.D., Renou-Wilson, F. & Strack, M. 2016. The role of waterborne carbon in the greenhouse gas balance of drained and re-wetted peatlands. *Aquat Sci* 78 (3):573-590.

Fasching, C., Ulseth, A.J., Schelker, J., Steniczk, G. & Battin, T.J. 2016. Hydrology controls dissolved organic matter export and composition in an alpine stream and its hyporheic zone. *Limnol Oceanogr* 61(2):558–571.

Gibson, H.S., Worrall, F., Burt, T.P. & Adamson, J.K. 2009. DOC budgets of drained peat catchments: implications for DOC production in peat soils. *Hydrol Process* 23 (13):1901–1911.

Hornberger, G.M., Bencala, K.E. & Mcknight, D.M. 1994. Hydrological controls on dissolved organic-carbon during snowmelt in the Snake River near Montezuma, Colorado. *Biogeochemistry* 25:147–165.

Kalbitz, K., Solinger, S., Park, J.H., Michalzik, B. & Matzner, E. 2000. Controls on the dynamics dissolved organic matter in soils: a review. *Soil Sci* 165(4):277–304.

Laudon, H., Berggren, M., Ågren, A., Buffam, I., Bishop, K., Grabs, T., Jansson, M. & Köhler, S. 2011. Patterns and dynamics of dissolved organic carbon (DOC) in Boreal Streams: The role of processes, connectivity, and scaling. *Ecosystems*. 14:880–893.

Moore, S., Gauci, V., Evans, C.D. & Page, S.E. 2011. Fluvial organic carbon losses from a Bornean blackwater river. *Biogeosciences* 8:901–909. doi:10.5194/bg-8-901-2011.

Pastor, J., Solin, J., Bridgham, S.D., Updegraff, K., Harth, C., Weishampel, P. & Dewey, B. 2003. Global warming and the export of dissolved organic carbon from boreal peatlands. *OIKOS* 100: 380–386.

Ryder, E., de Eyto, E., Dillane, M., Poole, R., Linnane, S. & Jennings, E. 2012. Impact of climate on export of DOC from a peatland catchment. Annual Meeting of Freshwater Biologists. Trinity College Dublin. 2nd of March 2012.

Stevenson, F.J. 1994. Humus Chemistry: Genesis, composition and reaction. Sec. Edition. JohnWilley & Sons Inc. NewYork. 496 p.

Turner, E.K., Worrall, F. & Burt, T.P. 2013. The effect of drain blocking on the dissolved organic carbon (DOC) budget of an upland peat catchment in the UK. *J Hydrol*. 479:169–179.

Soil amelioration on peat and its effect on methane (CH$_4$) emission and rice yield

H.L. Susilawati & A. Pramono
Division of Greenhouse Gas Emission and Absorption, Indonesian Agricultural Environment Research Institute, Pati, Central Java, Indonesia

P. Setyanto
Directorate of Vegetables and Medicinal Crop, Directorate General of Horticulture. South Jakarta, Indonesia

K. Inubushi
Graduate School of Horticulture, Chiba University, Matsudo, Japan

ABSTRACT: Agricultural land expansion into marginal areas, including peatlands, to meet increasing demands of agricultural product is a significant human intervention to the global environment because peatlands are a vital part of the global carbon cycling process. Increasing agricultural productivity of peatlands through land management strategies generally involve trade-offs to other objectives, such as achieving greenhouse gas (GHG) mitigation and other environmental issues. Synergistic options should be considered as the main objective to minimize the trade-offs. The aim of this study is to investigate whether the option of land management through soil amelioration can be used as a strategy to increase rice production, as well as mitigate methane (CH$_4$) emission in peat soils. The effects of different soil ameliorations on CH$_4$ emissions from peat soil under rice cultivation were studied using automatic closed chamber measurements. Twelve plots (approx. 25 m^2) were laid out and arranged using a Randomized Block Design (RBD). The data was collected from 6 consecutive rice cultivation periods. The results showed that CH$_4$ was emitted 4 – 42% lower when soil ameliorations were applied to the cultivated peat soils. The variations in CH$_4$ emissions between rice cultivation periods were large. Soil amelioration increased rice production by 2–15 %. Soil amelioration, e.g., zeolite, steel slag, manure, silicate fertilizer, compost, nitrification inhibitor, volcanic ash, and steel slag, can be recommended as a better option for increasing rice yield, as well as limiting CH$_4$ emissions from cultivated peat soils. Soil amelioration is vital for soil quality, agricultural production, as well as the reduction of CH$_4$ emissions from peat soil, but the usage of soil amelioration is strongly influenced by the economics of farming.

1 INTRODUCTION

Nowadays, the world faces an important threat of food security due to the impact of overpopulation and climate change. Rice (*Oryza sativa* L.) is an important staple crop in many developing countries and has become a major crop in many developed countries, consumed by more than half of the world's population. The consumption of rice is predicted to increase by about 24% in the next 20 years, and the world food production needs to be enhanced up to 70% to meet the demand for food (FAO, 2009; Patel et al., 2010). Rice is consumed and produced in China, India, Indonesia, Bangladesh, and Vietnam. In Japan, around 90% of the world's rice is produced and consumed (Maclean, 2002). Indonesia is the third-largest rice producing country in the world after China and India, as well as the world's biggest rice consumer. Although, as the third-largest rice producer, rice production in Indonesia is still not enough to fulfill its consumption. Consumption of rice in Indonesia is

around 139 kg capita^{-1} year^{-1}; it is higher than the average of the world's rice consumption (Jati, 2014). Figure 1 shows Indonesian rice production, total consumption, and harvested area during 1960-2018 (Ito, 2018). Since 1960, rice consumption in Indonesia has exceeded rice production. The demand for rice is estimated to rise, while the area for rice cultivation is predicted to decrease due to the increase in the human population. The Indonesian government is fully committed to attaining food self-sufficiency through rice intensification and opening new paddy field areas in marginal lands.

Agriculture is one contributor to three primary greenhouse gas (GHG) emissions: carbon dioxide (CO$_2$), methane (CH$_4$), and nitrous oxide (N$_2$O). National GHG emission, including land use, land-use change, and forestry (LULUCF), and peat fire sector for CO$_2$, CH$_4$, and N$_2$O, was estimated at approximately 1,844,329 Gg CO$_2$e in 2014 (MoE, 2017; Table 1). The agriculture sector contributed 6.15% from

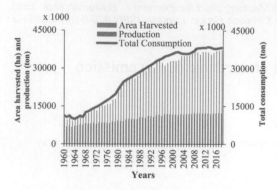

Figure 1. Area harvested, production and total consumption of rice in Indonesia during 1960–2018.

national GHG emission. According to Presidential Regulation no 61 year 2011, there are 2 targets for the agricultural sector to reduce GHG emission in Indonesia by 2020, i.e., business as usual (unconditional scenario) and with sufficient international support (conditional scenario) with the reduction of emission targets set around 26% and 41% below the national baseline emission level, respectively, while by 2030, it is targeted at 29% (unconditional) and 38% (conditional; Table 1).

Agriculture has potential options to reduce GHG emission through land management and agricultural farming practices with low GHG emissions. Agriculture can also be a sink for CO_2 through C sequestration into soil organic matter (SOM) and biomass products. Rice cultivation is one major source of a potent GHG emission from the agricultural sector, namely CH_4. Rice cultivation contributed about 11% of the global anthropogenic CH_4 emission (Ciais, 2013). More than 90% of CH_4 transportation from rhizosphere and oxygen diffusion into roots under anaerobic condition is released through the aerenchyma and intercellular space system of rice plants in roots, culm, leaf sheaths, and leaf blades, and the rest is emitted through the bubbles (Schütz et al., 1989). If the intensification and land use change to increase rice production is undertaken, without regard for the environment, it will substantially increase

GHG emissions from paddy fields. Many cultivation practices have been improved to minimize the trade-off between rice production and environmental issues, e.g., using high-yielding low emission crop varieties, balanced fertilization, and water saving irrigation.

In recent years, agricultural area has been reduced due to the conversion of agricultural land into residential and industrial areas and other functions. There is increased competition for usage of natural resources, e.g., water, land, energy, and other inputs into food production. Consequently, the other way to meet increasing demand of agricultural products is to look to new areas of potential land for agricultural activities. Tropical forest peatlands in Indonesia have largely been converted into agricultural and non-agricultural sectors. Natural and converted tropical peatlands are sources of CO_2, CH_4, and N_2O due to the high content of Carbon (C) and Nitrogen (N) (Arai et al., 2014; Inubushi et al., 2003). Excess water, low macro and micro-nutrients, high organic acid, and high acidity are the characteristics of peat soils that should be managed by water and soil managements if peat soils will be used for agricultural activities (Lobb, 1995). Soil ameliorants are needed to improve soil fertility to support better crop yield, as well as to reduce GHG emission. The effects of soil amelioration in mineral soil have been explored in many studies; however, few studies are related to soil amelioration in peat soil.

2 TROPICAL PEATLANDS

The coverage area of peatlands is less than 3.3% of the land surface (Hadi et al., 2001). The total peatlands area of 36 million ha can be found in the tropics and subtropics (Andriesse, 1988). It was estimated that tropical peatlands store 15-19% of the global peat carbon pool (Page et al., 2011). The largest peatlands area can be found in the Southeast Asia region. Indonesia has the largest area (14.9 million ha; 206,950 km^2, 87% of the total best estimate) in Southeast Asia (Table 2; Page et al., 2011; Ritung et al., 2011). Peats can be found across

Table 1. National GHG emission in 2004 and emission reduction target by sectors.

	GHG emission in 2004		Emission reduction by			
			2020		2030	
Sectors	Gg CO_2-e	%	26% (unconditional)	41% (conditional)	26% (unconditional)	41% (conditional)
Energy	602,458	32.67	4.95	4.71	37.97	36.61
IPPU	47,449	2.57	0.13	0.42	0.34	0.29
Agriculture	113,440	6.15	1.04	0.93	1.10	0.34
FOLU*	979,422	53.10	87.62	87.38	59.31	60.15
Waste	101,560	5.51	6.26	6.56	1.31	2.61
Total	1,844,329	100	100	100	100	100

Table 2. Estimation area of tropical peatlands in South East Asia.

Country	Area (km)	%	Maximum Peat Thickness (m)	Carbon storage (Gt)
Brunai	909	0.38	10	0.321
Indonesia	206,950	87.45	20	57.367
Malaysia	22,490	9.50	20	9.134
Myanmar	500	0.21	2	0.093
Papua New Guinea	5,000	2.11	10	1.384
Philippines	60	0.03	12	0.172
Thailand	638	0.27	3	0.032
Vietnam	100	0.04	2	0.013
Total	**236,647**	**100.00**		**68.516**

three main islands, i.e., Sumatra, Kalimantan, and Papua (West Papua) with total C stored of 41.1%, 33.8%, and 23%, respectively (Saharjo, 2011).

Peat is made up of partially decomposed dead organic material that has accumulated over thousands of years under permanently waterlogged and oxygen lacking conditions (Wösten et al., 2008). Decomposition of organic material, primarily plant material, during peat formation influences the chemical composition of the peat. Aboveground biomass is the main material of peat (Clymo, 1983). Mostly tropical peatlands are without moss cover (Rieley et al., 1996). The decomposition of organic matter is mainly influenced by high temperature and aerobic conditions of the tropical peatlands surface. The accumulation rates of tropical peatland range from approximately 4–5 mm/year^{-1} to 5–10 mm/year^{-1}, significantly faster than in most temperate and boreal peatlands, which are

approximately 0.5–1 mm year^{-1}, because tropical peatlands occur in consistently hot and often humid conditions (Gorham et al., 2003). The high porosity of the substrate makes peat hold a high water capacity that provides it with a water regulation function.

3 GREENHOUSE GAS EMISSION FROM DIFFERENT LAND USES IN PEATLAND

Peatlands play an important role in the global cycling of C because they can be net sinks of atmospheric CO_2 as well as a source of N_2O, CH_4, and CO_2 as a byproduct of organic matter decomposition. According to Ridlo (1997), peat and forest degradation in Indonesia contribute about 45% of total GHG emissions. GHG emission from peat soil is one target for C emission reduction, and the target is around 9.5-13% of GHG emissions from Indonesia by the year 2020 through peat management (Las and Surmaini, 2010).

Peat soil is a fragile ecosystem with important biological and hydrological functions. Demand to expand agriculture in peatlands resulted in water adjustment and soil improvement. Drainage of excess water resulted in a decrease of CH_4 emissions but an increase in soil decomposition, C losses, the emissions of CO_2 and N_2O from mineralization of the peat, and peat subsidence (Frolking et al., 2011). Drainage led to 0.5 to 4 cm per year of peat subsidence (Grønlund et al., 2008). Drainage of peatlands for agricultural or forest production is approximately 10–20% of total peatlands, and they emit 6% of global CO_2 emissions (FAO, 2013; Frolking et al., 2011). The CO_2, CH_4, and N_2O emission from different land uses in peat soil in Indonesia has been observed by many studies (Table 3). The emission rates from cultivated peat soils with non-paddy rice were dominated by CO_2 and N_2O,

Table 3. The emission of CO_2, CH_4 and N_2O from different land use in Indonesia.

Location	Land-use	Range (min-max) of			References
		CO_2	CH_4	N_2O	
		mg C m^{-2} hour^{-1}		µg N m^{-2} hour^{-1}	
Kalimantan	Abandoned upland crops field, abandoned paddy fields, secondary forest	110 to 180	0.068 to 0.22	-0.013 to -0.004	Inubushi et al., 2003
Sumatera	Drained forest, cassava field, upland paddy field, lowland paddy field	92.9 to 250.4	0.13 to 4.49	-5.88 to 68.21	Furukawa et al., 2005
Kalimantan	Secondary forest, paddy field, upland crops field, abandoned paddy field, abandoned upland, rice-soybean rotation field	0.01 to 0.03	-0.06 to 10.5	-0.001 to 0.584	Hadi et al., 2005
Sumatera	Sago palm	24 to 150	-0.038 to 0.99	nd	Watanabe et al., 2009

(Continued)

111

Table 3. (Cont.)

Location	Land-use	Range (min-max) of CO₂ mg C m⁻² hour⁻¹	CH₄	N₂O µg N m⁻² hour⁻¹	References
Kalimantan	Natural forest, regenerated forest, burned forest, grassland cropland	nd	nd	0.005 to 2.957	Takakai et al., 2006
Kalimantan, Sumatera	Ferns, sedges, pulp wood	5.8 to 59.3	0 to 0.27	-0.6 to 7.9	Jauhiainen and Silvennoinen, 2012
Kalimantan	Paddy, oil palm, vegetable	0.38 to 0.13	0.02 to 0.19	-7.78 to -52.34	Hadi et al., 2012
Kalimantan	Undrained natural forest, drained forest, burned forest, cropland	0.08 to 0.35	-0.02 to 0.36	7.31 to 13127	Arai et al., 2014
Kalimantan	Flooded forest, drained forest, flooded burnt site, drained burnt site	108 to 340	0.0084 to 5.75	-8.7 to 8.1	Adji et al., 2014

nd: no data

while the emission from cultivated peat soils with paddy rice were dominated by CH_4. The ranges of emission rates were different among the studies because they were influenced by temperature, maturity of the peat, land management, and type of crop.

The study of soil amelioration effects on rice productions and CH_4 emission was conducted at the research station of the Indonesian Agricultural Environment Research Institute (IAERI), Jakenan, Central Java, Indonesia. Peat soil samples of approximately 8 tons were transported from South Kalimantan to the IAERI research station. In this station, an automatic gas sampling device for CH_4 emission measurement consisted of 12 plexi-glass squared chambers with the size of 1 m x 1 m x 1 m (Figure 2). These chambers were individually placed in 12 experimental plots of 5 m x 5 m. Gas samplings were collected every 2 minutes and started at 06:00 in the morning and ended at 04:00 in the afternoon on each sampling day. Gas samples for CH_4 were determined by gas chromatography (GC). A GC is equipped with a flame ionization detector (FID) for CH_4 analysis.

4 SOIL AMELIORATION

Soil amendment or soil amelioration is the process of incorporating materials to modify soils to provide what the native or existing soils do not have naturally. Application of an amendment will largely depend on the existing soil and the traits of the soil that require alteration, for example incorporation of crop residues, usage of composts, application of calcium for increasing the nutrient holding capacity of highly sandy soil or repelling the negative effects of a saline soil near the coast.

Figure 2. Automatic closed chamber method.

The studies of soil amendments were mostly conducted to enhance soil fertility, to improve crop yield, as well as to reduce GHG emission, e.g., biochar, manure, steel slag fertilizer (Table 4). According to Ali et al., (2014) the mixture

Table 4. The emission of CH$_4$ and N$_2$O from soil ameliorants in different land use in field.

Type of ameliorants	Application rate (Mg ha^{-1})	Plant	Country	Soil types	Range of CH$_4$ mg C m^{-2} hour^{-1}	References
Compost	12	Paddy	Japan	Gley soil	5,07	Yagi and Minami, 1990
				Andosols	2,53	
Rice straw	6	Paddy		Gley soil	12,80	
				Andosols	4,27	
				Peat soil	21,73	
Biochar (wheat straw)	10-40	Paddy	China	Entic Halpudept	1.02-1.82	Zhang et al., 2012
Green manure	20	Paddy	India	Typic	2,99	Khosa et al., 2010
Rice straw compost	10			Ustochrept	0,80	
Wheat straw	10				7,08	
Farmyard manure	20				3,41	
Wheat straw	3.75-4.8	Paddy	China	Typic Epiaquepts	4.77-20.54	Ma et al., 2009
Steel slag	2-8	Paddy	China	No information	1.59-2.29	Wang et al., 2015
Urea, rice straw compost	0.22; 2, respectively	Paddy	Bangladesh	Clay loam	3,90	Ali et al., 2014
Urea, rice straw compost, silicate fertilizer	0.17; 2; 0.3, respectively				3,71	
Urea, sesbania, silicate slag	0.17; 2; 0.3, respectively				3,79	
Urea, azolla anabaena, silicate slag	0.17; 2; 0.3, respectively				3,55	
Urea, cattle manure compost, silicate slag	0.17; 2; 0.3, respectively				3,67	
Manure	4	Paddy	Indonesia	Peat soil	9,21	ICCTF 2011
Steel slag fertilizer (Pugam A)	0,75				9,39	
Steel slag fertilizer (Pugam T)	0,75				8,52	
Mineral soil	4				11,66	
Biochar	20-40	Maize	China	Inceptisols	-0.03 to -1.88	Zhang et al., 2016

application of silicate fertilizer, well-composted organic manures, and Azolla biofertilizer was a good option to minimize inorganic fertilizer, mitigate CH$_4$ emission, and enhance rice yield. Biochar application to soil is proposed to improve soil fertility, as well as sequester C, to mitigate climate change because biochar is made from biomass that has been pyrolyzed in a low oxygen condition, so it can transfer 50% of the carbon stored in plant tissue from the active to an inactive carbon pool (Lehmann et al., 2011). Hue (1992) showed that manure could reduce soil acidity and increase uptake of Calcium (Ca) in a tropical forage legume. According to (Khosa et al., 2010), incorporation of rice straw compost with high lignin content resistant to microbial decomposition could minimize CH$_4$ emission from rice fields with co-benefits of increased soil fertility and crop productivity. However, the results contradicted the study from (Yagi and Minami, 1990) that showed

the application of rice straw to the paddy fields significantly stimulated CH$_4$ emission because of it is positively correlated with the contents of readily mineralizable carbon. Dolomite is calcium magnesium carbonate with a chemical composition of $CaMg(CO_3)_2$ so it contains oxidants that act as electron acceptors to control CH$_4$ emissions. There is indigenous knowledge in cultivated peatland in Indonesia through combining minimum tillage known as *tajak*, *puntal*, *hambur* systems, followed by application of ash, salt, and manure (Noor, 2012).

Effects of different soil amelioration on CH$_4$ emission from peat soil under rice cultivation were studied using an automatically closed chamber method. Twelve plots (approximately 25 m^2) were arranged using randomized block design (RBD). The CH$_4$ emissions were collected from 6 consecutive rice cultivation periods. Table 5 showed the application of soil ameliorants

Table 5. CH_4 emission, rice yield and its index of soil ameliorant on peat cultivated paddy rice.

Ameliorant	CH_4 emission (ton CO_2eq ha^{-1} season^{-1})	CH_4 emission reduction (%)	Rice yield (ton ha^{-1})	Increase of grain yield (%)	Index of yield CO_2eq^{-1}
Without ameliorant	5.30 - 18.99	Baseline	3.08 - 5.16	Baseline	0.27 - 0.59
Dolomite	6.90 - 13.41	3 - 35	4.38 - 5.54	0.26 - 13.23	0.35 - 0.74
Zeolite	16,69	8	4,96	-	0,30
Steal slag	12,89	29	6,00	13,99	0,47
Rice straw	18,74	1	4,88	-	0,26
Animal manure	5.21 - 12.46	27 - 34	4.65 - 5.32	4.62 - 9.89	0.43 - 0.89
Silicate fertilizer	12,63	4	4,84	10,26	0,38
Compost	8,14	-	5,21	10,26	0,38
Peat fertilizer	7,50	14	6,16	17,20	0,82
Vulcanic Ash	4,96	43	6,09	16,28	1,23
Fe Fertilizer	6,93	20	4,55	-	0,66
NI	7,37	15	6,70	23,94	0,91
Biochar (Rice husk)	4,17	21	2,51	-	0,60
Biochar (*Eleocharis dulcis*)	4,90	8	2,44	-	0,50
Vulcanic Ash + Biochar (rice husk)	6,14	-	3,15	2,16	0,51
Vulcanic Ash + Biochar (*Eleocharis dulcis*)	5,25	1	2,27	-	0,43

reduced the total CH_4 emissions by approximately 1 - 43%, except application of compost and Volcanic Ash + Biochar (rice husk). A mechanism used to decrease CH_4 emissions is the addition of electron acceptors, which influences the sequential soil Eh reactions. The electron acceptors are ordered according to their Eh, and the substrate is used at lower concentrations by electron acceptors with a higher Eh (Lovley and Phillips, 1988). Some ameliorants contain electron acceptors, such as NO_3^-, Mn^{4+}, Fe^{3+}, and SO_4^{2-}, which decrease CH_4 production because of inhibitory and competitive effects with different microorganisms for common electron donors (Achtnich et al., 1995). Application of soil ameliorants, such as biochar, could provide better aeration that makes the soil more favorable for methanotrophs. The application of dolomite, steel slag, animal manure, silicate fertilizer, compost, peat fertilizer, volcanic ash, nitrification inhibitor, and Volcanic Ash + Biochar (rice husk) increased grain yield by approximately 0.26 – 23.94%, compared to no application of soil amelioration. The soil amelioration could increase yield because ameliorants improved soil quality to support the life of the plant. In this study, we use an index of yield per emission to minimize the trade-offs between rice production and CH_4 emissions. The higher values of the index from soil ameliorants, e.g., biochar, volcanic ash, peat fertilizer, dolomite, animal manure, and steel slag application, compared to no soil ameliorants means that application of soil ameliorants could give more advantages to produce more rice and mitigate GHG emission.

5 CONCLUSIONS

The option to fulfill the growing food demand is looking toward the areas of new arable land for agricultural activities, including peatland. However, utilization of peatlands causes disturbance to the ecosystem, so it is better to conduct production in degraded peatland. Cultivated peatland has received much attention due to poor nutrients for plant growth and contribution to GHG emission. The application of dolomite, steel slag, animal manure, silicate fertilizer, peat fertilizer, volcanic ash, nitrification inhibitor, and volcanic ash + biochar (rice husk) could reduce CH_4 emission and increase grain yield. Soil amelioration is vital for increasing rice production, as well as reducing CH_4 emission from peat soil, but the usage of soil amelioration is strongly influenced by the economics of farming. Sustainable agriculture means increasing production, as well as ecology adaptation to the environment.

REFERENCES

Achtnich, C., Bak, F. & Conrad, R. 1995. Competition for electron donors among nitrate reducers, ferric iron reducers, sulfate reducers, and methanogens in anoxic paddy soil. *Biol. Fertil. Soils* 19: 65-72. https://doi.org/10.1007/BF00336349.

Adji, F.F., Hamada, Y., Darang, U., Limin, S.H. & Hatano, R., 2014. Effect of plant-mediated oxygen supply and drainage on greenhouse gas emission from a tropical peatland in Central Kalimantan, Indonesia. *Soil Sci. Plant Nutr* 60: 216-230. https://doi.org/10.1080/00380768.2013.872019

Ali, M.A., Sattar, M.A., Islam, M.N. & Inubushi, K. 2014. Integrated effects of organic, inorganic and biological amendments on methane emission, soil quality and rice productivity in irrigated paddy ecosystem of Bangladesh: field study of two consecutive rice growing seasons. *Plant Soil* 378: 239-252.

Andriesse, J. 1988. Nature and Management Of Tropical Peat Soils. Rome: Food & Agriculture Organisation.

Arai, H., Hadi, A., Darung, U., Limin, S.H., Hatano, R. & Inubushi, K. 2014. A methanotrophic community in a tropical peatland is unaffected by drainage and forest fires in a tropical peat soil. *Soil Sci. Plant Nutr* 60: 577-585.

Ciais, P. 2013. Carbon and Other Biogeochemical Cycles. Climate Change 2013: The Physical Science Basis. Contribution of Working Group I to the Fifth Assessment Report of the Intergovernmental Panel on Climate Change (eds Stocker, TF et al.) Ch. 6. Cambridge University Press.

Clymo R.S. 1983. Peat. *In*: Ecosystems of the World, 4B. Mires: swamp, bog, fen and moor, Regional Studies. Elsevier, Amsterdam, the Netherlands, pp. 159–224.

FAO. 2009. How to feed the world in 2050 High level expert forum, issues brief. Rome, Italy. URL http://www.fao.org/fileadmin/templates/wsfs/docs/expert_paper/How_to_Feed_the_World_in_2050.pdf Accessed October 2013.

FAO. 2013. *FAO statistical yearbook, world food and agriculture*: 307. Rome: Food and Agriculture Organization of the United Nations.

Frolking, S., Talbot, J., Jones, M.C., Treat, C.C., Kauffman, J.B., Tuittila, E.S. & Roulet, N. 2011. Peatlands in the Earth's 21st century climate system. Environ. Rev. 19, 371–396.

Furukawa, Y., Inubushi, K., Ali, M., Itang, A.M. & Tsuruta, H. 2005. Effect of changing groundwater levels caused by land-use changes on greenhouse gas fluxes from tropical peat lands. *Nutr. Cycl. Agroecosystems* 71: 81-91.

Gorham, E., Janssens, J.A. & Glaser, P.H. 2003. Rates of peat accumulation during the postglacial period in 32 sites from Alaska to Newfoundland, with special emphasis on northern Minnesota. *Can. J. Bot* 81: 429-438.

Grønlund, A., Hauge, A., Hovde, A. & Rasse, D.P. 2008. Carbon loss estimates from cultivated peat soils in Norway: a comparison of three methods. *Nutr. Cycl. Agroecosystems* 81: 157–167. https://doi.org/10.1007/s10705-008-9171-5.

Hadi, A., Affandi, D.N., Bakar, R.A. & Inubushi, K. 2012. Greenhouse gas emissions from peat soils cultivated to rice field, oil palm and vegetable. *J. Trop. Soils* 17: 105-114.

Hadi, A., Haridi, M., Inubushi, K., Purnomo, E., Razie, F. & Tsuruta, H. 2001. Effects of land-use change in tropical peat soil on the microbial population and emission of greenhouse gases. *Microbes Environ* 16: 79-86.

Hadi, A., Inubushi, K., Furukawa, Y., Purnomo, E., Rasmadi, M. & Tsuruta, H. 2005. Greenhouse gas emissions from tropical peatlands of Kalimantan, Indonesia. *Nutr. Cycl. Agroecosystems* 71: 73-80.

Hooijer, Aljosja, Silvius, Marcel, Wösten, Henk, Page, Susan, Hooijer, A., Silvius, M., Wösten, H. & Page, S. 2006. PEAT-CO2. Assess. CO2 Emiss. Drained Peatl. SE Asia Delft Hydraul. Rep. Q 3943.

Hue, N.V. 1992. Correcting soil acidity of a highly weathered Ultisol with chicken manure and sewage sludge. *Commun. Soil Sci. Plant Anal*. 23: 241-264.

ICCTF (Indonesia Climate Change Trust Fund). 2011. Research and development of technology on sustainable peatland management to enhance carbon sequestration and greenhouse gas mitigation. Final presentation on greenhouse gas emission. (In Bahasa Indonesia).

Inubushi, K., Furukawa, Y., Hadi, A., Purnomo, E. & Tsuruta, H. 2003. Seasonal changes of CO_2, CH_4 and N_2O fluxes in relation to land-use change in tropical peatlands located in coastal area of South Kalimantan. *Chemosphere* 52: 603-608.

Ito, S. 2018. World Rice Statistics and Graphics: Indonesia URL: http://worldfood.apionet.or.jp/graph/num.cgi?byear=1960&eyear=2018&country=INDONESIA&article=rice&pop=0&type=e1

Jati, K. 2014. Staple food balance sheet, coefficient of variation, and price disparity in indonesia. *J. Adv. Manag. Sci*. Vol 2.

Jauhiainen, J. & Silvennoinen, H. 2012. Diffusion GHG fluxes at tropical peatland drainage canal water surfaces. *Suo* 63: 93-105.

Khosa, M.K., Sidhu, B.S. & Benbi, D.K. 2010. Effect of organic materials and rice cultivars on methane emission from rice field. *J. Environ. Biol* 31: 281-285.

Las, I. & Surmaini, W. 2010. Agricultural development technologies to reduce global warming. Paper on One Day Seminar for Alumni Gathering of Agricultural Faculty of Lambung Mangkurat University, Banjarbaru, March 11, 2010 (in Bahasa Indonesia).

Lehmann, J., Rillig, M.C., Thies, J., Masiello, C.A., Hockaday, W.C. & Crowley, D. 2011. Biochar effects on soil biota–a review. *Soil Biol. Biochem* 43: 1812-1836.

Lobb, D.A. 1995. Management and Conservation Practices for vegetable production on peat soils. East. Can. Soil Water Conserv. Cent. Gd. Falls N. B. Unpubl. Rep. 49.

Lovley, D.R. & Phillips, E.J. 1988. Manganese inhibition of microbial iron reduction in anaerobic sediments. *Geomicrobiol. J* 6: 145-155.

Ma, J., Ma, E., Xu, H., Yagi, K. & Cai, Z. 2009. Wheat straw management affects CH_4 and N_2O emissions from rice fields. *Soil Biol. Biochem* 41: 1022-1028.

Maclean, J.L. 2002. Rice almanac: Source book for the most important economic activity on earth. Int. Rice Res. Inst.

MoE. 2017. Third National Communication Under the United Nations Framework Convention on Climate Change. Directorate General of Climate Change, Ministry of Environment and Forestry, pp 219.

Noor M. 2012. Indigenous knowledge on peatland management. Proceeding of sustainability of peatland management. Indonesian Center for Agricultural Land Resources Research and Development (ICALRRD) Bogor, Indonesia. p 159. (In Bahasa Indonesia).

Page, S.E., Rieley, J.O. & Banks, C.J. 2011. Global and regional importance of the tropical peatland carbon pool. *Glob. Change Biol* 17: 798-818. https://doi.org/10.1111/j.1365-2486.2010.02279.x

Patel, D.P., Das, A., Munda, G.C., Ghosh, P.K., Bordoloi, J.S. & Kumar, M. 2010. Evaluation of yield and physiological attributes of high-yielding rice varieties under aerobic and flood-irrigated management practices in mid-hills ecosystem. *Agric. Water Manag* 97: 1269-1276.

Ridlo R. 1997. CO_2 emissions at peat reclamation for mega rice project in Central Kalimantan. *J. Alami* 2: 57-58 (in Bahasa Indonesia).

Rieley, J.O., Ahmad-Shah, A.A. & Brady, M. 1996. The extent and nature of tropical peat swamps. In: Maltby E, Immirzi CP, Safford RJ (Eds.): Tropical Lowland Peatlands of Southeast Asia. Proceedings of a Workshop on Integrated Planning and Management of Tropical Lowland Peatlands, Cisarua, Indonesia. IUCN, Gland, Switzerland. 294 pp.

Ritung, S., Wahyunto, & Sukarman. 2011. Peat land map of Indonesia, scale 1:250.000. Indonesian Center for Agro-climate Research and Development, Indonesian Agency for Agriculture Research and Development, Ministry of Agriculture, Jakarta, Indonesia.

Saharjo, B.H. 2011. Carbon baseline as limiting factor in managing environmental sound activities in peatland for reducing greenhouse gas emission. *Biodiversitas J. Biol. Divers.* 12.

Schütz, H., Seiler, W. & Conrad, R. 1989. Processes involved in formation and emission of methane in rice paddies. *Biogeochemistry* 7: 33-53.

Takakai, F., Morishita, T., Hashidoko, Y., Darung, U., Kuramochi, K., Dohong, S., Limin, S.H. & Hatano, R. 2006. Effects of agricultural land-use change and forest fire on N2O emission from tropical peatlands, Central Kalimantan, Indonesia. *Soil Sci. Plant Nutr.* 52: 662-674.

Wang, W., Sardans, J., Lai, D.Y., Wang, C., Zeng, C., Tong, C., Liang, Y. & Peñuelas, J. 2015. Effects of steel slag application on greenhouse gas emissions and crop yield over multiple growing seasons in a subtropical paddy field in China. *Field Crops Res* 171: 146-156.

Watanabe, A., Purwanto, B.H., Ando, H., Kakuda, K. & Jong, F.S. 2009. Methane and CO2 fluxes from an Indonesian peatland used for sago palm (Metroxylon sagu Rottb.) cultivation: Effects of fertilizer and groundwater level management. *Agric. Ecosyst. Environ* 134: 14-18.

Wösten, J.H.M., Clymans, E., Page, S.E., Rieley, J.O. & Limin, S.H. 2008. Peat–water interrelationships in a tropical peatland ecosystem in Southeast Asia. *Catena* 73: 212–224.

Yagi, K. & Minami, K. 1990. Effect of organic matter application on methane emission from some Japanese paddy fields. *Soil Sci. Plant Nutr* 36: 599-610.

Zhang, A., Bian, R., Pan, G., Cui, L., Hussain, Q., Li, L., Zheng, Jinwei, Zheng, Jufeng, Zhang, X. & Han, X. 2012. Effects of biochar amendment on soil quality, crop yield and greenhouse gas emission in a Chinese rice paddy: a field study of 2 consecutive rice growing cycles. *Field Crops Res* 127: 153-160.

Zhang, D., Pan, G., Wu, G., Kibue, G.W., Li, L., Zhang, X., Zheng, Jinwei, Zheng, Jufeng, Cheng, K. & Joseph, S. 2016. Biochar helps enhance maize productivity and reduce greenhouse gas emissions under balanced fertilization in a rainfed low fertility inceptisol. *Chemosphere* 142: 106-113.

Tropical Wetlands — Innovation in Mapping and Management — Sulaeman et al. (Eds)
© 2020 Taylor & Francis Group, London, ISBN 978-0-367-20964-3

Fresh water peatland in the Riau province, Sumatra

E. Suryani, E. Yatno & R.E. Subandiono
Indonesian Centre for Agricultural Land Resources Research and Development (ICALRRD), Jakarta, Indonesia

ABSTRACT: Indonesia has 14.93 million hectares of peat land, with the largest part of it, around 6.44 million hectares are in Sumatra. Of the 6.44, about 3.85 million hectares are in Riau, particularly ombrogenous peat and topogenous which are influenced by salty and fresh water. In the management of the land, it is important to know the nature and characteristics of each peat soil. The purpose of this study was to determine the nature and characteristics of freshwater peat soils in parts of Riau Province. A total of 31 representative profiles derived from freshwater topogenous peat in Siak, Kampar, Rokan Hilir, Rokan Hulu and Dumai Regencies were examined. In addition, 73 soil samples were analyzed to determine their physical, chemical and fertility properties. From the analysis, freshwater topogenous peatland has a hemic maturity indicated by an average fiber content of about 71%, and a thickness varying from shallow (50-<100 cm) to very deep (> 500 cm) with fine-textured mineral (silty clay) soil substrates. Soil classified as Hemic Organoso has a low BD (0.05 to 0.33 g cm^{-3}). However, peats in Kampar Regency have a relatively higher BD and ash content than other regions. The pH values range from 3.3 to 4.8. From the C/N ratio (15 to 69), indicating a requirement for nitrogenous fertilizer. Exchangeable cations (Ca, Mg, K, and Na) vary due to basic saturation (BS), though ranging from relatively low to very low. In addition, P_2O_5 levels (extracted 25% HCl) varied from very low to high (2 to 45 mg 100 g^{-1}).

1 INTRODUCTION

Peat soil is formed from organic sediment parent material, the result of the accumulation of plant/vegetation remnants, which decayed under anaerobic conditions. According to Hardjowigeno (1986), the formation of peat soil is a geogenic process- the formation of soil caused by deposition and transportation processes. This is in contrast to pedogenic processes, the formation of mineral soils. Peat soil is also known as Organosol (Soepraptohardjo, 1961; Subardja et al., 2016) or Histosols (2014, Soil Survey Staff).

Indonesia has 14.93 million hectares of peat land spread across Sumatra, Kalimantan, Papua and a small portion in Sulawesi (Ritung et al., 2011). In Sumatra, peatlands add up to around 6.44 million, 4.78 million in Kalimantan, 3.69 million in Papua, and approximately 23,844 in Sulawesi. These peatlands are found as ombrogenous and topogenous peats, both affected by fresh water and brackish water. According to Driessen and Sudjadi (1984), ombrogen peat is formed in an environment only affected by rainwater rather than tide or overflow. In contrast, topogenous peat is formed in the interior of coastal/river plains affected by tidal runoff or flooding. Topogen peat is relatively more fertile than ombrogen because of enrichments from the deposited materials. According to Driessen and Sudjadi (1984), freshwater topogenous peat is characterized by a level of exchangeable Ca (exch-Ca) higher than Mg (exch-Ca > exch-Mg). Contrastingly, water

in topogenous peat salty/brackish has exch-Mg levels higher than exch-Ca (exch-Mg >exch-Ca).

From Driessen and Sudjadi (1984), at least 487,950 ha, 45.95%, of the total area of peat land in Kampar, Siak, Rokan Hulu, Rokan Hilir regencies and Dumai City, Riau Province (1,062,154 ha) is considered freshwater topogenous peat. The rest is brackish topogenous and ombrogenous peats (Table 1). The distribution of peatlands in Kampar, Siak, Rokan Hulu, Rokan Hilir, and Dumai Regencies, Riau Province is presented in Figure 1.

The research area is traversed by many large and small rivers, some draining directly to the sea. The river's depth and water level fluctuate due to the intensity of rainfall or tides. It is possible to form topogenous peat which is relatively more fertile due enrichment from rivers and tides. The results of peatland mapping (BBSDLP, 2017) show freshwater topogenous peat in Kampar, Siak, Rokan Hulu, Rokan Hilir and Dumai Regencies, and Riau Province formed in back-swamps. The peat formed has a varying thickness, increasing from the riverbank to the inland swamp gradually. The dry land is covered by tertiary sedimentary rock material consisting of sandstone, claystone/mudstone, and shale.

Peat lands as part of the swamp ecosystem, play very significant roles. On the economic front, economic function, peat land is a source of income for some farmers. Moreover, it acts as a buffer for regional hydrology, helping to avoid flooding and drought. On an environmental basis, peat lands store

Table 1. Area and distribution of peatland in the studied area.

Location	Fresh water topogenous peat — Ha —	Other than topogenous peat*	Total
Kampar Regency	102,202	-	102,202
Siak Regency	130,138	253,193	383,331
Rokan Hilir Regency	174,999	216,981	391,980
Rokan Hulu Regency	31,903	19,535	51,438
Dumai city	48,708	84,495	133,203
Total	487,950	574,204	1,062,154

* Note: Topogenous brackish peat and ombrogenous peat (Driessen and Sudjadi, 1984).

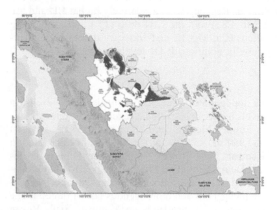

Figure 1. Distribution of peatland in Kampar, Siak, Rokan Hulu, Rokan Hilir Regencies and Dumai City.

very large carbon stocks with high emissions. It is also a native habitat of various types of rare plants such as ramin, jelutung, and swamp animals. More importantly, peatland functions as carbon storage (Agus et al., 2011). The management of peatlands in Kampar, Siak, Rokan Hulu and Rokan Hilir Regencies, and Dumai city requires careful planning and a lot of care to benefit local farmers and perform other functions. Therefore, the characteristics of peat soils, especially freshwater topogenous, are very important and should be considered.

This study provides information on the nature, characteristics and fertility levels of peat soils from Kampar, Siak, Rokan Hulu, Rokan Hilir Regencies and Dumai City, Riau Province. Currently, most of the peatlands are used for plantations, especially for oil palm, industrial timber (HTI), and rubber, though bushes still occupy a small portion of it. The information provided is expected to be of use to peatland managers in the area, especially in unopened peat lands in other locations with similar typologies.

2 MATERIAL AND METHODS

2.1 Climatic condition

The regencies of Kampar, Siak, Rokan Hulu, Rokan Hilir and Dumai City in Riau Province have wet climates with annual average rainfall ranging from 2,390 mm Siak Regency to 3,100 mm in Kampar Regency (Table 2). Average monthly rainfall ranges from 107 mm (July) to 364 mm (December). According to Schmidt and Ferguson (1951), the study area is classified as 'A' rain type and 'Afa' in the Koppen climate type. Rain type A has a Q value [the ratio of the number of dry months to wet months] is <14.3%. Wet periods (> 100 mm) occur throughout the year, there are no dry months (<60 mm). According to Oldeman et al., (1978), the study area included the B1 agroclimatic zone, a region with 7-9 wet months (> 200 mm) and dry months (<100 mm) <2 months respectively. The average annual temperature is 26.5°C to 27.2°C, while air humidity average 83%.

2.2 Landform and parent material

Freshwater topogenous peatlands in Kampar, Siak, Rokan Hulu, Rokan Hilir regencies and Dumai city, Riau Province are in the back swamp of the river and

Table 2. Average monthly rainfall of the research location.

Regencies	Monthly rainfall (mm)												
	Jan	Feb	Mar	Apr	Mei	Jun	Jul	Aug	Sep	Oct	Nov	Des	Total
Kampar	337	224	263	276	229	157	157	198	240	315	349	355	3,100
Rokan Hilir	259	201	251	256	206	159	148	183	261	311	320	305	2,860
Rokan Hulu	263	244	277	276	223	162	107	197	218	290	331	364	2,952
Siak	210	157	223	228	193	120	112	156	192	275	294	230	2,390
Kota Dumai	200	150	202	225	191	159	159	200	255	293	299	258	2,591

Source: id.climatedata.org/continent/asia/

its tributaries. A back swamp is a basin filled with organic material, or peat with varying thickness, gradually increasing from the edge of the river to the inland.

Soil parent material comes from deposits of organic matter with thicknesses ranging from 5 mm to > 500 cm in the middle of a basin. The peat maturity level is generally hemic to sapric. In the top layer, there are black burnt marks, about 5 cm thick, and in some locations, especially in the lower layers, there are inserts or mixed mineral soils thought to be the result of river runoff or floods.

2.3 *Soil sampling and analysis*

The 31 representative peat profiles were examined in the field. However, 73 soil samples from the representative profile were analyzed in the Bogor Soil Research Laboratory, for soil physical and chemical properties. The physical properties include Bulk Density or weight, ash, and fiber contents (LOI method). The chemical properties included: pH H_2O and pH-KCl (ratio 1: 2.5) levels of organic carbon (Walkley and Black methods), potential P_2O_5 and K_2O levels (25% HCl extraction), and P-available with Bray extract 1. The exchangeable cations (Ca, Mg, K, and Na) and cation exchange capacity (CEC) was determined in the extraction of ammonium acetate pH 7.0. The examination procedure referred to the Technical Guidelines for Analysis of Soil, Plant, Water, and Fertilizer Chemistry (Eviati and Sulaeman, 2012). Soil order was determined according to the National Soil Classification System (Subardja et al., 2016) and its equivalent according to the Soil Taxonomy System (Soil Surveys Staff, 2014) to the family level.

3   RESULT AND DISCUSSION

3.1 *Morphology and physical properties of peat*

Morphological properties like soil thickness and color, physical properties such as bulk density, ash, and fiber content in several representative profiles of freshwater topogenous peat are presented in Table 3. It is evident

the freshwater topogenous peat in the study area has varying thicknesses, ranging from shallow (50-<100 cm), medium (100-<200 cm), deep (200-<300 cm), very deep (300-<500 cm) and deepest (> 500 cm) with a level of maturity sapric (ripe) in the upper layer and hemic (half ripe) in the lower layer. The topsoil color is generally black (10YR 2/1) or dark gray (10YR 3/1), while the lower layers are dark reddish-brown (5YR 3/2) or very dark grayish brown (10YR 3/3). This is in line with Ambak and Melling, (2000) that tropical peat is generally reddish-brown to dark brown (dark) depending on the decomposition stage.

Bulk Density (BD) is one of the important properties in the use of peat soil. The weight of the peat soil contents in the study area ranged from 0.05 to 0.33 g cm-3. The weight of the top layer was higher than the lower layer, which was in harmony with Anshari et al. (2012; 2013); Dariah et al. (2011). The average weights of peat soil in the study area were; Lower Rokan 0.08 g cm-3, Rokan Hulu and Dumai City 0.09 g cm-3, Siak 0.15 g cm-3 and Kampar 0.22 g cm-3.

Peat soil has low bulk density, something that was evident in the studied area. The low bulk density of peat soil limits the use of agricultural machinery and even the choice of commodities to be developed (Ambak and Melling, 2000). According to Singh et al. (1986); Dent (1986) and Nugroho et al. (1997), peat is physically soft and has a low bearing capacity. Examples are often found in oil palm plants or other annual crops sloping growth, even falling due to weak foundation of the roots (Figure 3).

Ash content greatly varies, but it is usually >5%, characterizing peat soil formed as freshwater topogenous peat (Driessen and Sudjadi 1984). In Table 3, it can be seen that the upper layer's ash content is higher than what is in the lower layer. This is due to higher levels of weathering of organic matter in the layer (Anshari et al., 2012; 2013; Dariah et al., 2011), along with deposition of mineral materials through river runoff or flood. The average ash content of Kampar Regency is the highest (30.2%), followed by Siak District (16.1%), Dumai City (13.8%), Rokan Hilir Regency (8.2%) and Rokan Hulu Regency (5.6%). There is a positive correlation between bulk density

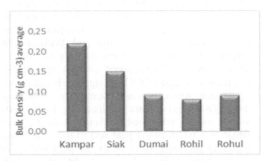 

Figure 2. Average bulk density (left) and ash content (right) of peat in the studied area.

119

Figure 3. Growth of oil palm plants on peat lands.

and ash content. This means bulk density increases with more ash content. The relationship between the two is explained in Figure 4, where the closeness is shown by the determination coefficient (R2) of 0.65.

Also, fiber levels vary widely, ranging from 38 to 92%, and on average, 71%. The average fiber content percentage indicates the level of peat maturity in the studied area is classified as hemic. Driessen and Sudjadi (1984) distinguish peat based on the level of decomposition or maturity as fibric peat, which is peat not decayed (raw). If squeezed it still contains fiber >75% (based on volume). Hemic peat (half ripe) contains fiber between 17 to 74%; while sapric peat is ripe and contains fiber <17%.

The utilization of hemic peat soil is quite beneficial to agriculture, but caution is needed. Natural peat soil has hydrophilic properties that can hold water up to 13 times its dry weight. According to Subagyo et al. (1996); Tadano et al. (1992), if peat is over-drying, the peat colloid get damaged, resulting in irreversible drying and changes to charcoal, failing to absorb nutrients and retain water as a result. In such conditions, according to Sabiham et al. (1999) and Salampak (1999), peat will lose available water, making it flammable.

Currently, the use of peatland for agriculture is restricted due to the issue of global climate change and environmental damage. Peatland management must refer to Permentan No. 14/2009, INPRES No.

10/2011 and Presidential Instruction No. 6 of 2013. But for areas with extensive peat lands, the implementation of the Presidential Decree is still a hard thing because the regional economy and its people are highly dependent on peat lands.

3.2 Chemical properties of peat soil

Among the important inherent properties of peat soils in the tropics are the parent material derived from wood, inundation, shrinking and subsidence (decrease in peat surface) due to irreversible drainage, very acidic pH, and low soil fertility. These characteristics are a limiting factor for the growth of agricultural products on peatland (Andriesse, 1988). Table 4 presents soil acidity, organic carbon, and alkaline content of some representative peat soils from the study area.

3.3 Soil reaction

Soil reaction (pH H_2O) can either be acidic to very acidic (pH 3.3 to 4.8). Peat soil is formed from the decomposition of organic matter, and therefore, more than 95% of it contains organic fractions, while the rest (less than 5%) are inorganic components. The level of acidity of peat soil is related to the content of organic acids (Andriesse, 1974); Miller and Donahue, 1990; Tan, 1993). When compared, the pH of KCl is lower than pH H_2O. - The second difference in pH (pH = pH KCl - pH H_2O) ranges from -0.3 to -1.2. This proves that the surface of colloidal peat soil is negatively charged (Tan, 1992), although the soil reaction is very acidic. According to Suhardjo and Widjaja-Adhi (1976), all the negative contents of peat soil are pH-dependent (pH-dependent charge) and originates from OH- and COOH- groups produced during the decomposition process. Stevenson (1994) suggests COOH- groups are produced by lignin compounds mostly contained in wood. It is the main parent material forming peat soil in the tropics and relates to the value of the CEC of peat soils. The results of the analysis show peat CEC of the study area ranges from moderate to very high (21.54 to

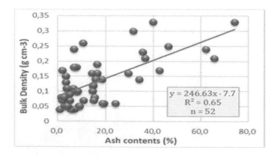

Figure 4. The correlation between ash content and bulk density of peat in the studied area.

Table 3. Thickness, color, BD, ash content, and the fiber content of several representatives peat in the studied area.

Profile	Horizon	Depth (cm)	Matrix color	Bulk density (g/cm³)	Ash content (%)	Fiber content	Level of maturity
TG 208/I	Oa	0-50	10YR 2/1	0.19	16.2	71	Sapric
II	Oe	50-100	7.5YR 3/4	0.10	8.2	71	Hemic
III	Oe	100-200	5YR 3/3	0.10	6.1	70	Hemic
TG 209/I	Oa	0-50	10YR 2/1	0.30	31.4	88	Sapric
II	Oe	50-250	7.5YR 3/3	0.14	15.7	75	Hemic
YG 138/I	Oa	0-50	10YR 2/1	0.33	39.7	75	Sapric
II	Oe	50-220	10YR 2/2	0.21	65.7	81	Hemic
III	Oe	220-360	10YR 2/2	0.21	65.7	75	Hemic
TG-205/I	Oa	0-50	10YR 2/2	0.18	7.8	70	Sapric
II	Oe	50-200	5YR 3/2	0.13	3.3	60	Hemic
TG-207/I	Oe	0-50	10YR 2/1	0.16	29.3	60	Hemic
II	Oe	50-200	7.5YR 3/4	0.10	7.3	79	Hemic
III	Oe	200-550	7.5YR 3/3	0.11	12.8	67	Hemic
YG 129/I	Oa	0-130	7.5YR 2.5/2	0.14	18.4	70	Sapric
II	Oe	130-280	7.5YR 3/3	0.26	10.3	75	Hemic
HF 26/I	Oe	0-50	10YR 2/1	0.08	2.5	56	Hemic
II	Oe	50-325	10YR 2/2	0.05	9.0	78	Hemic
HF 27/I	Oa	0-36	10YR 2/1	0.12	2.8	61	Sapric
II	Oe	36-80	10YR 2/2	0.10	8.2	33	Hemic
HF 1/I	Oa	0-15	10YR 2/1	0.12	14.5	71	Sapric
II	Oe	15-120	10YR 3/3	0.05	4.1	38	Hemic
HF 18/I	Oa	0-50	10YR 2/2	0.05	3.7	50	Sapric
II	Oe	50-100	7.5YR 2.5/2	0.03	25.4	71	Hemic
HI 3/I	Oa	0-100	10YR 3/2	0.07	5.1	50	Sapric
II	Oa	100-250	10YR 3/3	0.09	4.8	86	Sapric
HI 1/I	Oa	0-80	10YR 3/1	0.08	1.5	71	Sapric
II	Oe	80-190	10YR 3/3	0.05	8.7	64	Hemic
HF 12/I	Oa	0-30	10YR 3/1	0.12	14.7	57	Sapric
II	Oe	30-100	7.5YR 4/6	0.06	19.2	79	Hemic
HF 17/I	Oa	0-50	10YR 2/1	0.06	18.6	56	Sapric
II	Oe	50-70	7.5YR 2.5/2	0.06	23.9	63	Hemic

Note: 5YR 3/2-3/3 = dark reddish brown, 7.5YR 2.5/2 = very dark gray, 7.5YR 3/3 = dark brown, 7.5YR 3/4 = dark brown, 7.5YR 4/6 = strong brown, 10YR 2/1 = black, 10YR 2/2 = very dark brown, 10YR 3/1 = dark gray, 10YR 3/2 = very dark grayish brown, 10YR 3/3 = very dark grayish brown.

104.13 me 100 g⁻¹). The average is also very high (59.5 3 me 100 g⁻¹).

3.4 C-organic

Organic carbon levels range from 13.62 to 55.66%, and generally, the upper layers are lower than the lower layers. According to Soepraptoharjo (1961); Subardja et al. (2016); and Soil Survey Staff (2014) peat soil is formed from deposits of organic matter, with a thickness of 50 cm or more and contain organic carbon of at least 12% (dry weight). From this definition, all the properties presented in Table 4 meet the requirements of peat soil. If observed, the levels of organic carbon negatively correlate with the bulk density and ash content. This means that the

bulk density and ash content decreases with increasing levels of organic carbon (Figure 5).

Figure 5 (left) shows the bulk density decreases with increasing levels of organic carbon, evidenced by the determinant coefficient ($R^2 = 0.45$). The level of organic carbon is shown by the mass of carbon relative to the dry period of peat soil. Furthermore, carbon content and weight determine its mass per unit volume, also called carbon density. The carbon density value, multiplied by the volume of peat soil (peatland area multiplied by its depth) equals the total carbon stored in the area.

Similar to the bulk density, the ash content decreases with increasing levels of organic carbon, shown by the determinant coefficients ($R^2 = 0.85$) in Figure 5 (right). Ash content is influenced by the

Table 4. Soil acidity, C-organic, and exchangeable cations of representative peat at the studied area.

Profile	Horizon	Depth cm	pH (1:5) H₂O	KCl	C-organic %	Exchangeable cations Ca me/100 g	Mg	K	Na	Total	CEC	BS %	Ca/Mg ratio
TG 208/I	Oa	0-50	3.7	2.7	40.41	4.55	0.53	0.07	0.22	5.37	45.75	12	8.58
II	Oe	50-100	3.7	2.8	44.20	5.86	0.57	0.11	0.34	6.88	41.95	16	10.28
III	Oe	100-200	3.9	3.1	42.87	6.61	0.56	0.14	0.44	7.75	35.12	22	11.80
TG 209/I	Oa	0-50	4.6	4.3	34.81	2.76	0.35	0.07	0.16	3.34	30.17	11	7.89
II	Oe	50-250	4.7	4.3	48.25	3.98	0.35	0.06	0.08	4.45	30.37	15	11.31
YG 138/I	Oa	0-50	4.1	3.7	23.99	5.12	0.54	0.11	0.19	5.96	51.30	12	9.48
II	Oe	50-220	4.8	4.3	13.62	4.22	0.42	0.02	0.12	4.78	24.95	19	10.05
III	Oe	220-360	4.7	4.3	42.45	4.21	0.56	0.07	0.18	5.02	32.51	15	7.52
TG-205/I	Oa	0-50	3.6	3.1	40.14	5.48	0.93	0.09	0.35	6.85	76.63	9	5.89
II	Oe	50-200	3.6	3.0	46.60	8.60	1.58	0.16	0.38	10.72	72.37	15	5.44
TG-207/I	Oe	0-50	3.9	3.2	39.61	14.73	2.51	0.08	0.16	17.48	65.77	27	5.87
II	Oe	50-200	3.8	3.1	55.66	10.68	2.53	0.18	0.44	13.83	54.30	25	4.22
III	Oe	200-550	3.8	3.0	49.06	11.86	1.83	0.09	0.23	14.01	54.34	26	6.48
YG 129/I	Oa	0-130	3.5	2.7	42.22	5.24	1.82	0.41	0.28	7.75	53.97	14	2.88
II	Oe	130-280	4.1	2.9	36.42	3.30	0.64	0.32	0.39	4.65	43.30	11	5.16
HF 26/I	Oe	0-50	3.3	2.4	52.84	6.64	2.24	0.18	0.81	9.87	80.75	12	2.96
II	Oe	50-325	3.3	2.3	43.77	6.36	2.13	0.35	0.30	9.14	64.45	14	2.99
HF 27/I	Oa	0-36	3.7	2.6	50.71	8.12	3.87	0.18	0.15	12.32	82.51	15	2.10
II	Oe	36-80	3.6	2.6	43.43	1.44	0.98	0.16	0.20	2.78	69.04	4	1.47
HF 1/I	Oa	0-15	3.7	2.8	43.65	7.33	2.38	0.11	0.49	10.31	90.11	11	3.08
II	Oe	15-120	4.0	3.0	53.48	6.69	6.13	0.14	0.97	13.93	102.21	14	1.09
HF 18/I	Oa	0-50	3.6	2.7	51.75	4.93	3.06	0.12	0.38	8.49	71.05	12	1.61
II	Oe	50-100	3.7	2.8	25.29	3.35	2.42	0.15	0.17	6.09	51.83	12	1.38
HI 3/I	Oa	0-100	3.7	2.8	40.38	5.84	3.38	0.24	0.65	10.11	82.17	12	1.73
II	Oa	100-250	3.5	2.6	47.56	3.92	1.97	0.15	0.49	6.53	75.00	9	1.99
HI 1/I	Oa	0-80	3.3	2.5	55.61	5.22	2.14	0.20	0.44	8.00	96.65	8	2.44
II	Oe	80-190	3.7	2.8	48.99	1.31	2.24	0.25	1.04	4.84	71.25	7	0.58
HF 12/I	Oa	0-30	3.7	2.8	41.10	8.87	3.55	0.48	0.18	13.08	54.08	24	2.50
II	Oe	30-100	3.3	2.4	39.63	3.31	1.96	0.28	0.36	5.91	43.57	14	1.69
HF 17/I	Oa	0-50	3.4	2.5	45.34	6.36	1.69	0.21	0.43	8.69	64.55	13	3.76
II	Oe	50-70	3.5	2.7	35.94	6.15	1.56	0.17	0.71	8.59	56.82	15	3.94

Note: Oa = sapric peat, Oe = hemic peat.

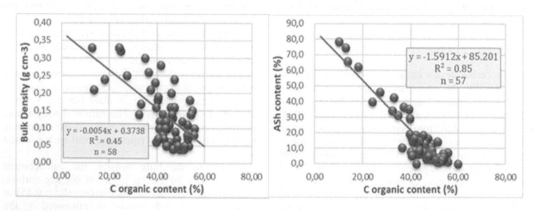

Figure 5. The correlation between C-organic with bulk density (left), and the correlation between organic C with ash content (right).

level of peat decomposition. The further the level of decomposition, the higher the ash content. Conversely, the lower the level of weathering, the lower the ash content. The analysis shows that the average ash content in sapric peat soil is 20.25%, followed by 13.94% hemic peat and 3.63% fibric peat.

3.5 Bases content

Exchangeable bases such as exch-Ca vary from very low to high (1.31 to 14.73 me/100 g), while exch-Mg varies from very low to high (0.22 to 7.23 me/100 g). Higher levels of exch-Ca than exch-Mg are shown by the ratio Ca/Mg > 1, showing the peat soil formed is freshwater topogenous (Driessen and Sudjadi, 1984). Exch-K levels range from very low to moderate (0.02 to 0.48 me/100 g), while exch-Na varies from very low to very high (0.02 to 10.70 me/100 g). The variety of cations contents causes' base saturation (BS) also varies from 4 to 28%, although it is still relatively low to very low. Driessen and Sudjadi (1984) distinguish peat land from ombrogenous peat and topogenous peat. Ombrogenous peat is formed in an environment only affected by rainwater rather than tide or overflow. Contrastingly, topogenous peat was formed in the interior of coastal/river plains affected by tidal runoff/flooding. Generally, topogenous peat is relatively more fertile than ombrogenous peat because it gets a lot of enrichment from the deposited material.

3.6 Peat fertility

Naturally, peat has a relatively low fertility rate. The nutrient content of peat is largely determined by the environment of its formation. Based on its fertility level, Driessen and Sudjadi (1984) distinguish between ombrogenous and topogenous peats affected by freshwater and saltwater/brackish water. Topogenous peat is relatively more fertile than ombrogenous peat. Table 5 presents the nutrient content of some representative peat soils.

3.7 Nitrogen (N)

Nitrogen (N) is the primary macronutrient mostly needed by plants, though its availability is very low in the soil due to its high mobility. In general, N is distinguished from N-organic and N-inorganic. The plants absorb N in N-inorganic form as ammonium (NH^{4+}) and nitrate (NO^{3-}). Apart from fertilizers, the two ions are derived from the decomposition of organic matter (Benbi and Richter, 2002), where availability is largely determined by the condition and amount of organic matter in the soil (Cookson et al., 2005). Weathering of organic matter involves microorganisms which utilize carbon compounds contained in organic matter as an energy source. This causes the carbon content of organic matter to decrease, lowering the C/N ratio.

In Table 5, it is evident the C/N ratio ranges from 15 to 69, indicating some of the organic carbon weathered and released several N. This is shown by the total N-content which varies from very low to very high (0.04 to 2.04%). According to Radjagukguk (1997), although total N-analysis is very high, N availability for plants on peat soil is low because it is an organic N-form. Therefore, N. fertilization is needed depending on the requirements of the plants.

3.8 Phosphorus (P)

Phosphorus (P) comes second to N among the most absorbed primary macro nutrients by plants. P_2O_5 levels (extracted 25% HCl) on peat soil in the study area varied greatly from very low to high (2 to 45 mg 100 g-1). According to Stevenson (1984), most of the P elements in peat soils are in the form of P-organic, with 2% nucleic acid, 1% phospholipid, and 35% inositol phosphate. The rest have not been identified. The release of inositol phosphate is very slow compared to other esters.

Ivanoff et al. (1998) on peat soils containing high mineral materials suggested that mineral materials reduce mobility and degradation of P. Freshwater topogenous peat soil in the studied area received mineral enrichment from river runoff or floods which prevented P from being washed away. This is evident from the high content of P_2O_5 (extracted Bray 1) in the upper layer compared to the lower layer (in line with the high ash content in the upper layer which is thought to have originated from enriching mineral materials).

3.9 Potassium (K)

Potassium (K) is the third primary macro element after N and P needed by plants. The K element is absorbed in the form of K+ (Tisdale et al., 1990). Based availability, K can be divided into the following classes: (a) K not available because it is in the form of source minerals K, (b) K available in forms that can be exchanged in soil solutions, and (c) K available in non-exchangeable forms because it is fixed by colloidal soil.

The K content in peat soil is low. This is because the peat has a little K absorption capacity, and since most K is in the soil solution, it is easily washed away (Andriesse, 1997). The K content of peat soil in the study area ranged from very low to moderate (2 to 25 mg 100 g-1), though the average was low (10 mg 100 g-1). Therefore, K fertilizer is needed to meet the nutritional requirements of plants.

3.10 Peat soil classification

Soils in the research area are formed from deposits of organic matter, which is the result of the accumulation of plant/vegetation remains decayed under anaerobic conditions. In an anaerobic state, the decomposition of organic matter is slow, though over time there is a buildup of sufficiently thick organic matter to form the soils called peat or Organosol (Soepraptoharjo, 1961; Subardja et al., 2016) or Histosols (Soil Survey

Table 5. N content, C/N ratio, potential P_2O_5 dan K_2O content, and available P_2O_5 content of representative peat soil in the studied area.

Profile	Horizon	Depth (cm)	N content (%)	C/N ratio	P_2O_5 (mg 100 g^{-1})	K_2O	P_2O_5 (ppm)
TG 208/I	Oa	0-50	1.36	30	11	5	53
II	Oe	50-100	1.26	35	9	9	42
III	Oe	100-200	1.01	42	5	9	25
TG 209/I	Oa	0-50	1.26	28	24	7	20
II	Oe	50-250	1.20	39	29	5	30
YG 138/I	Oa	0-50	0.62	39	30	6	20
II	Oe	50-220	0.38	36	26	3	15
III	Oe	220-360	0.76	56	22	6	11
TG-205/I	Oa	0-50	1.13	36	26	10	10
II	Oe	50-200	0.98	48	6	10	15
TG-207/I	Oe	0-50	1.18	34	22	10	50
II	Oe	50-200	1.46	38	19	16	48
III	Oe	200-550	1.70	29	11	9	24
YG 129/I	Oa	0-130	1.46	31	18	9	49
II	Oe	130-280	1.09	50	13	20	36
HF 26/I	Oe	0-50	1.24	43	7	10	13
II	Oe	50-325	0.76	55	15	18	33
HF 27/I	Oa	0-36	1.60	32	22	9	41
II	Oe	36-80	0.99	44	10	8	32
HF 1/I	Oa	0-15	1.73	25	13	5	24
II	Oe	15-120	1.14	47	11	9	12
HF 18/I	Oa	0-50	1.02	51	11	7	12
II	Oe	50-100	0.48	53	3	8	6
HI 3/I	Oa	0-100	1.04	39	31	18	16
II	Oa	100-250	0.90	53	4	8	11
HI 1/I	Oa	0-80	1.21	46	9	10	17
II	Oe	80-190	0.91	53	12	9	10
HF 12/I	Oa	0-30	0.73	56	21	20	55
II	Oe	30-100	0.62	64	19	13	43
HF 17/I	Oa	0-50	1.10	41	17	11	29
II	Oe	50-70	0.61	59	8	9	19

Staff, 2014). Peat soil is classified according to the National Soil Classification (Subardja et al., 2016) and its equivalent according to Keys to Soil Taxonomy (Soil Survey Staff, 2014). The results of field observations are supported by the analysis of soil samples in the laboratory. This indicated that the peat soil formed at the research site is classified into three; Organosol Fibric, Organosol Hemic, and Organosol Sapric. Table 6 presents the classification of peat soil in the studied area.

3.11 Organosol Fibric

Organosol Fibric is peat soil with a fibric maturity level (raw) in the control section (0-160 cm). In the upper layer (0-30 cm), it has a hemic maturity level, while the underlying layer is fibric. Peat thickness varies from medium (100-<200 cm) to very deep (300-<500 cm) with soft gray sandy clay (5Y 6/1) soil substratum. It is equivalent to Hemic Haplofibrists and Typic Haplofibrists (Soil Survey Staff, 2014).

3.12 Organosol Hemic

Organosol Hemic is a peat soil with hemic maturity level (half-ripe) in the control section (0-130 cm). The top layer (0-30 cm) has a sapric maturity level, and hemic maturity level in the bottom layer (30-200 cm). Some of these soil have mineral deposits in the bottom layer, indicating the enrichment of mineral materials. Peat thickness varies from shallow (50-100 cm) to very deep (> 500 cm) with a light gray silty clay substratum (5Y 5/1). This peat is equivalent to Terric Haplohemists, Sapric Haplohemists, and Typic Haplohemists (Soil Survey Staff, 2014).

Table 6. Peat soil classification in the studied area.

National Soil Classification (Subardja *et al.* 2016)		Keys to Soil Taxonomy (Soil Survey Staff 2014)	
Type	Kind	Group	Subgroup
Organosol	Organosol Fibric	Haplofibrists	Hemic Haplofibrists
			Typic Haplofibrists
	Organosol Hemic	Haplohemists	Terric Haplohemists
			Sapric Haplohemists
			Typic Haplohemists
	Organosol Sapric	Haplosaprists	Terric Haplosaprists
			Hemic Haplosaprists
			Typic Haplosaprists

3.13 Organosol Sapric

Organosol Sapric has a sapric maturity level in the control section (0-130 cm), and its thickness varies from shallow (50-100 cm) to very deep (> 500 cm). Under the sapric peat layer, there is peat with hemic maturity level. Underlying this peat is soft, pale gray (5Y 6/1) silty clay substratum. This soil is equivalent to Terric Haplosaprists, Hemic Haplosaprists, and Typic Haplosaprists (Soil Survey Staff, 2014).

4 CONCLUSION

Riau Province has peatland of 3.85 million hectares with around 1.06 million located in Kampar, Siak, Rokan Hulu, Rokan Hilir and Dumai City Regencies. Of this area, 487,950 ha, 45.95%, is topogenous freshwater peat, indicated by the ratio of Ca/Mg > 1 and ash content > 5%. The following presents the properties and characteristics of freshwater topogenous peat in the studied area:

1. Freshwater topogenous peat in Kampar, Siak, Rokan Hulu, Rokan Hilir and Dumai Regencies has a hemic level of maturity indicated by the average fiber content around 71%. The soil is classified as Organosol Hemic, with its thickness varying from shallow (50- <100 cm) to very deep (> 500 cm) with light gray (silty clay) substratum (5Y 5/1). At the subgroup level, the soil is classified as Terric Haplohemists, Sapric Haplohemists, and Typic Haplohemists.
2. Bulk density ranged from 0.05 to 0.33 g cm^{-3}, and was higher at the top than the lower layer. The average bulk density (0.22 g cm^{-3}) and ash content (30.2%) of Kampar is higher than in

other regencies. Ash content is related to the level of peat maturity. Generally, where the sapric peat has a higher ash content than hemic and fibric peats. This indicates that the peat soil of Kampar Regency is more mature than other regencies.
3. pH H$_2$O ranged from acidic (pH 3.3) to very acidic (pH 4.8) due to high organic acids. Organic carbon ranges from 13.62 to 55.66%, and the upper layer is lower than the lower one. Based on the C/N ratio (15 to 69), some part of organic carbon has weathered and released a number of N. This is seen from N-total varying from very low (0.04%) to very high (2.04%), though, it is still in the form of organic N. Therefore, to meet N nutrient requirements, N-fertilization is recommended.
4. Exchangeable cations such as Ca, Mg, K, and Na vary from very low to very high. Similarly, base saturation (BS) varies, although still relatively low to very low. The P$_2$O$_5$ levels (extracted 25% HCl) also varied greatly from very low to high (2 to 45 mg 100 g^{-1}). The presence of river runoff carrying mineral materials prevents P leaching, explaining why the content of P$_2$O$_5$ (extracted Bray 1) in the upper layer is higher than in the lower layer. With the low K level (10 mg 100 g^{-1}), K fertilization is need.

ACKNOWLEDGMENT

We express our gratitude to the Center for Agricultural Research and Development for facilitating this research throughout the 2017 FY. We also thank Mr. Ir. Hikmatullah, M.Sc and the team that has provided the data needed in compiling write this report.

REFERENCES

Agus, F., K. Hairiah dan A. Mulyani. 2011. Pengukuran Cadangan Karbon Tanah Gambut. Petunjuk Praktis. World Agroforestry Centre-ICRAF, SEA Regional Office dan Balai Besar Penelitian dan Pengembangan Sumberdaya Lahan Pertanian (BBSDLP), Bogor, Indonesia. 58 p.

Ambak, K., and J. Melling. 2000. Management practices for sustainable cultivation of crop plants on tropical peatland. In Proceedings of the International Symposium on Tropical Peatlands. Bogor, Indonesia, 22-23 November 1999. Hokkaido University & Indonesian Institute of Sciences. pp. 119-134 (2000).

Andriesse, J.P. 1997. Lecture Note on The Reclamation of Peat swamps and Peat In Indonesia. Faculty of Agriculture University of Bogor. Lecture 4.

Andriesse, J.P. 1988. Nature and Management of Tropical Peat Soil. Soil Researches Management and Conservation Service. FAO Land and Water Development Division. Rome.

Andriesse, J.P. 1974. Tropical Peats in South East Asia. Dept. of Agric. Res. Of the Royal Trop. Inst. Comm. 63. Amsterdam 63 p.

Anshari G., E. Gusmayanti, M. Afifudin and G. O. Widhanarto. 2012. A study of carbon balance in customary peat forest (hutan nung) in Taman Nasional Sentarum. Final Report. Pontianak: Universitas Tanjungpura.

Anshari G., E. Gusmayanti and M. Afifudin. 2013. Assessing degradation of tropical peat domes and dissolved organic carbon (DOC) export from the Belait, Mempawah and Lower Kapuas Kecil River in Borneo. PEER Annual Technical Report 2012-2013. Center for Wetlands Peoples and Biodiversity (CWPB), Pontianak: Universitas Tanjungpura.

BBSDLP 2017. Identifikasi dan karakterisasi lahan gambut mendukung One Map Policy (Laporan Akhir RPTP). Balai Besar Litbang Sumberdaya Lahan Pertanian. Badan Litbang Pertanian.

Benbi, D.K., and J. Richter. 2002. A critical review of some approaches to modelling nitrogen mineralization (Published online: 11 April 2002). Biol Fertil Soils (2002) 35:168–183.

Cookson, W.R., D. Abaye, P. Marschner, and D.V. Murphy. 2005. The contribution of soil organic matter fractions to carbon and nitrogen mineralization and microbial community size and structure. Soil Biology and Biochemistry 37(9):1726–1737.

Dariah A., E. Susanti dan F. Agus. 2011. Simpanan karbon dan emisi CO2 lahan gambut, In. Nuraida, N. L., A. Mulyani, dan F. Agus, eds. Pengelolaan Lahan Gambut Berkelanjutan. Bogor: Balai Penelitian Tanah.

Dent, F.J. 1986: Southeast Asian coastal peats and their use–An overview. In Proceedings Second International Soil Management Workshop, Thailand/Malaysia. Classification, Characterizations and Utilization of Peat Land, Ed. H. Eswaran et al., p. 27-49.

Driessen, P.M., and M. Sudjadi. 1984. Soils and specific problems of tidal swamps. Workshop on Research Priorities in Tidal Swamp Rice. P143-1160. IRRI, Los Banos, Philippines.

Eviati dan Sulaeman. 2012. Petunjuk Teknis Analysis Kimia Tanah, Tanaman, Air, dan Pupuk. Edisi 2. Badan Penelitian dan Pengembangan Pertanian, Kementerian Pertanian.

Hardjowigeno, S. 1986. Genesis dan Klasifikasi Tanah. Jurusan Tanah, Fakultas Pertanian IPB. Bogor.

Ivanoff, D.B., K.R. Reddy, and S. Robinson. 1998. Chemical fractionation of organic phosphorus in selected Histosols. J. Soil Sci. 163(1):36-45.

Miller, M.H. dan R.L. Donahue. 1990. Soils. An Introduction to Soils and Plant Growth. Prentice Hall Englewood Cliffs. New Jersey. 768 p.

Nugroho, K., G. Gianinazzi and IPG Widjaja Adhi. 1997. Soil hydraulic properties of Indonesia peat. In: J. O. Riely and S.E. Page (eds) Biodiversity and Sustainability of Tropical Peat and Peatland. Samara Publ. Ltd. Cardigan. UK, pp 147-156.

Oldeman, L.R., Irsal L., M. Darwis. 1978. An agroclimatic map of Suamatera, scale 1:3,000,000. Contr. Res.Inst. of Agric., Bogor. Bulletin No. 60.

Radjagukguk, B. 1997. Peat soil of Indonesia: Location, classification, and problems for sustainability. pp. 45-54. In J.O. Rieley and S.E. Page (Eds.).Biodiversity and

Sustainability of Tropical Peat and peatland. Proceedings of the International Symposium on Biodiversity, Environmental Importance and Sustainability of Tropical Peat and Peatlands, Palangkaraya, Central Kalimantan 4-8 September 1999. Samara Publishing Ltd. Cardigan. UK.

Ritung, S., Wahyunto, Nugroho K., Hikmatullah, Suparto dan Tafakresnanto, C. 2011. Peta Lahan Gambut Indonesia Skala 1:250.000. Balai Besar Litbang Sumberdaya Lahan Pertanian. Badan Litbang Pertanian, Kementan, Bogor.

Sabiham, S., T.B. Prasetyo, dan S. Dohong. 1997. Phenolic acid in Indonesia peat. pp. 289-292. In Rieley and Page (Eds.). Biodiversity and Sustainability of Tropical Peat and Peatland Samara Publishing Ltd. Cardigan UK.

Salampak, 1999. Peningkatan Produktivitas Tanah Gambut yang Disawahkan dengan Pemberian Bahan Amelioran Tanah Mineral Berkadar Besi Tinggi. Disertasi Program Pascasarjana, Institut Pertanian Bogor.

Schmidt F.H. and J.H.A. Ferguson, 1951. Rainfall Types Based on Wet and Dry Period Ratios for Indonesia with Western New Guinea, Verh. No. 42. Kementerian Perhubungan, Jawatan Meteorologi dan Geofisika. Jakarta.

Singh, G., Pau, T.Y., Rajah, P.C.V., and Lee, F.W. 1986: Experiences on the cultivation and management of oil palms on deep peat in United Plantation Berhad. In Proceedings Second International Soil Management Workshop, Thailand/Malaysia. Classification, Characterizations and Utilization of Peat Land, Ed. H. Eswaran et al., p. 77-194.

Soepraptohardjo, M. 1961. Jenis-jenis Tanah di Indonesia. Lembaga Penelitian Tanah. Bogor.

Soil Survey Staff. 2014. Keys to Soil Taxonomy. 12nd Ed. USDA Natural Resources Conservation Service. Washington DC.

Stevenson, F.J. 1994. Humus Chemistry. Genesis, Composition, and Reactions. John Wiley and Sons.Inc. New York. 443 p.

Subagyo, H., D.S. Marsoedi dan A.S. Karama. 1996. Prospek Pengembangan Lahan Gambut untuk Pertanian. Seminar Pengembangan Teknologi Berwawasan Lingkungan untuk Pertanian Lahan Gambut. Bogor, 26 September 1996.

Subardja, D, S. Ritung, M. Anda, Sukarman, E. Suryani, dan R.E. Subandiono. 2016. Petunjuk Teknis Klasifikasi Tanah Nasional. Balai Besar Penelitian dan Pengembangan Sumberdaya PertaniaN, Badan Litbang Pertanian, Bogor. Edisi 2016.

Suhardjo, H. and I P.G. Widjaja-Adhi. 1976. Chemical characteristics of the upper 30 cm of peat soils from Riau. ATA 106. Bull. Soil Res. Inst. Bogor.

Tadano, T., Pantanahiran, W., and Nilnond, C. 1992: Inhibitory effect of canal water drained from a tropical deep peat soil on the elongation of rice roots. Soil Sci. Plant Nutr., 38, in press.

Tan, K.H. 1993. Principles of Soil Chemistry. Marcel Dekker, Inc. New York. 362pp.

Tisdale, S.C., W.L. Nelson and J.O. Beaton, 1990. Soil Fertility and Fertilizer Elements Required in Plant Nutrition. pp: 52–92. 4th ed. Maxwell Macmillan Publishing, Singapore.

Tropical Wetlands — Innovation in Mapping and Management — Sulaeman et al. (Eds)
© 2020 Taylor & Francis Group, London, ISBN 978-0-367-20964-3

Characteristics of peat soils and management implication for agricultural development in Asahan regency, North Sumatra

E. Yatno, E. Suryani & Suratman
Indonesian Center for Agricultural Land Resources Research and Development (ICALRRD), Jakarta, Indonesia

ABSTRACT: Understanding the natural behavior and change after the reclamation of peatland for agricultural use as well as the consideration of local wisdom are the major factors to achieve sustainable peatland management. Some peat soils in oil palm plantation areas in the Asahan Regency, North Sumatra were sampled to determine their physio-chemical characteristics and management strategies. The results showed the peat soil thickness to vary from shallow with 65 mm to deep with 250 cm overlaying the substratum of fine textured (clay, silty clay) mineral soils. The degree of peat maturity varied between hemic to sapric at the top soils, and hemic to fibric in the subsoils. The peat soils are very acid with pH in H_2O of 3.5-4.4, have a very high organic C content of 34-54%, a very high total N of 0.92-1.72%, a very high exchangeable capacity of 50-109 cmolc/kg, and a very low base saturation of 8-19%. Therefore, proper water management and soil nutrients balancing through fertilization should be applied to achieve effective peatland management for agricultural development.

1 INTRODUCTION

Indonesia has a large peatland area of 14.93 million ha distributed in Sumatra, Kalimantan, Papua and few hectarages in Sulawesi Islands and was also observed to be covering a large area of 24,232 ha in the Asahan Regency in North Sumatra (BBSDLP, 2011). Most of the peatlands were used for oil palm plantation. However, there is no sufficient information on the characteristics of the peatland. Therefore, it is important to conduct a study on these characteristics and developed peatland management.

Peat soil is defined as the soil formed from organic sediments of parent material which are obtained from the accumulation of decayed plant/vegetation remnants under anaerobic conditions. According to Hardjowigeno (1986), this is a geogenic process involving deposition and transportation, which is in contrast to the formation of mineral soils through pedogenic processes. However, peat soil is also known as Organosol (Soepraptohardjo, 1961; Subardja et al., 2016) or Histosols (Soil Survey Staff, 2014).

Peat soils have very different characteristics compared to the mineral soils. Physically, they have low bearing capacity due to the low bulk density of 0.05-0.3 g/cm3 (Nugroho et al., 1997). Moreover, over-drainage causes subsidence, which consequently leads to irreversible drying and a decrease in water retention. Its chemical properties generally include very acidic soil

reaction with a pH of 3.0-4.5, very high organic carbon contents, and cation exchange capacity (Sabiham et al., 1999 and Salampak, 1999).

Peatland is currently used for the cultivation of food and horticultural crops as well as oil palm and rubber plantations. However, as stated earlier, over-drainage could decrease water retention capacity and leads to soil nutrient leaching. Therefore, the management of peatland requires strong planning and high level of attention.

The objective of this study was to characterize chemical characteristics of peat soils and the implication of their management on sustainable agricultural development in the Asahan Regency, North Sumatra Province.

2 MATERIALS AND METHODS

2.1 Description of the study area

The study area included the oil palm plantation areas located in the southeastern part of the Asahan Regency, North Sumatra Province, Indonesia. Four representative soil profiles, AP 2, EY 3, EY 6, and EY 7, were formed from organic materials and used in this study. They were all located in the peat dome and freshwater topogenous peat landform ranging from flat to nearly flat topography with a slope less than 3 %, and elevation between 8 and 14 m above sea level (asl). The location and description of the studied profiles are presented in Table 1 and Figure 1.

Table 1. Location and description of the studied soils.

Profiles	Location	Coordinate	Elevation m asl	Landform	Relieve	Slope (%)
AP 2	Alang Bombon, Aek Kuasan	2°42'6"N 99°45'42" E	14	Peat Dome	Nearly Flat	2
EY 3	Perbangunan, Sei Kepayang	2°47'44" N 99°50'33" E	8	Peat Dome Edge	Nearly Flat	2
EY 6	Bangun Baru, Sei Kepayang	2°50'32"N 99°54'15"E	9	Fresh Water Topogenous Peat	Flat	1
EY 7	Bangun Baru, Sei Kepayang	2°48'49"N 99°54'13"E	9	Fresh Water Topogenous Peat	Flat	1

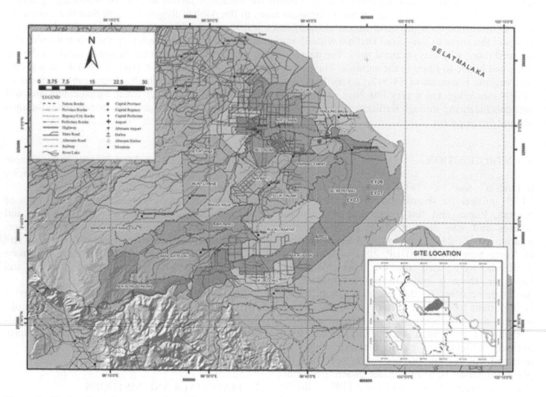

Figure 1. Site location of representative soil profiles.

2.2 Methods

The morphological characteristics of the profiles were described in the field following the Soil Survey Manual (Soil Survey Division Staff, 1993 and Sukarman et al., 2017). All the soils were classified as a family level according to Keys to Soil Taxonomy (Soil Survey Staff, 2014) and National Soil Classification (Subardja et al., 2016).

Furthermore, laboratory analysis was conducted on the soil chemical properties with the pH measured with a glass electrode in water using a soil:solution ratio of 1:5. Organic carbon was determined by the Walkley-Black wet combustion method, while total Nitrogen was measured using the Kjeldahl method. Moreover, the P_2O_5 and K_2O contents were determined by using the HCl 25% extraction method while the cation exchange capacity (CEC) was determined through saturation with 1 M NH_4OAc at pH 7.0. In addition, the exchangeable Al was determined using the KCl extraction method. The analysis procedure used was in reference to the Soil Survey Laboratory Methods Manual (Burt, 2004) and the Technical Guidelines for Analysis of Soil, Plant, Water, and Fertilizer Chemistry (Eviati and Sulaeman 2012).

3 RESULTS AND DISCUSSION

3.1 *Soil morphological characteristics*

The results showed the peat thickness to vary from shallow with 65 mm to deep with 250 cm overlying the substratum of fine textured (clay, silty clay) mineral soils with EY 6 having the shallowest thickness. The topsoil color is generally very dark brown (7.5YR 2.5/2) to dark reddish brown (5YR 3/2), while the lower layers are very dark brown (7.5YR 2.5/2) and reddish brown (7,5YR 3/2). In agreement with this, Ambak and Melling (2000) also found tropical peat to be generally reddish brown to dark brown (dark) depending on the decomposition stage. Furthermore, the degree of peat maturity was also found to vary from hemic (half ripe) to sapric (ripe) in the top soils, and hemic (half ripe) to fibric (unripe) in the subsoils as shown in Table 2.

3.2 *Soil chemical properties*

All the peat soils were characterized by very acidic soil reaction (pH H_2O 3.5-4.4) as shown in Table 3. This was associated with the high exchangeable H of 1-10 cmolc/kg found mainly in the upper layers of the peat soils. Furthermore, no significant relationship was found between peat soil with sapric and hemic maturities. The reaction of the substratum layers was also found to range from very acidic to slightly acidic

(pHH_2O 4.4-5.8). The level of acidity of peat soil is related to the organic acids content (Andriesse, 1974; Miller and Donahue, 1990); Tan, 1993). However, this reaction may have a negative effect on plant growth and production. Therefore, the soil needs high input to achieve optimal soil productivity.

The studied soil was also found to have a very high organic C content of 34-54% and total Nitrogen of 0.92-1.72%. According to Soepraptoharjo (1961), Subardja et al. (2016), and Soil Survey Staff (2014), peat soil is soil formed from deposits of organic matter with a thickness of 50 cm or more and containing organic carbon of at least 12% (dry weight). Moreover, no significant difference was observed with the organic C content between the soil peat with sapric, hemic, and fibric maturities, and content found can increase the exchangeable cation capacity.

Furthermore, the cation exchange capacity (CEC) can provide a figure the soil with the capability to retain and release cation nutrients from the exchange complex and it was found to vary between 50 to 109 cmolc/kg. Moreover, the peat soils with sapric maturity generally have a higher capability to retain and release soil nutrients than hemic and fibric maturities.

All the samples were found to have generally low to medium P_2O_2, and K_2O, and no significant difference was observed between the value for soil peat with sapric, hemic, and fibric maturities as shown in Table 3. However, the upper layers of profile EY 3 and EY 6 showed a medium value.

Table 2.   Morphological characteristics of the studied soils.

Horizon	Depth (cm)	Colour	Peat Maturity	Substratum Soil Texture	Consistency
Profile AP 2 (Hemic Haplosaprists)					
Oa	0 – 70	5YR 3/2	sapric	-	non-sticky
Oc1	70 – 185	7.5YR 3/2	hemic	-	non-sticky
Oe2	185 – 250	7.5YR2.5/2	hemic	-	non-sticky
Oi	250 – 320	7.5YR 2.5/1	fibric	-	non-sticky
Cg	320 – 350	2.5Y 4/2	-	clay	sticky
Profile EY 3 (Hemic Haplosaprists)					
Oa1	0 – 18	7.5YR 2.5/1	sapric	-	non-sticky
Oa2	18 – 65	7.5YR 3/1	sapric	-	non-sticky
Oe	65 – 105	7.5YR2.5/2	hemic	-	non-sticky
Cg1	105 – 121	7.5YR5/2	-	clay loam	sticky
Cg2	121 – 150	2.5Y 4/2	-	silty clay loam	sticky
Profile EY 6 (Typic Haplosaprists)					
Oa1	0 – 25	7.5YR 2.5/2	sapric	-	non-sticky
Oa2	25 – 65	7.5YR 3/2	sapric	-	non-sticky
Cg1	65 – 90	7.5YR 4/2	-	-	non-sticky
Cg2	90 – 120	2.5Y 4/2	-	clay	sticky
Profile EY 7 (Sapric Haplohemists)					
Oa	0 – 30	7.5YR 2.5/2	sapric	-	non-sticky
Oe1	30 – 70	7.5YR 2.5/3	hemic	-	non-sticky
Oe2	70 – 95	7.5YR 2.5/2	hemic	-	non-sticky
Cg	95 – 120	2.5Y 6/1	-	silty clay loam	sticky

Table 3. Chemical properties of the studied soils.

Horizon	Depth	pH H₂O	Org C	Total N	HCl 25 % P₂O₅	K₂O	CEC	BS	Al³⁺	H⁺
	cm		%		mg/100 g		cmolc/kg	%	cmolc/kg	
Profile AP 2 (Hemic Haplosaprists)										
Oa	0 – 70	3.7	51.65	1.24	4	9	92	11	1.34	10.46
Oe1	70 – 185	3.8	54.01	1.72	25	8	109	12	1.75	8.52
Oe2	185 – 250	3.6	48.61	0.96	3	16	87	8	3.53	9.73
Oi	250 – 320	4.1	52.61	1.17	7	44	90	15	3.86	6.18
Cg	320 – 350	4.8	0.54	0.09	19	0	4	22	0.00	
Profile EY 3 (Hemic Haplosaprists)										
Oa1	0 – 18	4.0	49.67	1.28	33	17	74	14	0.94	2.78
Oa2	18 – 65	4.1	53.46	1.52	12	11	89	19	0.80	2.73
Oe	65 – 105	4.4	34.23	1.01	10	8	50	15	3.88	1.64
Cg1	105 – 121	5.7	6.14	0.18	4	18	31	38	0.00	0.08
Cg2	121 – 150	5.8	4.34	0.14	4	19	26	48	0.00	0.06
Profile EY 6 (Typic Haplosaprists)										
Oa1	0 – 25	3.8	40.98	1.13	29	4	59	12	2.94	2.25
Oa2	25 – 65	3.8	52.22	0.92	14	7	71	14	4.99	2.75
Cg1	65 – 90	4.4	11.60	0.29	10	7	26	23	6.59	1.16
Cg2	90 – 120	5.3	6.14	0.17	4	10	29	32	0.58	0.63
Profile EY 7 (Sapric Haplohemists)										
Oa	0 – 30	3.5	35.77	1.48	20	5	59	13	3.16	2.72
Oe1	30 – 70	3.9	51.72	1.58	13	4	89	14	1.59	3.68
Oe2	70 – 95	4.1	48.22	1.29	7	8	74	15	1.26	1.44
Cg	95 – 120	5.4	2.78	0.09	5	32	23	44	0.36	0.39

Remarks: CEC = Cation Exchange Capacity. BS = Base Saturation.

Moreover, the base saturation value of all samples was very low, as observed in the 8-19% obtained. This shows they all have low soil fertility and no significant difference was found between the maturities. On the contrary, the values obtained for the soil substratum layers are higher than those for the peat soil layers. This is associated with the high content of extractable Al and H and poor or low supply of exchangeable bases in the peat soil layers, as shown in Table 3.

3.3 Soil classification

Soils in the studied area are formed from deposits of organic matter obtained from the accumulation of decayed plant/vegetation remnants under anaerobic conditions. Furthermore, the decomposition of organic matter was slow due to this condition, and it builds up to form peat or Organosol (Soepraptoharjo, 1961; Subardja et al., 2016) or Histosols (Soil Survey Staff, 2014).

Moreover, most of the samples studied including AP 2, EY 3, and EY 6 were classified at Geat Group level as Haplosaprists (Soil Survey Staff, 2014) or Organosol Sapric according to National Soil Classification (Subardja et al., 2016) while EY 7 was classified as Sapric Halpohemists of Organosol Hemic due

to its thin nature with degree of sapric maturity as well as hemic domination.

3.4 Land management for sustainable agricultural use

The utilization of hemic peat soil is quite potential for agriculture, but adequation caution is required. The soil naturally has hydrophilic properties with the ability to hold water up to 13 times its dry weight. According to Subagyo et al. (1996) and Tadano et al. (1992), over-drying peat damages its colloid and leads to irreversible drying which changes it to charcoal and makes it lose the ability to absorb nutrients and retain water. In such conditions, according to Sabiham et al. (1999) and Salampak (1999), peat loses available water and becomes flammable.

At present, the use of peatland for agriculture is restricted due to the issue of global climate change and environmental damage, as manage with reference Permentan No. 14/2009, INPRES No. 10/2011 and Presidential Instruction No. 6 of 2013. However, the implementation of this Decree faces a lot of challenges, except for some areas with extensive peatlands, because the regional economy and its people are highly dependent on the land.

Land clearing of peatland for food or estate crops without proper water management strategy can destroy the ecosystem. and increase greenhouse gas emission. Therefore, appropriate strategies should be employed for agricultural development.

Peatland reclamation for agricultural use starts with the creation of drainage channel, land clearing, and land preparation. However, the drainage channels serve two functions and the first is to efficiently discard water excess while the second is to control soil water surface to achieve optimum condition for plant growth. Moreover, the clearing also has an impact on the peat soil physical and chemical properties.

Furthermore, peatlands cultivated for oil palm plantation or other agricultural use generally experience compaction and subsidence due to soil drainage. Moreover, the substratum mineral layer may contain pyrite materials or sand, which influences the peat soil fertility. Therefore, proper water management is required in managing peatland for agricultural use for optimal benefit through the reduction of fire and pyrite oxidation.

4 CONCLUSIONS

The degree of peat maturity in the oil palm plantation areas in the Asahan Regency varied from hemic to sapric at the top soils and hemic to fibric in the subsoils. The peat soils were characterized by very acid soil reaction, very high organic C content, total N, exchangeable capacity, and exchangeable H. as well as a very low base saturation. Therefore, it is recommended that the peatland should be managed carefully through the application of proper water management and balanced soil nutrients fertilization.

REFERENCES

Ambak, K. & Melling, J. 2000. Management practices for sustainable cultivation of crop plants on tropical peatland. In Proceedings of the International Symposium on Tropical Peatlands. Bogor, Indonesia, 22-23 November 1999. Hokkaido University & Indonesian Institute of Sciences. pp. 119-134 (2000).

Andriesse, J.P. 1974. Tropical Peats in South East Asia. Dept. of Agric. Res. Of the Royal Trop. Inst. Comm. 63. Amsterdam 63 p.

BBSDLP. 2011. Peta Lahan Gambut Indonesia Skala 1:250.000. Bogor: Balai Besar Litbang Sumberdaya Lahan Pertanian. Badan Litbang Pertanian.

Burt, R. 2004. Soil survey laboratory method manual. Soil Survey Investigation Report No. 42. Version 4. USDA Natural Res. Conserv. Service. National Survey Center. 700p.

Eviati & Sulaeman. 2012. Petunjuk Teknis Analysis Kimia Tanah, Tanaman, Air, dan Pupuk. Edisi 2. Badan Penelitian dan Pengembangan Pertanian, Kementerian Pertanian.

Hardjowigeno, S. 1986. Genesis dan Klasifikasi Tanah. Bogor: Jurusan Tanah, Fakultas Pertanian IPB.

Miller, M.H. & Donahue, R.L. 1990. Soils. An Introduction to Soils and Plant Growth. Prentice Hall Englewood Cliffs. New Jersey. 768 p.

Nugroho. K., Gianinazzi, G. & Widjaja Adhi, I.P.G. 1997. Soil hydraulic properties of Indonesia peat. In: J. O. Riely and S.E. Page (eds) Biodiversity and Sustainability of Tropical Peat and Peatland. Samara Publ. Ltd. Cardigan. UK. pp 147-156.

Sabiham. S. 1999. Studies on Peat in the Coastal Plains of Sumatra and Borneo. Physiography and Geomorphology of the Coastal Plain. South Asean Studies. Kyoto Univ. 26 (3): 308-335.

Salampak. 1999. Peningkatan Produktivitas Gambut Tanah Gambut yang Disawahkan Dengan Pemberian Bahan Amelioran Tanah Mineral Berkadar Besi Tinggi. Disertasi S3 Program Pasca sarjana IPB. Bogor.

Soepraptohardjo, M. 1961. Jenis-jenis Tanah di Indonesia. Bogor: Lembaga Penelitian Tanah.

Soil Survey Division Staff. 1993. "Soil Survey Manual". USDA Handbook No.18. United States Department of Agriculture. Washington DC.

Soil Survey Staff. 2014. "Keys to Soil Taxonomy". 12th ed. Natural Resources Conservation Service. USDA. Washington DC.

Subagyo, H., Marsoedi, D.S. & Karama, A.S. 1996. Prospek Pengembangan Lahan Gambut untuk Pertanian. Seminar Pengembangan Teknologi Berwawasan Lingkungan untuk Pertanian Lahan Gambut. Bogor, 26 September 1996.

Subardja. D.S., Ritung, S., Anda, M., Sukarman, Suryani, E. & Subandiono, R.E. 2016. Petunjuk Teknis Klasifikasi Tanah Nasional. Edisi revisi 2016. Balai Besar Penelitian dan Pengembangan Sumberdaya Lahan Pertanian (BBSDLP). Badan Penelitian dan Pengembangan Pertanian. Kementerian Pertanian. Bogor. 45 hlm.

Sukarman, Ritung, S., Anda, M. & Suryani, E. 2017. Pedoman Pengamatan Tanah di Lapangan. Badan Penelitian dan Pengembangan Pertanian. IAARD Press. Jakarta. 136 hlm.

Tadano, T., Pantanahiran, W., & Nilnond, C. 1992: Inhibitory effect of canal water drained from a tropical deep peat soil on the elongation of rice roots. Soil Sci. Plant Nutr., 38, in press.

Tan, K.H. 1993. Principles of Soil Chemistry. Marcel Dekker, Inc. New York. 362pp.

Part E. Tidal land management

Water plant and iron oxidizing bacteria for phytoremediation of water waste with passive treatment in acid sulphate soil

W. Annisa & Y. Lestari
Indonesian Swampland Agricultural Research Institute (ISARI), Kebun Karet Loktabat Banjarbaru, Indonesia

K. Napisah
Indonesian Center for Agricultural Land Resources Research and Development, Bogor, Indonesia

ABSTRACT: Paddy grown on acid Sulphate soil is subjected to Fe toxicity. Water management is not only the key but also the only practical way of reclaiming acid sulphate soil with special emphasis on leaching and flushing. The purpose of this study was to examine the effective formula for improving wastewater quality in sulphate soil in an environmental friendly manner. The experiment was conducted within a factorial design with three factors, the first one being the management of acid water by mean phytoremediation using two different water plants; *Eleocharis dulcis* and *Eleocharis reflekta*. The second factor was the inoculant with carrier and involved the following: without inoculant, carrier wood charcoal without inoculant carrier charcoal husk without inoculant, inoculant with wood charcoal, and inoculant with charcoal husk. The third factor was the management of water, specifically returned and leaching. The result showed a phytoremediation plant which lowers acid levels by raising the pH (by 23%) is *Eleocharis dulcis*. The application of inoculant with wood charcoal into inundated acid sulphate soil increased soil and water waste pH up to 5.16 and 5.73 respectively, which was higher the option without inoculant (soil and water waste pH 3.98 and 4.38 respectively). The management of water waste with returned decreased sulphate concentration from 12.04% to 5.47%.

1   INTRODUCTION

The reduction process in acid sulphate field takes place with the help of anaerobic bacteria and organic matter. In the flooded soil, organic matter is a substrate for Fe^{3+} reducing microbes. According to Satawathananont et al. (1991), when the redox potential value decreased to -50 mV of Fe^{2+}, the concentration increased. Water has a significant impact on acid sulphate. For instance, the quality of water in the inlet on sulphate sour soil is influenced by the tidal dynamics of the river. From Subagyo et al. (1999), the washing of toxic materials was efficient with the presence of fresh water, both from rain and the tide. The results of Hanhart & Ni (1993) on the water quality in the Mekong delta of Vietnam in drainage channels at 41 and 85 days after rice planting were worse than in the inlets indicated by pH values between 2.6-2.9. Besides, Aluminum dissolved 57-148 mg. kg-1 and total acidity ranged from 11 to 26 mol L-1. Contrastingly, in the inlet the pH of water ranged from 6.3 to 6.8, the soluble aluminum varied between 1.3 and 7.2 mg.kg-1, and the total acidity was 0.15-0.56 mol.L-1. This indicates that the toxic elements are washed and carried to the drainage, lowering the water quality as a result. There are several parameters affecting the Water Quality Criteria (KTA) for irrigation. For the water quality is good

if the range of pH value is between 6 and 8.5, BOD and COD concentrations of 20 mg/L, and 100 mg/L respectively, iron content of 10 mg/L and sulfate 7 mg/L. These parameters are basic indicators of the quality of environmentally friendly waste water which may be reused for irrigation.

Some aquatic plants have the ability to absorb or filter out toxic elements existing in the water. For instance, Purun tikus (*Eleocharis dulcis*) grows predominantly in acid sulphate soil environment and may be used as an absorbing agent of iron due to a symbiosis relationship between plants and microorganisms. In addition, Thiobacillus ferrooxidans is a bacterium with an iron oxidase which may metabolize metal ions such as ferrous iron. It grows on oxidized soil or where reduction occurred (continuous flooding for 1 month) (Mariana et al., 2012). From the reaction of microorganisms to oxygen, Thiobacillus ferrooxidans bacteria are microaerophils (mandatory aerobic organisms that continue to grow well on low oxygen content). It has the ability to oxidize either Fe^{2+} to Fe^{3+} or the reduced sulfur compounds and utilize the oxidant as its energy source. The use of Thiobacillus ferroxidan bacteria in improving the quality of waste water not only increases rice production but also reduces methane (CH_4) emissions from the cultivation. This relates to the role of SO_4^{2-} as an electron acceptor which inhibits CO2 reduction to CH₄. Methane thermodynamically (CH₄) is formed after most iron ferric (Fe^{3+}) is reduced to ferrous iron (Fe^{2+}). Soil microbes have a mechanism of causing changes in the mobility of metallic elements, making them either more difficult or easier for absorption. The changes in the mobility of metallic elements by microbes are grouped into two: redox changes of inorganic metals and changes in metallic form from inorganic to organic and vice versa. Specifically, this change is the process of methylation and demylation. Through the oxidation of metals such as iron, microbes can obtain energy. On the other hand, the reduction of metals may take place through the process of decimilation in anaerobic respiration when microbes use metals as electron acceptor terminals (Atlas and Bartha, 1998; Stolz and Oremland, 1999; Niggemyer et al 2001; Irfan DP 2006). The purpose of this study was to examine the effective formula for improving water waste quality in sulphate sulfate soil with environmental friendliness.

2 METHODOLOGY

This study was conducted in the greenhouse of Indonesian Swampland Agriculture Research Institute (ISARI). It used a randomized factorial design of two factors repeated 3 replications. A closed system was used in water management.

The first factor examined was the formula of inoculant as:

1. Control (without inoculant)
2. Inoculant A without carrier
3. Inoculant B without carrier
4. Inoculant A with carrier of wood charcoal husk
5. Inoculant A with carrier of husk charcoal
6. Inoculant B with carrier of wood charcoal husk
7. Inoculant B with carrier of husk charcoal
8. Carrier of wood charcoal without inoculant
9. Carrier of husk charcoal without inoculant

Figure 1. Desain of simulation water management for rice in acid sulphate soil.

The second factor was plant of phytoremediation including:

1. Rat purun (*Eleocharis dulcis*) plant
2. Sea urchins (*Eleocharis reflaxta*) plant

The Rat Purun rod and sea urchin were first grown two weeks before the research commenced to allow sufficient time for them to breed. Plants harvested as phytoremediation to be grown in the waste channel (greenhouse simulation) had the same age, fresh leaves and height of 20-30 cm in purun, though the piglets had the root length of 10-15 cm. Without the process of drying, the acid sulphate soil taken in the field at the selected site was put into a pot of 9 kg and chalked according to the optimum dosage. Basal fertilizers were applied to all the treatments according to recommended doses in tidal swamp land. The commended dosage for basal fertilizer: urea is SP-36: KCl namely: 200 kg.ha^{-1}: 100 kg. ha^{-1}: 100 kg.ha^{-1}. Urea fertilizer is usually applied in 2 stages; at planting time and when the rice is 4 weeks old, SP-36 and KCl are given once at planting time. After incubation for 1x24 hours, there was new planting of Rice seedlings with Variety of Inpara 2. The plants were maintained to the vegetative maximum. They were washed periodically every two weeks using river water. The leaching water from the rice plants was collected in a tub treated with water plants as phytoremediation material. Among the basic things observed included the height of rice plants and the number of tillers every two week. There was intensive observation of wastewater every two weeks on pH, DHL and EH.

3 RESULTS AND DISCUSSIONS

3.1 *Characteristics of the acid sulphate soil*

Based on Soil Survey Staff (2010), sulphate soil belongs to the Typic Sulfaquent subgroup because of the pH> 3.5, but after oxidation pH decreased to <3.5. The pyrite oxidation process produced Fe^{2+}, SO$_4^{2-}$ and H$_2$ which resulted in a decrease in pH. According to Dent (1986), the pH value <2.5 or 3 after application of H$_2$O$_2$ indicates strong sulphate acidity. The pyrite oxidation process with H$_2$O$_2$ was described by Singer & Stumm (1970) in the reaction equation:

$$FeS_2 + 7H_2O_2Fe^{2+} + 2SO_4^{2-} + 2H^+ + 6H_2O \quad (R1)$$

In Table 1, the iron content of ferrous (Fe^{2+}) and the one for sulfate (SO$_4^{2-}$) was high at 269.9 ppm and 319.8 ppm. There was Aluminum saturation of 9.77% in the high root zone. According to Bremen and Buurman (2002), the solubility of aluminum increase sharply in very acid conditions. This was

Table 1. Chemical properties of acid sulphate soil.

Variabel	pH	Org- C %	N- tot %	K-soluble (cmol(+)/kg)	CEC	Al- soluble	H-soluble	P Bray 1 (ppm P)	Fe soluble (ppm)	SO$_4$ soluble (ppm)
Soil	3.56	3,67	0,25	0,35	46,9	9,77	6,08	10,59	269,9	319,8

Figure 2. Soil sampling and characteristic acid sulphate soil.

the case in the iron content of ferrous (Fe^{2+}) in the high root zone of 1005.0 mg.kg^{-1} and related to the high pyrite content.

3.2 Soil quality

Based on the results, this land has a very sour soil acidity followed by Al saturation, pyrite content and high iron content and sulfate.

Figure 3. Greenhouse activities.

Soil quality at simulation in greenhouse on the paddy field and on drainage channel in the picture below. The results showed the soil pH increased during the initial period of observation, decreased and then continued to increase until the fourth period of surveillance. The decrease in the second phase of observations is attributed to the process of oxidation in the presence of microbes in the drainage canal. It

is inversely proportional to soil pH on drainage channels which continues to decline until the fourth period of observation. The highest soil pH values were seen in the treatment of inoculant A without carrier and B inoculant without carrier which reached value of 5.02. This is attributed to the speed of adaptation and the effectiveness of higher inoculants due to the direct contact with roots.

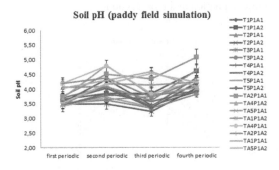

Figure 4. Soil pH periodically every two weeks after planting (paddy field simulation).

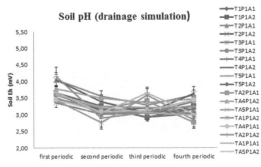

Figure 5. Soil pH periodically every two weeks after planting (drainage simulation).

3.3 Water waste quality

The quality of wastewater in acid sulphate is affected by the biogeochemical processes occurring in the soil. According to Zedler (2003), the inputs on agricultural cultivation have a positive effect on the quality of waste water. The facts show inoculant with charcoal husk increase the pH of linden water

Figure 6. Soil Eh periodically every two weeks after planting (paddy field simulation).

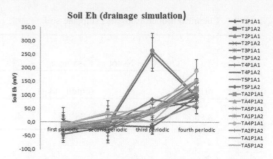

Figure 9. EH soil periodically every two weeks after planting.

to a value of 5.03 with a potential redox value (Eh) of 98.8 mV and 100.9 mV (Figure 8 and Figure 9). This is due to the oxidation process which is effective in the presence of microbes. According to Atlas and Bartha (1993), bacteria resistant to heavy metals have survival mechanisms through bioaccumulation, bioprecipitation and bioreduction.

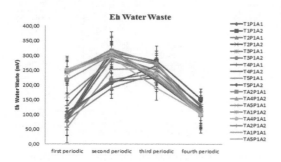

Figure 7. Eh water waste periodically every two weeks after planting.

Figure 10. The growth of paddy in greenhouse.

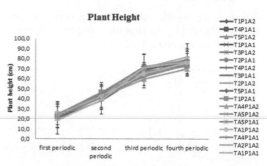

Figure 11. Plant height periodically every two weeks after planting (drainage simulation).

4 CONCLUSION

The Improved quality of wastewater was shown in the treatment of Inoculant with carrier of wood charcoal husk. The phytoremediation plant increased wastewater pH from 3.75 to 5.03 with redox potential of 98.8 mV. The height of the rice plant was 78 cm at fourth phase of observation.

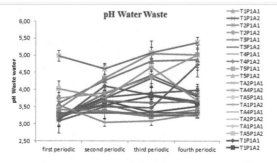

Figure 8. pH of wastewater every two weeks after planting.

3.4 The growth of rice

The height of the plants at the drainage channel with inoculant was better than without inoculant. The highest plant height to the maximum vegetative phase in the treatment of inoculant B without carrier was 82 cm (Figure 11).

REFERENCES

Atlas, R. M. dan Bartha, R. 1998. Microbial Ecology Fundamentals dan Applications. Benjamin Cummings Publishing Company Inc., California: 65.

Bremen, N.V. & Buurman, P. 2002. Soil Formation. Second Edition. Kluwer Academic Publishers. New York. Boston. Dordrecht. London. Moscow.

Dent, D.L. 1986. Acid Sulphate Soils:A Baseline for Research and Development. International Institute for Land Reclamation and Improvement – ILRI, Wageningen.

Hanhart, K and D.V. Ni. 1993. Water Management of Rice Fields at Hoa An, Mekong Delta, Vietnam. 161–176. *In* Dent, D.L., and M.E.F. Van Mensvorort (Ed). Selected Papers of The Ho Chi Minh City Symposium on Acid Sulphate Soils. ILRI Pub. 53. International Institute for Land Reclamation and Improvement. Wageningan. Netherlands.

Irfan D. Prijambada. 2006. The role of microorganisms in phytoremediation of heavy metal polluted soil. *In* Ariyanto, D.P, Marwoto, Supriyadi, Maryani, Saptawati, L., Rosariastuti, M.M.A.R., dan Mulyani, S. (eds.) Improved Role of Microbiological Research & Technology in the Field of Food Safety & Health. Proceedings of the Annual Scientific Meeting of Indonesian Microbiological Society (in Bahasa Indonesia). Surakarta, August 26th -27th, 2006.

Mariana, Z.T., F. Razie and M. Septiana. 2002. Population of iron and sulfur oxidizing bacteria due to flooding and drying in acid sulphate soils of South Kalimantan. Agrosains. 15 (1): 22–27.

Niggemyer A., Spring S., Stackebrandt E., Rosenzweig R. F. (2001). Isolation and characterization of a novel As(V)-reducing bacterium: implications for arsenic mobilization and the genus Desulfitobacterium. Appl. Environ. Microbiol.,67: 5568–5580.

Stolz Ronald S, J.F and Oremland. 1999. Bacterial respiration of arsenic and selenium. EMS Microbiology Reviews, Volume 23, Issue 5, 1 October 1999, Pages 615–627, https://doi.org/10.1111/j.1574-6976.1999.tb00416.x.

Satawathananont, W.H. Patrick and P.A Moore. 1991. Effect of Controlled Redox Conditions on Metal Solubility In Acid Sulphate Soil. Plant and Soil 133: 281–290.

Singer, PC., and Stumm W. 1970. Acidic mine drainage the rate determining step. *Science* 167: 1121.

Soil Survey Staff. 2010. Keys to Soil Taxonomy. Eleventh Edition. Handb. 336, Natural Resources Conservation Service-USDA.

Subagyo, K., H. Suwardjo., A. Abas., I.P.G. Widjaja-Adi. 1999. The Effect of Washing, Chalk and K fertilization on the Chemical Properties of Soil, Water Quality and Rice Results in Acid Sulphate Soil in Unit Tatas, Central Kalimantan. Preaching of Soil Research and Fertilizers 12: 35–47.

Zedhler, J.P. 2003. Wetlands at your service: reducing impacts of agriculture at the watershed scale. Frontiers in Ecology and the Environment. Vol.1 Issue 1. March 2003. P. 65–72. First published: 01 March 2003. https://doi.org/10.1890/1540-9295(2003)001[0065: WAYSRI]2.0.CO;2. Cited by: 228.

Tropical Wetlands — Innovation in Mapping and Management — Sulaeman et al. (Eds)
© 2020 Taylor & Francis Group, London, ISBN 978-0-367-20964-3

Management optimization technology of acid sulphate tidal swampland for improving farmers income (case study of Sidomulyo Village Tamban Catur District Kapuas Sub-district)

Y.R. Darsani & W. Annisa
Indonesian Swampland Agricultural Research Institute (ISARI), Kebun Karet Loktabat Banjarbaru, Indonesia

ABSTRACT: This research is aimed at obtaining management optimization technology for adaptive tidal swamp land in order to increase farmers income. This research was conducted in a land management area of 20 ha with the cooperation of 20 farmers in Sidomulyo Village, TambanCatur District, Kapuas Sub-district, in Central Kalimantan in 2017. The technologies implemented include (1) water management, (2) soil tillage, (3) varieties, (4) seed age, (5) planting systems, (6) fertilization, (7) weeding, (8) pest control, and (9) harvest method. The study was conducted in two planting seasons; Rainy Season and Dry Season, 2017. Using the farming who didn't cooperate as a control, data were collected from daily notes and surveys, as well as analyzed to balance both cost and revenue. The results showed that the application of tidal swampland management optimization technologies packages along with the Planting Index being IP 200, could increase rice productivity by 303%, making the rice-rice planting pattern not just economically profitable but also feasible to be developed on a wide scale. Farmer's household income increased by IDR 16,549,247/household/year or 44.35% when compared to the previous year's income.

1 INTRODUCTION

Swampland is marginal land and has the potential to be developed into agricultural land. According to Nugroho and Suriadikarta (2010), the area of tidal swampland in Indonesia is estimated to be about 20.12 million hectares, consisting of 2.07 million ha potential land, 4.23 million ha of acid sulfate land, 10.89 million hectares of peat land and 0.44 million ha of saline. Of this total, 8.54 million ha has the potential for agriculture.

Agricultural development on acid sulfate tidal swampland faces obstacles which include high acidity. In acidic conditions, the solubility of iron (Fe), Aluminum (Al), and Manganese (Mn) increase to the extent of poisoning the plants in an effort to remove acids derived from Fe-free ions, So4 and Al^{3+} and Mn which are at poisonous levels to plants (Alwi, 2011). In addition, farmers find it difficult to implement new technology or adopt them because of the limited resources available such as capital, labor, and the technical know-how needed for the mastery of the technology adopted is still low. Imam (2007) further stated that the failure of technological innovations is due to errors in identifying problems, the need for technological innovation, and also the absence of mentoring programs. This has made it difficult to show the results of the innovation. Farmers rejection of new technology is not only due to conventional farmers but also because the new technologies cannot be integrated into the real conditions the farmers are facing. The low application of cultivation technology can be seen in the large gap in production potential between the results of research and the field results obtained by farmers. Understanding and mastery of the new technology packages that are poorly understood by farmers but expected to be used as a blanket solution, whereas it is not just a solution to a piece of the problem, but is also one of the causes, as said by Mashar (2000) in Haryono et al. (2012). Therefore, specific site studies are needed and carried out on farmer's land so as to be able to come up with the best technology for that site. The Farmers will bring the technology into the balance when faced with the options of increased income or the risk of failure (Byerlee and Collinson, 1980; Adjid, 1985). This was buttressed by Bunnch (2001), stating that the adoption of technology can run fast if the technology is able to increase farmers income by at least 50-150%.

As a result of this, Indonesian Swampland Agricultural Research Institute (ISARI) and other institutions have carried out a lot of research, including a) land management, b) water management, c) crop management, d) management of amelioration and nutrients, and e) weed control (Alihamsyah et al., 2004).

Opportunities for successful utilization of potential can be done by solving problems in swampland. These solutions are found by carrying out studies on the application of integrated technologies with specific environments, including: (1) water management, (2) soil tillage, (3) varieties, (4) Seed age, (5) cropping systems, (6) fertilization, (7) weeding, (8) pest control and, (9) how to harvest with the Planting Index to IP 200. The expectation is to provide optimal and profitable results, not only in terms of business but also on the sustainability of its resources.

This study aims to obtain optimization technology for adaptive acid sulfate tidal swampland management in order to increase farmers income.

2 METHODOLOGY

The study was conducted on farmlands involving 20 farmers (on-farm research) on an area of about 20 ha which was designed with one water management unit based on hydrological units and overflow types, with rice-rice crops commodity in 2017. Land arrangement and technological touches were carried out in all areas in the form of (1) water management, (2) soil tillage, (3) varieties, (4) seed age, (5) planting system, (6) fertilization, (7) weeding, (8) pest control, and (9) harvesting method (Table 1). The touch of technology applied to 20 ha farmer area was as a result of intensive research on superimposed activities in 2015 and 2016.

Table 1. Rice cultivation technology applied in tidal swampland.

Applied Technology	Planting Season 1 (Rainy season 2016/2017)	Planting Season 2 (Dry season 2017)
Water Management	Two Way Water System	Two Way Water System
Soil Tillage	Mechanization (hand tractor/OTS)	Mechanization (hand tractor/OTS)
Varieties	Argo Pawon	Argo Pawon
Seedling Age	15-22 DAS	15-22 DAS
Planting System	Mechanization (Indo JarwoTransplanter 2:1)	Mechanization (Indo Jarwo Transplanter 2:1)
Fertilization		
Urea (kg/ha)	DSS	DSS
SP 36 (kg/ha)	DSS	DSS
KCl (kg/ha)	DSS	DSS
Lime	DSS	DSS
Cow manure (t/ha)	1 t/ha	0.5 t/ha
Pest Control	IPT	IPT
Harvest Method	Mechanization	Mechanization

Note:
DAS: Day after seedling
DSS: Decision Support System (DSS)
IPT: Integrated Pest Control

2.1 Rice

Tillage was carried out by using a hand tractor, with the plowing done in wet condition, and then raked lengthwise and crossed until the soil was sufficiently muddy. Tillage was carried out no later than 15 days before the transfer of seeds, during which Dolomite and manure were given according to recommended doses. A Nursery was made outside the plantation business area to save time and the seedlings were maintained in the nursery for 15 to 20 days. The actual rice planting was done with a rice planting machine called jajar legowo 2:1 which was named Indo Jarwo Transplanter 2:1. Besides accelerating planting time, it also reduced planting costs. Jajar legowo planting method was able to produce a total population of 213,300 plants/hectare or 33.31% more than 25 cm x 25 cm tiles planting method, which resulted in a plant population of only 160,000/ha. N fertilizer was given 2 times, at the age of 7 days after planting (HST) 1/3 part, then 2/3 parts were given with the provision of perceiving BWD (Leaf Color Chart from IRRI). While P and K fertilizers were given at the time of planting by spreading evenly across the surface of the land. Rice harvesting was carried out by using a mini combine harvester which was a rice harvester machine designed specifically for the condition of Indonesian rice fields.

Observations were made on the yield of milled dry grains (t GKG/ha) by taking tiles from three different places from each of the farmers involved. Rice farming data collection was carried out by recording immediately at the time the activity was carried out (daily notes) with the cooperating farmers while collecting data from non-cooperators was with the use of survey methods through interviews and structured questionnaires. With the number of cooperators at 20 people, 20 non-cooperators were randomly selected to serve as the control. Data collected were the number of seeds, amount of fertilizer, number of pesticides, herbicides and its prices, amount of labor used and wage levels, yields and prices, sources of income and expenditure of cooperators and non-cooperators.

Then the collected data were tabulated and analyzed. Financial analysis was carried out to determine business feasibility, breakeven, and sensitivity of rice farming. Farming is considered feasible if the Gross B/C value is greater than one. The formulation of Gross B/C is (Kasijadi and Suwono, 2001; Nurmanaf, 2005):

$$\text{Gross B/C} = \frac{P \times Q}{Bi}$$

Where:
P = Price of product (Rp/kg)
Q = Production result (kg/ha)
Bi = Production cost i-th (Rp/ha)

The economic feasibility of the recommended technology is indicated by the value of Marginal Benefit Cost Ratio or MBCR (Banta & Surya, 1984).

3 RESULTS AND DISCUSSIONS

3.1 *Farmers characteristics*

Characteristics of cooperating farmers in Sidomulyo village is in Table 2. The average age of cooperating farmers in Sidomulyo Village was 48.20 years and uncooperating farmers were 49.0 years. The age range was of 30-74 years. The educational level of cooperators was 8.3 years and that of non-cooperators was 6 years.

The level of education was dominated by those that graduated from elementary school to junior high school. The cooperators and non-cooperators had sufficient education. However, increasing skills in the field of technological components for crop and livestock cultivation is still very much needed.

The average duration of farming experience for cooperators was 24.3 years and non-cooperator was 21.9 years. The experience of the two groups of farmers could be said to be quite long so that this experience could be used as a driving factor for the success of rice farming in tidal land. Also, the length of the farming experience affected the adoption of farming technology; longer experience makes it easier to understand innovation and also results in higher participation in agricultural development (Wahyunindyawati et al., 1999).

Labor is one of the factors of production needed in managing a farm and is a very important contribution to agricultural production. According to Leknas (1977) in Gunawan et al. (1979), a worker is considered to be full-time when devoting his time to work as follows: (a) Men aged > 15 years = 35 hours per week, (b) Young men 15 years and women > 15 years = 20 hours per week. Based on this concept, the availability of labor in the form of man-days for cooperating farmers in Sidomulyo Village tidal swampland was 568.75 man-days per household/ year. Whereas for non-cooperative farmers, 433 man days/household/year. The potential of workers who can participate in farming on tidal swamp land, 591 man days/household/year. (Rina et al., 2014). The availability of labor in the family is enough to contribute to the wide use of land. A farmer's family workforce is distributed into farming, non-farming and leisure activities. The results showed that farmers were more interested in doing activities as laborers in both agriculture and non-agricultural fields because the level of wages available from rice farming was still low (Rina and Mawardi, 2013). Therefore, it is necessary to create incentives so that farmers want to work intensively in agriculture. The area of land owned by cooperating farmers was 2.9 ha on average or ranges from 0.37 to 7 ha and the area of land cultivated averages 2.5 ha or ranged from 0.37 to 7 ha or only 86.2% was cultivated. While the area of land ownership of non-cooperator farmers was 2.6 ha, cultivated by 2.0 ha or only 76.9% had been cultivated.

3.2 *Capital availability*

Capital limitations are an obstacle for innovations. The capital used by farmers in farming is usually derived from savings which are the acquisition of income after deducting food and non-food expenditure per year which can either come from farming and non-farming. The availability of cooperating farmer capital in 2017 was an average of IDR 11,647,895 per household/year and that of non-cooperator, IDR 1,583,784 per household/year. The amount of surplus income of cooperating farmers compared to non-cooperating farmers indicated that by increasing cropping intensity, the availability of farmers capital could also be increased.

3.3 *Analysis of farming commodity*

3.3.1 *Rice*

Rice production obtained by cooperative farmers in the 2016/2017 rainy season had a peak of 5.17 tons/ha, with the lowest being 2.89 tons/ha while in the 2017 dry season the highest was 6.72 tons/ha and the lowest was 3.17 tons/ha. Production in the dry season was higher than the rainy season due to the bird pest attack in the rainy season. Comparison of farming costs and revenue analysis of rice-rice cropping patterns in cooperating rice-*bero* farmers and non-cooperating farmers in Sidomulyo Village,

Table 2. Characteristics of cooperative farmers in Sidomulyo Village, TambanCatur District, Kapuas Sub-district, 2017.

No	Description	Cooperator		Non-Cooperator	
		Average	Range	Average	Range
1.	Age (year)	48.2	30 -75	49.0	34-74
2.	Education (year)	8.3	6 -12	6.0	6-9
3.	Experience (year)	24.3	6 -50	21.9	10-44
4.	Productive labor (people/ household/ year)				
	- Man	1.7	1-3	1.2	1-2
	- Woman	1.4	1-3	1.2	1-2
5.	Land ownership (Ha)	2.9	0.37-7	2.6	0.6-6
6.	Cultivated area (Ha)	2.5	0.37-7	2.0	0.6-6

Source: Primary data

Table 3. Analysis of rice farming costs and revenues of cooperator and non-cooperator farmers in Sidomulyo Village, TambanCatur, Kapuas Sub-district, 2017.

| No | Description | Cooperator | | | | Non - Cooperator | |
| | | Rainy Season 2016/2017 | | Dry Season 2017 | | Dry Season 2017 | |
		Physical unit	Value Rp.000	Physical unit	Value Rp.000	Physical unit	Value Rp. 000
1.	Production	4.033 kg	17.341	5.075 kg	21.822	2.252 kg	13.512
2.	Production Cost		11.330		11.254		10.013
	Saprodi		3.043		2.365		1.595
	Seed	25 kg	225	25kg	225	12 kg	84
	Urea	50 kg	100	50 kg	100	113 kg	227
	Ponska	150 kg	900	150 kg	900	134 kg	804
	Lime	500 kg	500	500 kg	500	57 kg	57
	Organic fertilizer	500 kg	400	-		-	
	Herbicide	4,0 ltr	240	4,0 ltr	240	3,8 ltr	228
	Drugs		350		310		140
	Plastic		318.35		90		45
	Tax		10		-		10
	Labor	67,8 man-days	8.286	68,9 man-days	8.889	75.8man-days	8.418
3.	Profit		6.011		10.568		3.498
4.	R/C		1,53		1,94		1,40
5.	MBCR	2.04					

Source: Primary Data
Note: The price of superior rice is IDR 4300

Tamban Catur District, Kapuas Sub-district (Table 3). Table 3 showed that the cooperative farmers' rice production increased by 303% compared to non-cooperators' rice production, and likewise the profit value of rice farming (rice-rice cropping pattern) Agro Pawon rice varieties grown in 2016/2017 rainy season and Argo in 2017 dry season amounted to IDR 16,580,050/ha (cooperating farmers) and the pattern of local rice-bero or rice varieties Siam amounting to IDR 3,498,750/ha (non-cooperating farmers).Rice-rice cropping patterns (introduced patterns) provided a profit increase of IDR 13,081,300/ha or 3.74 times compared to the benefits of local rice farming of non-cooperating farmers (farmers patterns). The rice-rice cropping pattern proved to be worth developing on a wider scale (MBCR> 2).

3.3.2 Cattle

Maintenance of cattle by groups (17 people) of Sidodadi farmer.10 cows was assisted by the Kapuas Animal Husbandry Department. Farmers take care of the cows in turn with 7 people in charge of looking for grass, 1 person cleaning cages and 2 people feeding each cow for 1 hour per day. Analysis of cattle farming for 1 year (Table 4).

Table 4 showed that profits obtained from cattle in groups for 1 year were IDR 63,600,000 or IDR 3,741,176.5/household/year. The

cultivation of livestock in groups was quite efficient. The value of this profit could still be increased by increasing the volume of livestock kept.

Table 4. Analysis of cattle farming as many as 10 heads for 1 year in Sidomulyo Village, 2017.

No.	Description	Physical	Price (Rp/unit)	Value (Rp)
1.	Production: Cattle	11 cows	-	157.750.000
	Feces	5000 kg	750	3.750.000
2.	Production cost			97.900.000
	Calf	10 calves	7.000.000	70.000.000
	Rope	1 roll	100.000	100.000
	Medicines	1 bottle	200.000	200.000
	Cage depreciation	1 year	600.000	600.000
	Labor	270 man-days	100.000	27.000.000
4.	Profit			63.600.000
5.	R/C			1,61

Note: The cattle's price is range from IDR 14.000.000 to IDR 15.500.000.

3.4 Farmer household income

The household incomes of cooperators and non-cooperators that was sourced from agriculture and other means are presented in Table 5.

Table 5 showed that the income of cooperative farmer households in 2017 was Rp. 53,858,167,-/HH/YR and non-cooperator Rp. 30,962,350,-/HH/YR. The income of cooperator farmers came from agriculture (rice, livestock and perennial crops/vegetables) 69.22% and non-agriculture 30.78%. While household income from non-cooperator farmers comes agriculture from 66.23% and non-agricultural 33.77%.The contribution of livestock to the income sources of cooperator farmers was 16.47% (Rp. 8,867,576,-/HH/YR) and non-cooperators 22,99% (Rp. 6,807,650,-/HH/YR). If the income of cooperative farmer households compared to non-cooperators there was an increase of Rp. 22,895,817,-/HH/YR or 73.95%.

Table 5. The income of cooperators and non-cooperators in Sidomulyo Village, Tamban Catur District, Kapuas Sub-district, 2017.

No.	Description	Cooperator		Non-Cooperator	
		Value (IDR/HH/YR)	%	Value (IDR/HH/YR)	%
1.	Agriculture	37.278.168	69,22	20.507.350	66,23
	-Rice	26.337.925	48,90	12.587.200	40,65
	-Cattle	8.867.576	16,47	6.807.650	21,99
	-Other crops	2.072.667	3,85	1.112.500	3,59
2.	Non Agriculture	16.579.999	30,78	10.455.000	33,77
	Sum (1+2)	53.858.167	100,00	30.962.350	100,00

The increase in income was followed by an increase in expenditure. Expenditure of cooperative farmers for food Rp. 23,341,906,-/HH/YR and non-food Rp. 18,868,366,-/HH/YR or total expenditure Rp. 42,210,272,-/HH/YR. While non-cooperator farmers amounted to Rp. 29,378,566,-/HH/YR consisting of food Rp. 14,675,000,-/HH/YR and non-food Rp. 14,703,566,-/HH/YR. Even though the expenditure was quite large for cooperative farmers, there was still a surplus of income of Rp. 11,647,895,-/HH/YR and non-cooperator Rp. 1,583,784,-/HH/YR (Table 6).

3.5 Farmer's income performance

Table 7 showed that cooperative farmer household income in 2015 was IDR 30,049,800/household/year and in 2016, it was IDR 37,308,920/household/year resulting in an increase of IDR 7,259,120/household/year or 24.16%. Similarly, the increase in farm income from 2015 to 2016 amounted to 13.21%. The increase was still low due to the low production of rice achieved by farmers, as can be seen from the decreasing contribution of rice to farmers income by 19.65%. However, the contribution to income by live-stock farming increased by 158.47% in 2016 followed by an increase in expenditure. Expenditures for food and non-food items increased by 7.74%, but despite increased spending, there was still an increase in the income surplus of 278.88% compared to 2015. Furthermore, the cooperative farmer's household income in 2017 was IDR 53,858,167/household/year compared to income in 2016 IDR 37,308,920/household/year resulting in an increase of IDR 16,549,247/household/year or 44, 35%. This increase was still low due to the new cropping intensity of 200% and a decrease in income contribution from livestock from IDR 12,673,636/household/year or 34% in 2016 to IDR 8,867,576/household/year or 16.47% in 2017. Nevertheless, in 2017 there was a balance of Rp. 11,647,895/household/year which could be used by farmers as capital stock.

Table 6. Income and expenditure of household cooperator and non-cooperator in Sidomulyo Village, TambanCatur District, Kapuas Sub-district, 2017.

No.	Farmer	Income (IDR/HH/YR)	Expenditure (IDR/HH/YR)		Income surplus (IDR/HH/YR)
			Food	Non Food	
1.	Cooperator	53.858.167	23.341.906	18.868.366	11.647.895
2.	Non-Cooperator	30.962.350	14.675.000	14.703.566	1.583.784

Source: Primary Data

Table 7. Income and expenditure of cooperative farmer households in Sidomulyo Village, 2015-2017.

No.	Description	Income (IDR/HH/YR)	Expenditure (IDR/HH/YR)	Income surplus (IDR/HH/YR)
	2015[1]			
1.	Rice	19.346.500	-	-
2.	Cattle	4.903.300	-	-
3.	Vegetable and Fruits	800.000	-	-
4.	Non-Agriculture	5.000.000	-	-
	Sum	30.049.800	28.231.330	1.818.470
	2016[2]			
1.	Rice	15.544.830	-	-
2.	Cattle	12.673.560	-	-
3.	Vegetable and Fruits	140.530	-	-
4.	Non-Agriculture	8.950.000	-	-
	Sum	37.308.920	30.419.082	6.889.838
	2017[3]			
1.	Rice	26.337.925	-	-
2.	Cattle	8.867.576	-	-
3.	Vegetable and Fruits	2.072.667	-	-
4.	Non-Agriculture	16.579.999	-	-
	Sum	53.858.167	42.210.272	11.647.895

Source: 1) Subagio and Rina (2016), 2) Subagio et al (2016), 3) Primary Data.

4 CONCLUSION

The innovation technologies of rice cultivation in acid sulfate tidal swamps were: (1) water management, (2) soil tillage, (3) varieties, (4) seed age, (5) planting system, (6) fertilization, (7) weeding, (8) pest control, and (9) integrated harvesting methods. These innovations helped to increase productivity by 303% compared to non-cooperator rice production. Therefore, making the rice-rice pattern economically profitable and feasible to develop. Farmer's household income increased by IDR16, 549,247/household/year or 44.35% compared to the previous year's income.

REFERENCES

Adjid, D.A. 1985. Participation pattern of rural community and development of planned agriculture: cases of overlay farming group in special intensification (insus) of rice in West Java. Dissertation. Bandung: Padjadjaran University.

Alihamsyah, T., Prayudi, B., Sulaiman, S., Ar-Riza, I., Noor, I. & Sarwani, M. 2004. 40 Years of Balittra. Future Research and Development Programs. Balittra. Research and Development Agency. Agriculture Department.

Alwi, M. 2011. Weathered Pyrite and Jarocyte Inactivation Through Leaching and Utilization of Biofilter in Acid Sulfate Soil. Dissertation. Bogor Agricultural Institute. 170 pages.

Banta, G.R. & Surya, A.R.J. 1984. *Economic Analysis of New Technologies in Basic Procedures for Agro economic Research*. Los Banos: IRRI.

Byerlee D. & Collinson, M. 1980. *Planning Technologies. Appropriate to farmers: Concepts and Procedures*. Mexico: CyMMYT.

Bunnch, Roland. 2001. *Two corn cobs: Guidelines for Agricultural Development based on People. Second Edition*. Jakarta: Indonesian Torch Foundation.

Gunawan, M., Nurmanaf, A.R. & Sawit, M.H. 1979. Supply and Demand of Works in the Agricultural Sector. Report on the Rural Dynamics Study Project. The Agro Economic Survey in collaboration with the Ministry of Agriculture's Planning Bureau.

Haryono, Subagyono, K. & Sunandar, N. 2012. Role and strategy of agricultural research and development in increasing production and productivity of food. In M. Muhaemin et al (ed). *Proceedings of the National Seminar on Food Self-Reliance 2012 "Increasing Competitiveness and Added Value of Local Resource-Based Agricultural Products"*. Collaboration with the Faculty of Industrial Technology UNPAD with West Java BPTP and the Regional Research Council of West Java Province.

Imam, S.H.M. 2007. Success tips for integrated farming management. *Proceedings of the National Seminar on Agricultural Innovation and Institutional Agriculture in Efforts to Increase Community Empowerment, Yogyakarta, 24-25 Agustus 2007*. BP2TP.

Kasijadi, F. & Sawono. 2001. Application of assemblies of application of technology in increasing competitiveness of rice farming in East Java. *Journal of Assessment and Development of Agricultural Technology*. 4 (1) January 2001. Bogor: Research and Development Center for Agricultural Social-Economic.

Nugroho, K. & Suriadikarta, D.A. 2010. Production capacity of food in swamp field. In Sumarno & N. Suharta (Eds.). *Land Resource Analysis Towards Sustainable Food Security*: 71-87. Jakarta: Agricultural Research Agency.

Nurmanaf. 2005. National Farmers Panel (Patanas). The Dynamics of Household Socio-Economic and Rural Communities: Analysis of Agricultural Profitability and Dynamics of Prices and Agricultural Wages. Bogor: Final Report. Research and Development Center for Agricultural Social-Economic.

Rina, Y. & Mawardi. 2013. Availability of labor and employment opportunities in areas Ex-Development of million hectares of peatlands in Central Kalimantan. *Jurnal Agroscientiae* 20(1): 1-8. South Kalimantan: Scientific Journal of Agricultural Sciences, University of Lambung Mangkurat.

Rina, Y., Noorginayuwati & Subagio, H. 2014. Socio-economic aspects of rice cultivation in tidal swamp land. In Dedi Nursyamsi et al (ed.). *Technology for Tidal Swamp Land Innovation Supports Food Sovereignty*: 275-299. IAARD PRESS.

Subagio, H., Annisa, W. & Noor, M. 2016. Model of Tidal Farming Based on Food Crops and Livestock. Balittra Research Results for Budget Year 2016. Research and Development Center for Agricultural Land Resources. Swampland Agricultural Research Institute. 48 pg.

Subagio, H & Rina, Y. 2016. institutional status and farming system in tidal swamp land (case of Sidolmulyo Village, Tamban Catur Subdistrict, Kapuas Regency). In Muslimin et al (Ed.). *National Seminar Proceedings of the Specific Location Agricultural Technology Innovation Supporting Sustainable Food Sovereignty.* Pg.542-542550.

Wahyunindyawati, Kasijadi, F., Purbiati, T. & Soemarsono, S. R. 1999. Adoption survey of salak vegetatative nursery technology in production centers of Bali and East Java. *Journal Hort* 9(3): 235-242.

Decreasing pyrite content and acidity on potential acid sulphate soil of South Kalimantan by leaching-oxidation

Y. Lestari
Indonesian Swampland Agriculture Research Institute (ISARI), Banjarbaru, Indonesia

S.N.H. Utami
Agriculture Faculty, Gadjah Mada University, Yogyakarta, Indonesia

D. Nursyamsi
Indonesian Center for Agricultural Land Resources Research and Development (ICALRRD), Jakarta, Indonesia

ABSTRACT: Pyrite (FeS_2) is a secondary mineral attribute of acid sulphate soil that is highly insoluble and can produce a high concentration of ferrous ions and sulfuric acid (H_2SO_4) which leads to extreme acidity of soil and water (pH<4). Leaching is one of the techniques that can be used to reduce acidity in acid sulphate soils. The objective of this research was to determine the effect of leaching and oxidation on decreasing pyrite content and acidity of a potential acid sulphate soil. The study was conducted in a greenhouse in Indonesian Swampland Agriculture Research Institute (ISARI), Banjarbaru. The experiment used a Randomized Complete Design (RCD) with 9 replications. The treatments were wetting of the soil followed by drying for 7 and 14 days, in a continuous cycle for 6 months. The results showed that pyrite content of the acid sulphate soil decreased faster with 14 days of treatment compared to the 7 days treatment. The decrease was 80.68% and 70.08%, for the 14 and 7 days treatment, respectively. The quantity of titration acidity (sums of available H^+ and Al^{3+}) decreased by 14.90 $cmol^{(+)}kg^{-1}$ and 15.23 $cmol^{(+)}kg^{-1}$ respectively. It can be concluded that the longer the pyrite is exposed to air and followed by leaching, the more pyrite was oxidized and titration acidity can be removed.

1 INTRODUCTION

Large parts of the lowland areas in Southeast Asia were submerged in seawater some 4300 years ago due to the rise in sea level. This caused coastal plains of the region to be pyritized (Shamshuddin et al. 2017). Pyrite forms when sulphate (SO_4^{2-}) from seawater and iron (Fe^{3+}) are reduced to sulfides (S^{2-}) and (Fe^{2+}). The reaction was catalyzed by the microbes in decaying organic matters in an anaerobic condition. Therefore, it can be deduced that the reaction between sulfides (S^{2-}) and ferrous (Fe^{2+}) ion form pyrite (FeS_2) (Enio et al., 2011).

In Indonesia, the actual and potential acid sulphate soil covers about 6.71 Mha. The potential ones are characterized by sulfur in the form of sulfide (reduced, not oxidized), typically pH>6, total S $\geq$ 0.2%, incubated soil pH drop more than 0.5 units when compared to the field under anaerobic condition. While the actual ones are characterized by field pH <4 as a result of pyrite oxidation and, sometimes, due to the presence of jarosite. When soils are oxidized, there is a sulphurous smell and extensive iron stains on any drain or pond surfaces (Yli-Halla et al., 2012).

According to Maria et al. (2002), soil has the potential of containing sulphuric acid if pyrite is oxidized, and this result in what is known as potential acid sulphate. However, the oxidation of one mole of pyrite by O_2 will produce 2 moles of H^+, 1 mole of Fe^{2+} and 2 moles of SO_4^{2-}, but if Fe^{3+} acts as an oxidizer, it will release 16 moles of H^+, 15 moles of Fe^{2+} and 2 moles of SO_4^{2-}.

Pyrite levels in Indonesian swamps are up to 7%, and if oxidized, they can produce acid up to 252 $cmol^{(+)}kg^{-1}$. To neutralize this acidity, requires 400 t ha^{-1} of lime. Utilization of potential acid sulphate soils will only be successful if the pyrite degradation has subsided and characterized by leachate EC <0.5mS and pH <4 (Maas, 2000).

Naturally, the degradation of pyrite in acid sulphate soils takes a long time but can be accelerated by reducing soil moisture. This will increase the diffusion of oxygen into the soil layer to oxidize the pyrite in it (Cook et al., 2004). Based on the results of a lysimeter experiment, pyrite will degrade faster with lower moisture content (Maas et al., 2001). According to Mustafa (2010), soil tillage by hoeing 20 cm deep increases soil surface area and aids pyrite oxidation. The rate can also be accelerated by continuously repeating the wet conditions of the soil (Maas et al., 2000). Drying of acid sulphate soil for a long period results in more oxidation of pyrite and produces high Fe and SO_4^{2-} (Khamidah, 2002).

Furthermore, aeration through the surjan farming system has the ability to reduce pyrite content by 70% over 14 months (Maas, 1989).

Another important step in decreasing pyrite content is by reducing soil acidity and leaching toxic substances. Based on XPS, pyrite oxidation produces coated materials such as Ferric(oxy) hydroxysulphates and oxyhydroxides within the 3-5 nm thick outermost layer of the reached surface (Yigqun Ma and Chuxia Lin, 2013). The formation of these materials reduces the oxidation area and can be removed through leaching. This process is also useful for the maintenance of the soil moisture by making sure it is not cohesive. This is necessary because if the soil turns dry and become cohesive at high levels of acidity, it will be difficult to degrade the pyrite because it would have been isolated in the soil matrix. Furthermore, in addition to maintaining moisture, the water also functions as a reactant.

Therefore, the objective of this research was to determine the effects of decreasing pyrite content and acidity by leaching-oxidation on a potential acid sulphate soil.

2 METHODOLOGY

The experiment was conducted at the Greenhouse of Indonesian Swampland Agricultural Research Institute (ISARI) between August 2014 and January 2015. The potential acid sulphate soil used was taken from Jelapat, Barito Kuala, South Kalimantan on S 03° 14' 16.1", E 114° 31' 02", at a depth of 35-65 cm and classified as Typic Sulfaquent. The treatment was arranged in a completely randomized design with nine replications and dried by the sun for (i) 7 days and (ii) 14 days. The process of leaching-oxidation was conducted by sun-drying the potential acid sulphate soils and leaching the oxidation product repeatedly.

For the process of leaching-oxidation, the soil and sieved air-dried soil with rice husk mixed in a 2:100 ratio based on dry weight (0.2:10 kg). After it has been well mixed, the soil was packed into sacks for drying-leaching by immersing them in a large basin of a local well containing approximately 25 litre of water for three days. The soil was then drained and dried for seven days. This process was conducted for six months.

Some properties of the acid sulphate soils were observed before and after the process, and they include pH (H_2O) 1:2.5, EC (1:2.5), available-H, available-Al and the levels of pyrite. The leachate was also observed for pH (H_2O), EC and sulphate concentration.

The pH was analyzed by using a glass electrode pH meter HORIBA Model 9625 and EC through the use of EC-meter glass electrode WTW Cond Model 3110. Analysis of H^+ and Al^{3+} was conducted by titration through KCl 1M extractor and SO_4^{2-} by spectrophotometer Model Spectronic 20 (λ=432 nm). Pyrite contents were analyzed by H_2O_2 method and measured using spectrophotometer Model Spectronic 20 (λ = 494 nm). The overall data were analyzed statistically through the application of SAS software 9.1.3.

3 RESULT AND DISCUSSION

3.1 Initial soil chemical properties

The chemical properties of potential acid sulphate soils before and after the experiment are presented in Table 1. It was observed that the difference in soil pH was larger before and after the process. Conversely, drying of the soil was observed to have increased EC value. According to Akpan-Idiok and Esu (2012), the field pH of potential acid sulphate soil is near neutral until alkaline. It was found that drying results in strong acid soil reaction due to pyrite oxidation, which produces H^+, Fe^{2+}, and SO_4^{2-}. Therefore, oxidation of ferro (Fe^{2+}) to ferri (Fe^{3+}) gave an additional proton and the reaction between the proton and sulphate produces sulfuric acid which increases soil acidity (Vuai et al., 2003). The results of the study by Kawahigashi (2008) showed that after the Hydraquent Sulfaquent (Mekong Delta, Vietnam) taken from the Gr horizon was dried at 40°C for 11 days (air dry conditions), soil pH decreased by 0.8 units from 3.5 to 2.7. This is due to the oxidation of pyrite, which produces sulfuric acid and then releases large amounts of metal due to cation exchange or weathering of minerals such as Al^{3+}. Furthermore, Al^{3+} hydrolysis produces several protons which causes the soil solution to become very acidic. According to Havlin et al. (2014), the reaction is as follows:

$$Al^{3+} + H_2O = Al(OH)^{+2} + H^+$$
$$Al(OH)^{+2} + H_2O = Al(OH)_2^+ + H^+$$
$$Al(OH)_2^+ + H_2O = Al(OH)_3^0 + H^+$$
$$Al(OH)_3^0 + H_2O = Al(OH)_4^- + H^+$$

Table 1. Chemical properties of the acid sulphate soil from Jelabat, before and after drying.

Chemical properties	Acid sulphate soil before drying (sulphidic material)	Acid sulphate soil after drying
pH H_2O	4.81	2.97
EC (mS.cm^{-1})	0.99	8.62
Available-H (cmol$^{(+)}$kg^{-1})	4.39	7.60
Available-Al (cmol$^{(+)}$kg^{-1})	11.38	25.30
Pyrite content (%)	4.39	2.64

After drying, Al-available and H-available of acid sulphate soil were 7.60 cmol$^{(+)}$kg^{-1} and 25.30 cmol$^{(+)}$ kg^{-1} respectively and these concentrations were higher than before drying. This strong acidity was influenced by the solubility and availability of some nutrients in the soil (Goles and Kyuma, 1997). Therefore, it can be deduced that the drying of potential acid sulphate soils leads to pyrite oxidation and produces extreme acidity. Furthermore, a very low pH causes the destruction of the clay mineral lattice to release silicates and Al^{3+} since its weathering has been observed to be occurring at pH of 3.5 (Shamsuddin and Auxtero, 1991). In addition, at low pH, there is also hydrolysis of Al-hydroxide to produces Al^{3+}. The pKa of Al is 5 (Rosilawati et al., 2015). Therefore, it can be concluded that the quantity of Al-available is due to the weathering of clay minerals or hydrolysis of Al-hydroxide.

The results showed that pyrite content (2.64%) of the dried soil is lower than for the sulfidic material (4.39). Pyrite is a secondary mineral that is very insoluble in water (Kusel, 2003) and it can be reduced in acid sulphate soils by the oxidation process. Drying increases aeration to make the diffusion of oxygen better. This, consequently, enhances the oxidation process. Therefore, drying potential acid sulphate soils has the ability to reduce the pyrite content.

3.2   pH and EC of leachate

The pH of soil solution is a measure of the concentration of H$^+$ ion. However, a high concentration of H$^+$ is reflected in low pH. As shown in Figure 1, the pH value of leachate fluctuates at every observation. At the beginning of the leaching, the pH value decreased for both 7- and 14-days sun drying treatment.

Meanwhile, after the first four days, it started increasing. The value observed at the beginning could be attributed to the accumulation of H$^+$ due to pyrite oxidation. According to Vuai et al. (2003), oxidation of pyrite in acid sulphate soil produces a high concentration of strong acid (H$_2$SO$_4$). The increase in the pH of leachate after the 4th leaching is due to the loss of H$^+$. Also, changes in soil sulphate levels are related to the pH (Michael et al., 2016). Therefore, it can be concluded that more leaching caused the reduction of soil sulphate concentration and leachate water pH.

The EC of soil solution is a measure of the total concentration of the dissolved ion. Figure 1 shows that the EC of the leachate decreases with frequent leaching. This can be associated with the removal of more acidic and basic cations serving as conductors of electricity from the soil.

3.3   SO$_4$$^{2-}$ of leachates

SO$_4$$^{2-}$ is a product of pyrite oxidation found to be stable under acidic conditions (Subagyo, 2006). From the 2nd to 8th leaching, the concentration of SO$_4$$^{2-}$ accumulation was observed to be higher in the

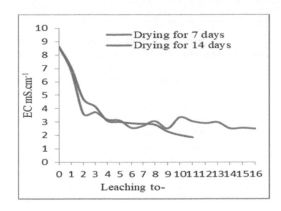

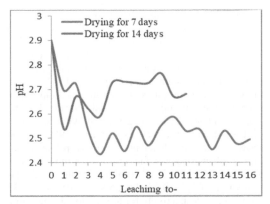

Figure 1. The effects of leaching oxidation by sun drying for 7 and 14 days after leaching on pH and EC of acid sulphate soil leachate.

14 days treatment than the 7 days one. However, on the 10th and 11th cycle, the variations were small (Figure 2). According to Maas (2003), drying acid sulphate soil for too long can make the soil hard and difficult to reverse when rewetted. Therefore, if the soil becomes too dry at a high level of acidity, it will be difficult to degrade the pyrite due to its protection in the soil matrix.

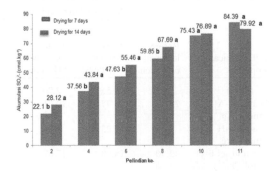

Figure 2. The effects of leaching-oxidation by sun drying for 7 and 14 days after leaching on the release of SO$_4$$^{2-}$ (cmol$^{(+)}$kg^{-1} accumulation in leach to-2, 4, 6, 8, 10 and 11.

3.4 Changes in soil chemical properties

Chemical properties of acid sulphate soils before and after leaching-oxidation for six months or 16 times (for 7 days drying treatment) and 11 times (for 14 days treatment) of drying-wetting cycle are presented in Table 2. It was observed that the soil reaction (pH) becomes more acidic for the two treatments. As discussed by Loganathan (1987), soil acidity is caused by high availability of H^+, Fe^{3+} and Al^{3+} and reduction of basic cations such as Ca, Mg, K, and Na. In this research, acidity from pyrite oxidation was able to fill the exchange site and leached out basic cations. Moreover, the low pH after leaching-oxidation was because of the reduced content of available H^+ and Al^{3+} rather than basic cations.

The initial soil available-H^+ of 2.97 $(cmol^{(+)} kg^{-1})$ was decreased to 0.37 $cmol^{(+)} kg^{-1}$ and 0.64 $cmol^{(+)} kg^{-1}$ for the 7- and 14-days sun-drying treatment, respectively. This is due to the removal of pyrite oxidation products such as Fe^{2+}, Al^{3+}, and SO_4^{2-} from the soil by leaching as well as nutrients from plants such as N and P and available basic ones such as K, Na, Ca and Mg. This is in agreement with the results of the Maas laboratory experiment (1989) which showed that using leaching to remove pyrite oxidation product also has the ability of removing Na^+, K^+, Ca^{2+}, Mg^{2+} Fe^{2+} Al^{3+} cations with the accumulation value of 5.75, 0.77, 2.50, 8.94, 28.05, and 11.38 $cmol^{(+)}kg^{-1}$, respectively for leach 1 to 23. The same results were obtained by Anda et al. (2002) through the use of leaching processes as part of the back swamp management which was observed to cause the loss of N nutrients and available basic ones such as K, Na, Ca and Mg. The concentration of NH_4^+, NO_3^-, K^+, Na^+, Ca^{2+} and Mg^2 in the waste-water of the Dadahup secondary canal were found to be 0.14-0.12 me L^{-1}, 0.02 me L^{-1}, 0.05-0.11 me L^{-1}, 0.15-0.44 me L^{-1}, 0.52-0.74 me L^{-1}, 0.81-2.56 me L^{-1} respectively. Furthermore, the waste-water in Palingkau secondary canal contained NH_4^+, NO_3^-, K^+, Na^+, Ca^{2+} and Mg^{2+} at 0.24-0.27 me L^{-1}, 0.01 me L^{-1}, 0.12-0.16 me L^{-1}, 0.29-0.31 me L^{-1}, 0.64-0.81 me L^{-1}, and 1.45-1.48 me L^{-1} respectively. Field experiments conducted also showed that open raised bed system (in contact with groundwater) decreases P_2O_5 by 100-180 $\mu g\ g^{-1}$ (Maas, 1989).

The sums of available H^+ and Al^{3+} decreased from the initial 32.90 $cmol^{(+)}\ kg^{-1}$ to 18.00 $cmol_{(+)}\ kg^{-1}$ for 7 days and 17.67 $cmol^{(+)}kg^{-1}$ for 14 days. This can be explained by the removal of soil acidity by leaching. According to Golez and Kyuma (1997), periodic leaching was found effective in reducing soil acidity.

3.5 Decreasing pyrite content

Continuous sun drying and leaching for 14 days caused a reduction of pyrite content faster than the 7 days treatment, as shown in Figure 3. This is mainly due to the lower soil moisture content and increased diffusion of oxygen. The speed of pyrite oxidation was found to increase with the drying time. According to Singh and Ardejani (2003), the pyrite oxidation rate is influenced by the diffusion coefficient for oxygen transport, soil particle size, and pyrite content. However, in this study, the soil particle size and pyrite content were the same. Therefore, it can be concluded that the pyrite oxidation rate was influenced by the diffusion of oxygen t.

Figure 3 shows that the pyrite content was almost constant after the 12th leaching for the 7 days treatment and 9th leaching for the 14 days treatment. Maas (1989) explained that pyrite resists oxidation if it is well-crystalline or isolated in the soil matrix.

Table 2. The effect of leaching-oxidation on the characteristics of potential acid sulphate soil.

Soil chemical properties	Before leaching-oxidation	After leaching-oxidation	
		Drying for 7 days treatment (16 times)	Drying for 14 days treatment (11 times)
pH (H₂O)	2.97	2.55	2.70
Available-Al (cmol $^{(+)}$kg^{-1})	25.30	8.77	10.61
Available-H (cmol $^{(+)}$kg^{-1})	7.60	7.23	6.96
Titration acidity (cmol$^{(+)}$kg^{-1})	32.90	18.00	17.67
Pyrite content (%)	2.64	0.79	0.51

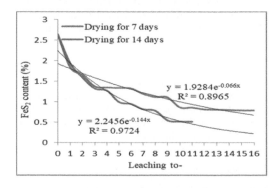

Figure 3. The effects of drying for 7 and 14 days on the pyrite degradation.

4 CONCLUSION

The leaching-oxidation process conducted on an acid sulphate soil for 6 months by drying-wetting cycles decreased pyrite content by 1.85% for 7- and 2.13% for 14-days drying treatment. The speed of reduction was found to be faster in the 14 days drying treatment. Titration acidity (available Al^{3+} and H^+) decreased faster in the 7 days treatment than the 14 days by 14.90 $cmol^{(+)}kg^{-1}$) and 15.23 $cmol^{(+)}kg^{-1}$, respectively.

REFERENCES

Anda, M., Siswanto, A.B., Eko, R. & Subagyo, H. 2002. Properties of soil and water of a 'reclaimed' tidal backswamp in Central Kalimantan, Indonesia. *Soil Sci. Soc. Am. J.*

Cook, F.J., Dobos, S.K., Carlin, G. D. & Miliar, G.E. 2004. Oxidation of pyrite in acid sulphate soils: in situ measurement and modeling. *Soil Research* 42(6): 499-507.

Enio, M.K., Shamsuddin, J., Fauziah, C.I. & Husni, M.H. A. 2011. Pyritization of the Coastal Sediments in the Kelantan Plains in the Malay Peninsula during the Holocene. *Am J. Agri. & Biol. Sci* 6(3): 393-402.

Golez, N.V. & Kyuma, K. 1997. Influence of pyrite oxidation and soil acidification on some essential nutrient elements. *Aquacultural Engineering* 16(Issues 1-2): 107-124 (Abstract).

Havlin, J.L., Tisdale, S.L., Beaton, J.D. & Nelson, W.L. 2014. *Soil Fertility and Fertilizers. An Introduction to Nutrient Management. Eight Edition.* New Jersey: Pearson Prentice Hall.

Kawahigashi, M., Do, N.M., Nguyen, V.B. & Sumida, H. 2008. Effects of drying on the release of solutes from acid sulphate soils distributed in the Mekong Delta, Vietnam. *Soil Science and Plant Nutrition* 54:495-506.

Khamidah. 2002. Perubahan Beberapa Sifat Kimia Tanah Sulfat Masam Delta Telang, Musi Banyuasin, Sumatera Selatan Akibat Pengeringan dan Pencucian. Program Studi Ilmu Tanah-S1. Jurusan Tanah Fakultas Pertanian. Institut Pertanian Bogor.

Kusel, K. 2013. Microbial cycling of iron and sulfur in acidic coal mining lake sediments. *Water, Air and Soil Pollution* 3: 67-90.

Loganathan, P. 1987. *Soil Quality Consideration in the selection of sites for aquaculture.* United Nations Development Programe. Food and Agriculture organization of United Nations. Nigerian Institute for Oceanography and Marine Research Project RAF/82/009.

Maas, A. 1989. Genesis, Classification and Reclamation of Potential Acid Sulphate Soils in South Kalimantan, Indonesia. PhD. Thesis. State University of Ghent, Belgium.

Maas, A., Sutanto, R. & Purwadi, T. 2000. Pengaruh air laut terhadap oksidasi pirit dan tahana hara tanah sulfat masam. 2000. *Jurnal Ilmu Tanah dan Lingkungan* 2 (2):41-45.

Maas, A., Sutanto, R. & Purwadi, T. 2001. Ameliorasi Tanah Sulfat Masam Potensial untuk Budidaya Tanaman Pangan yang Dikelola Dengan Sistem Mekanisasi. Laporan Penelitian Hibah Bersaing VII/4 Perguruan Tinggi Tahun Anggaran 2001. Fakultas Pertanian. Lembaga Penelitian Universitas Gadjah Mada.

Maas, A. 2003. *Peluang dan Konsekuensi Pemanfaatan Lahan Rawa Pada Masa Mendatang.* Pidato Pengukuhan Jabatan Guru Besar pada Fakultas Pertanian. Universitas Gadjah Mada.

Maria, E., Freeman, C. & David, R. 2012. Impact of pH and redox potential changes on acidic sulphate soils. *Anual Conference of the Canadian Society for Civil Engineering.*

Michael, P.S., Fitzpatrick, R. W. & Reid, R.J. 2016. The important of soil carbon and nitrogen for amelioration of acid sulphate soils. *Soil Use and Management* 32: 97-105.

Mustafa, A., Rachmansyah & Anugriati. 2010. Distribusi kebutuhan kapur berdasarkan nilai S_{POS} tanah untuk tambak sulfat masam di Kabupaten Mamuju propinsi Sulawesi Barat. *Proseding Forum Inovasi Teknologi Akuakultur 2010.*

Rosilawati, A.K., Shamsuddin, J. & Fauziah, C.I. 2014. Effects of incubating an acid sulphate soil treated with various liming materials under submerged and moist conditions on pH, Al and Fe. *African Journal of Agricultural Research.* 9(1): 94-112.

Shamsuddin, J. & Auxtero, E.A. 1991. Soil solutions composition and mineralogy of some active acid sulphate soils in Malaysia as affected by laboratory incubation with lime. *Soil Science* 152(5): 365-375.

Shamsuddin, J., Panwar, Q.A., Alia, F.J., Shazana, M.A. R.S., Radzia, O. & Fauziah, C.I. 2017. Formation and Utilisation of Acid Sulphate Soils in Southeast Asia for Sustainable Rice Cultivation. *Pertanika J. Trop. Agric. Soil* 40(2): 225-246.

Singh, R.N. & Ardejani, F.D. 2003. Evaluation of factor affecting pyrite oxidation and subsequent pollutant generation in backfilled open cut coal mines. 8th International Congress on Mine Water & the Environment, Johannesburg, South Africa. 173-186.

Subagyo, H. 2006. Klasifikasi dan penyebaran lahan rawa. Dalam D.A., Suriadikarta, U., Kurnia, H.S., Mamat, W., Hartatik & D., Setyorni (Penyunting). *Karakteristik dan pengelolaan lahan rawa.* Bogor: Balai Besar Penelitian dan Pengembangan Sumberdaya Lahan pertanian. Badan Penelitian dan Pengembangan Pertanian. Departemen Pertanian.

Vuai, S.A., Nakamura, K. & Tokuyama, A. 2003. Geochemical characteristics of runoff from acid sulphate soil, in the northern area of Okinawa Island, Japan. *Geochemical Journal* 37: 579-592.

Yingqun Ma & Chuxia Lin. 2013. Microbial Oxidation of Fe^{2+} and Pyrite Exposed for Flux of Micromolar H_2O_2 in Acidic media. *Scientific Report 3.*

Yli-Halla, M. M., Räty & Puustinen, M. 2012. Varying depth of sulfidic materials: a challenge to sustainable management. *7th International Acid Sulphate Soil Conference in Vaasa, Finland 2012.*

Relationship between soil Fe concentration and agronomic traits of rice in acid sulfate soils

I. Khairullah
Indonesian Swampland Agriculture Research Institute (ISARI), (IAARD), Banjarbaru, South Kalimantan, Indonesia

D. Indradewa, A. Maas & P. Yudono
Faculty of Agriculture, Gadjah Mada University, Yogyakarta, Indonesia

D. Nursyamsi
Indonesian Center for Agricultural Land Resource Research and Development (ICALRD), (IAARD), Banjarbaru, South Kalimantan, Indonesia

ABSTRACT: Iron toxicity is the main problem in using acid sulfate soils for growing rice, as it decreases its agronomic traits. The experiment aimed (1) to study the effects of Fe concentrations on some agronomic traits of 15 improved rice varieties and (2) to find varieties that are tolerant to iron toxicity in acid sulfate soils of tidal swampland. The experiment was conducted in the greenhouse of Indonesian Swampland Agricultural Research Institute, Banjarbaru, Indonesia, and made use of Factorial Randomized Completely Design with three replications. The first factor was 15 improved rice varieties, and the second was four Fe concentrations of acid sulfate soils (229.44; 236.34; 564.15, and 1277.50 mg kg^{-1} Fe). Furthermore, observation variables included plant growth (plant height, tiller number, root and straw dry weight), grain yield, harvest index, and iron toxicity symptoms. The results showed that an increase in Fe concentrations of acid sulfate soils led to a decrease in rice agronomic traits (maximum tiller number, plant height, dry weight of roots and straw, grain yield and harvest index), but an increase in iron toxicity symptoms. Varieties Inpara-1 and Inpara-2 revealed the smallest decrease in rice agronomic traits and exhibited the lightest iron toxicity symptoms, whereas IR64 showed the greatest decrease in the traits and the most severe symptoms. Inpara-1 and Inpara-2 were identified as the varieties tolerant to iron toxicity in the acid sulfate soils of tidal swampland.

1 INTRODUCTION

Tidal swamplands considered as sub-optimal lands compared to irrigated wetland, have significantly contributed to the increase in rice production. It is divided into acid sulfate, potential, peatland, and saline soils, where acid sulfate soils and peatlands dominate. Agricultural development in acid sulfate soils was constrained by problems such as, water management, soil fertility and acidity (Maas et al., 1992), high in Fe, Al, and Mn, deficiency in P, low base cations, and high salinity (Ikehashi and Ponnamperuma, 1978; Ponnamperuma, 1977). The specific technology for soil and water management, as well as rice varieties adaptive and resistant to iron toxicity is a better option to increase the growth of rice in acid sulfate soils.

The main problem with acid sulfate soils is the pyrite layer (FeS$_2$). In waterlogged conditions, an increase in soil pH causes a reduction from Fe^{3+} to Fe^{2+}. Therefore, this is an indication that the concentration of Fe^{2+} increases at a high rate in soil solution. This phenomenon occurs particularly on actual acid sulfate soils flooded by rain or high tide (Widjaja-Adhi et al., 2000). Also, it is important to note that Fe^{2+} concentration within the range of 300-400 mg kg^{-1} can be toxic to rice plants (Ikehashi and Ponnamperuma, 1978; Yoshida et al., 1971).

Iron toxicity is a physiological disease of rice plant associated with an excess of dissolved Fe (Tanaka and Yoshida, 1970), salinity, P deficiency, low bases (Ikehashi and Ponnamperuma, 1978), nutrient stress, and low pH (Benckiser et al., 2005), as well as plant physiological conditions (Ottow et al., 1989). Iron toxicity was first reported (Ponnamperuma et al., 1955) to be one of the main problems of rice production in several countries in Asia, Africa, and South America (Van Breemen and Moormann, 1978; Yoshida, 1981, De Datta et al., 1994). However in Indonesia, it occurs at the wetlands of West Java, Sumatera, Kalimantan, and at tidal swampland in Sumatera, Kalimantan and Irian Jaya (Ismunadji et al., 1989; Puslitbangtan, 1991; Ottow et al., 1982; KEPAS, 1985).

The growth and grain yield of rice on acid sulfate soils were strongly influenced by the iron toxicity, and were phenotypic indicators of some agronomic traits of rice plants. The decrease in grain yield due to 30-100% iron toxicity depends on a number of factors which include rice resistance (Virmani, 1977), intensity of Fe toxicity (Cai et al., 2003; Majerus et al., 2007), and soil fertility status (Audebert and Sahrawat, 2000). The decrease in grain yield grown on iron toxic wetland in

Cihea, West Java reached 52% compared to healthy plants (Ismunadji et al., 1973). However, in the case of heavy iron toxicity condition in Belawang, South Kalimantan Province, only 160kg ha^{-1} of rice could be produced (Noorsyamsi and Sarwani, 1989).

This experiment aimed (1) to study the effects of increasing the Fe concentrations of acid sulfate soils on rice agronomic traits (maximum tiller number, plant height, dry weight of roots and straw, grain yield, harvest index) and iron toxicity symptoms of 15 improved rice varieties; (2) and to discover the varieties that are tolerant to iron toxicity on acid sulfate soils of tidal swampland.

2   MATERIALS AND METHODS

2.1   Soil sample

This experiment made use of acid sulfate soils from A, B, and C tidal types in Barito Kuala District, South Kalimantan Province. Type A included typic sulfaquent, raw, and a reduced inner pyrite layer. Type C included typic sulfaquent, mature, and the pyrite layer is shallow as well as oxidized. As many as 360 soil samples were taken at the depth 0-20 cm of acid sulfate soil in reductive condition. Subsequently, the soils were put into 360 pots for experiment.

2.2   Experimental design

The experiment was conducted in the greenhouse of Indonesian Swampland Agricultural Research Institute (ISARI) in Banjarbaru, South Kalimantan Province in 2010. Factorial Randomized Completely Design was also used with three replications. Furthermore, treatment consists of two factors; the first were 15 rice varieties (eight rice varieties of tidal swampland and seven irrigated rice varieties). These varieties were: Margasari, Lambur, Batanghari, Punggur, Indragiri, Mendawak, Inpara-1, Inpara-2 (as tidal swampland rice varieties), IR64, Ciherang, Cibogo, Ciapus, Inpari-1, Inpari-6, and Inpari-10 (as irrigated rice varieties). The second factors were four Fe concentrations of acid sulfate soils, i.e. 229.44; 236.34; 564.15; and 1277.50 mg kg^{-1} Fe. These concentrations were extracted using NH$_4$Cl.

21-day old seedlings were planted in each pot, two seedlings per pot. Afterwards, basal fertilizers of N, P, and K were applied at the rate of 90 kg ha^{-1} N, 60 kg ha^{-1} P$_2$O$_5$, and 60 kg ha^{-1} K$_2$O. The N fertilizer was applied three times, the first at seven days after transplanting (DAT) together with all P and K fertilizers. The second application was at 28 DAT, and the third at primordia phase. The fertilizers were applied by burying in the soils, while supplementary N fertilizer was in a spreading manner.

Flooding was carried out manually during vegetative phase, and was maintained at 3cm. The water used in flooding the plants was taken from the acid sulfate soils from which soil samples were taken. It was taken from high tide and collected in large plastic drums. Also, plant protection was intensively carried out to protect from pests, diseases, and weeds. Subsequently, harvesting was done at yellow ripe grain phase.

Variables were observed for plant agronomic traits, i.e. maximum tiller number, plant height, root and straw dry weight, grain yield, and harvest index. They were all observed at maximum vegetative phase, except grain yield and harvest index. To determine the variety resistance to iron toxicity, there was an inspection for symptoms at maximum vegetative phase using the scoring system of Standard Evaluation System for Rice (IRRI, 1996).

2.3   Data analysis

To determine the response of varieties to soil Fe concentration treatments, the data was analyzed for variance, followed by F test 5% level. If the F test turned out to be significant, the analysis was continued using Duncan's Multiple Range Test (DMRT) 5%. Furthermore, to determine a variety's tolerance towards iron toxicity simple linear regression analysis was conducted. The data analyzing program used SAS software for Windows version 9.

3   RESULTS AND DISCUSSION

3.1   Rice growth

The parameter of rice growth covers maximum tiller number, plant height, root and straw dry weight. All variables were influenced by the interaction of varieties and Fe concentration, except maximum tiller number. In average, the highest maximum tiller number was shown by Inpara-1, followed by Inpara-2, and Margasari. However, IR64 had the least, and was not significantly different from Ciherang and Inpari-6 (Table 1).

The increase in Fe concentrations led to the decrease in maximum tiller number. The most maximum tiller numbers were shown by low Fe concentrations (229.44 mg kg^{-1}), whereas the least tiller numbers were obtained by high soil Fe concentrations (1277.50 mg kg^{-1}). The decrease in maximum tiller number from 229.44 mg kg^{-1} to 1277.50 mg kg^{-1} reached -57.4%. However, that of all varieties decreased with increased Fe concentration. Inpara-1 showed the lowest decrease, followed by Inpara-2, whereas IR64 had the highest increase (Table 1).

Furthermore, an increase in Fe concentrations led to a decrease in plant height for all varieties. Nevertheless, at low (229.44 mg kg^{-1}) and high Fe concentrations (1277.50 mg kg^{-1}), the highest plant was Margasari, while the shortest was IR64. However, at Fe concentrations from 229.44 to 236.34 mg kg^{-1}, there were decreases in the heights of Margasari, Lambur, IR64, Ciapus, Inpari-1, and Inpari-6. Most of the varieties also suffered this decrease at Fe concentrations from 236.34 to 564.15 mg kg^{-1}, except Inpara-1 and Inpara-2. Furthermore, the increase in Fe

Table 1. Maximum tiller number of 15 rice varieties at four Fe concentrations.

Varieties	Fe Concentration (mgkg^{-1})				
	229.44	236.34	564.15	1277.50	Mean
Margasari	15.3	13.0	9.7	8.0	11.5 b
Lambur	14.0	13.0	8.7	6.7	10.6 c
Batanghari	12.7	10.7	5.7	4.3	8.3 e
Punggur	14.0	11.7	7.3	6.0	9.8 d
Indragiri	13.7	12.0	8.7	6.3	10.2 cd
Mendawak	14.3	12.3	8.0	6.7	10.3 cd
Inpara-1	15.3	13.7	10.7	9.3	12.3 a
Inpara-2	15.0	13.7	9.3	8.3	11.6 b
IR64	13.0	10.3	4.3	2.7	7.6 f
Ciherang	12.0	10.7	5.7	4.0	8.1 ef
Cibogo	12.7	10.7	6.3	4.7	8.6 e
Ciapus	12.3	11.3	6.0	4.7	8.6 e
Inpari-1	12.7	10.7	6.3	4.7	8.6 e
Inpari-6	12.0	10.7	5.3	4.0	8.0 ef
Inpari-10	12.7	10.7	5.7	4.7	8.4 e
Mean	13.4 w	11.7 x	7.2 y	5.7 z	(-)

Number followed by the same letter in a column or a row was not significantly different (DMRT 5%).

concentration from 564.15 to 1277.50 mg kg^{-1} led to a decreasing in plant height of most varieties, except Batanghari, Ciherang and Ciapus (Table 2).

The variety with the heaviest dry roots at Fe concentration 229.44 mg kg-1 was Inpara-2, which was not significantly different from Inpara-1. However, the heaviest dry root at high Fe concentrations (1277.50 mg kg-1) was showed by Inpara-1, followed by Inpara-2, whereas the lightest was that of IR64 (Table 3).

The root dry weight of all varieties decreased due to an increase in Fe concentrations from 229.44 to

Table 2. Plant height (cm) of 15 rice varieties at four Fe concentrations.

Varieties	Fe Concentration (mg kg^{-1})								
	229.44		236.34		564.15		1277.50		Mean
Margasari	117.3	a	107.0	bc	100.3	d-g	94.0	h-k	104.7
Lambur	99.3	d-h	90.7	k-p	81.7	rst	76.7	t-x	87.1
Batanghari	100.3	d-g	97.3	d-j	88.0	n-q	70.0	y	88.9
Punggur	96.0	e-l	93.0	h-n	80.0	stu	74.3	u-y	85.8
Indragiri	101.3	c-f	97.7	d-j	89.3	m-p	79.0	s-v	91.8
Mendawak	99.0	d-i	93.0	h-n	89.7	l-p	75.0	u-y	89.2
Inpara-1	97.7	d-j	94.0	g-n	88.7	nop	79.7	stu	90.0
Inpara-2	95.3	f-m	90.0	l-p	85.0	p-s	76.3	t-y	86.7
IR64	96.7	e-k	89.0	m-p	71.3	xy	62.0	z	79.8
Ciherang	102.0	cde	98.3	d-j	78.3	t-w	72.7	wxy	87.8
Cibogo	99.0	d-i	98.3	d-j	80.7	r-u	73.0	v-y	87.8
Ciapus	111.3	b	103.3	cd	86.3	o-r	81.7	rst	95.7
Inpari-1	98.3	d-j	92.7	i-o	82.7	q-t	70.3	xy	86.0
Inpari-6	100.3	d-g	92.3	j-o	80.0	stu	70.3	xy	85.8
Inpari-10	97.0	d-k	94.3	g-n	79.7	stu	71.7	xy	85.7
Mean	100.7		95.4		84.1		75.1		(+)

Number followed by the same letter in a column or a row was not significantly different (DMRT 5%).

154

Table 3. Root dry weight (g) of 15 rice varieties at four Fe concentrations.

Varieties	Fe Concentration (mg kg^{-1}) 229.44		236.34		564.15		1277.50		Mean
Margasari	6.90	bc	5.47	h-k	3.57	qr	2.60	s	4.63
Lambur	6.90	bc	5.57	ghi	3.87	pq	3.10	rs	4.86
Batanghari	5.60	ghi	5.00	j-m	1.23	v-y	0.83	xy	3.17
Punggur	5.67	ghi	4.97	klm	2.03	t	0.93	wxy	3.40
Indragiri	6.27	def	5.83	e-i	3.50	qr	2.10	t	4.43
Mendawak	6.47	cd	4.77	mn	2.67	s	1.83	tu	3.93
Inpara-1	7.07	ab	6.47	cd	6.03	d-g	5.30	i-l	6.22
Inpara-2	7.50	a	6.33	de	6.03	d-g	4.20	op	6.02
IR64	5.97	d-h	4.40	no	0.87	wxy	0.23	z	2.87
Ciherang	5.50	g-j	4.40	no	1.23	v-y	0.73	y	2.97
Cibogo	5.57	ghi	4.50	mno	1.70	tuv	0.90	wxy	3.17
Ciapus	5.73	f-i	4.67	mno	1.40	uvw	0.90	wxy	3.18
Inpari-1	5.67	ghi	4.93	lmn	1.87	tu	1.10	wxy	3.39
Inpari-6	5.43	h-l	4.47	mno	1.30	v-x	0.70	y	2.98
Inpari-10	5.73	f-i	4.47	mno	2.70	s	0.93	wxy	3.46
Mean	6.13		5.08		2.67		1.76		(+)

Number followed by the same letter in a column or a row was not significantly different (DMRT 5%).

236.34 mg kg-1, except that of Indragiri. At Fe concentrations from 236.34 to 564.15 mg kg-1, root dry weight of all varieties decreased, except Inpara-1 and Inpara-2. At Fe concentrations from 564.15 to 1277.50 mg kg-1, only Lambur, Ciherang, and Ciapus did not decrease in roots dry weight.

At low Fe concentration (229.44 mg kg-1), the heaviest straw dry weight was obtained by Margasari, and was not substantially different from Lambur, Batanghari, Mendawak and Inpara-2. However, at high Fe concentrations (1277.50 mg kg-1), Inpara-1 showed the heaviest straw dry weight, which was not substantially different from Inpara-2, Margasari, and Lambur, whereas IR64 had the lightest straw dry weight. The increase in Fe concentrations from 229.44 to 236.34 mg kg-1 resulted in a decrease in straw dry weight for all varieties, except Indragiri. Furthermore, all varieties decreased in straw dry weight following an increase in Fe concentrations from 236.34 to 564.15 mg kg-1. Also, increased Fe concentrations from 564.15 to 1277.50 mg kg-1 led to the decrease of straw dry weight in all varieties, except Lambur, Punggur, Inpara-1, and Ciherang (Table 4).

Increasing the Fe concentration resulted in a decrease in plant height, dry weight of roots and straw of all varieties, and this could be seen from the slope of linear regression equation (Table 5). The smallest slope was shown by Inpara-1 (for plant height, root and straw dry weight) followed by Inpara-2, whereas the largest was shown by IR64. The negative correlation between Fe concentration and plant height, root and straw dry weight were observed. This therefore translates that the higher Fe concentration, the greater reduction in plant height, and dry weight of roots and straw. Inpara-1 showed the lowest decrease, followed by Inpara-2, whereas IR64 had the highest decrease (Figure 1 and Attachment Table 5). Therefore, this data suggests that high Fe concentration negatively influenced rice growth. In this case, Inpara-1 and Inpara-2 were more tolerant to high Fe concentration compared to other varieties.

The decreases in plant growth following an increase in Fe concentrations of all varieties were linear and negatively correlated. This was caused by the inhibition of plant growth due to high Fe concentration causing the plants to suffer from iron toxicity. Under this toxic condition, plant growth and tillering ability are either blocked or stalled, and many leaves change color to reddish

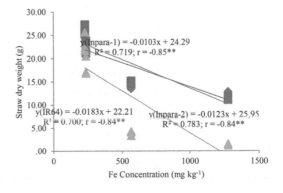

Figure 1. Relationship between Fe concentration and straw dry weight of Inpara-1, Inpara-2, and IR64.

155

Table 4.   Straw dry weight (g) of 15 rice varieties at four Fe concentrations.

Varieties	Fe Concentration (mg kg⁻¹)									Mean
	229.44		236.34		564.15		1277.50			
Margasari	27.13	a	22.53	e-i	14.00	no	11.80	pqr		18.87
Lambur	25.17	a-d	22.77	e-i	12.30	opq	10.57	qrs		17.70
Batanghari	25.30	abc	17.40	m	7.00	tu	3.53	x		13.31
Punggur	24.63	b-e	21.73	hij	10.40	qrs	8.73	st		16.38
Indragiri	23.07	d-i	21.13	ijk	11.03	pqr	8.80	st		16.01
Mendawak	26.83	a	23.83	c-h	13.03	nop	9.87	rs		18.39
Inpara-1	24.60	b-e	22.20	ghi	14.00	no	12.50	opq		18.33
Inpara-2	26.13	ab	22.87	e-i	14.83	n	11.57	pqr		18.85
IR64	23.23	c-h	18.13	lm	3.90	wx	1.43	y		11.68
Ciherang	22.37	f-i	19.40	klm	4.30	vwx	2.43	xy		12.13
Cibogo	24.57	b-f	19.80	jkl	6.80	tu	4.33	vwx		13.88
Ciapus	24.03	b-g	19.73	jkl	6.57	u	4.03	vwx		13.59
Inpari-1	23.00	d-i	18.77	lm	6.10	uv	3.30	xy		12.79
Inpari- 6	23.50	c-h	19.00	lm	4.50	vwx	2.43	xy		12.36
Inpari-10	23.47	c-h	19.03	lm	5.97	uvw	3.23	xy		12.93
Mean	24.47		20.56		8.98		6.57			(+)

Number followed by the same letter in a column or a row was not significantly different (DMRT 5%).

Table 5.   Linear regression equation for plant growth of 15 rice varieties at four Fe concentration.

Varieties	Plant height	Root dry weight	Straw dry weight
Margasari	$y = -0.0173x + 114.67$	$y = -0.0034x + 6.60$	$y = -0.0124x + 26.01$
	$R^2 = 0.711$; $r = -0.84**$	$R^2 = 0.719$; $r = -0.85**$	$R^2 = 0.701$; $r = -0.84**$
Lambur	$y = -0.0175x + 97.155$	$y = -0.003x + 6.58$	$y = -0.0127x + 25.03$
	$R^2 = 0.713$; $r = -0.85**$	$R^2 = 0.705$; $r = -0.84**$	$R^2 = 0.707$; $r = -0.84**$
Batanghari	$y = -0.0276x + 104.83$	$y = -0.0042x + 5.61$	$y = -0.0169x + 23.08$
	$R^2 = 0.945$; $r = -0.97**$	$R^2 = 0.702$; $r = -0.84**$	$R^2 = 0.700$; $r = -0.84**$
Punggur	$y = -0.0192x + 96.90$	$y = -0.0042x + 5.80$	$y = -0.0137x + 24.28$
	$R^2 = 0.774$; $r = -0.88**$	$R^2 = 0.806$; $r = -0.90**$	$R^2 = 0.703$; $r = -0.84**$
Indragiri	$y = -0.0196x + 103.13$	$y = -0.0038x + 6.59$	$y = -0.0126x + 23.29$
	$R^2 = 0.816$; $r = -0.90**$	$R^2 = 0.871$; $r = -0.93**$	$R^2 = 0.744$; $r = -0.86**$
Mendawak	$y = -0.0201x + 100.78$	$y = -0.0042x + 5.61$	$y = -0.0147x + 26.87$
	$R^2 = 0.879$; $r = -0.94**$	$R^2 = 0.702$; $r = -0.84**$	$R^2 = 0.768$; $r = -0.88**$
Inpara-1	$y = -0.0155x + 98.92$	$y = -0.0014x + 7.03$	$y = -0.0103x + 24.29$
	$R^2 = 0.800$; $r = -0.89**$	$R^2 = 0.807$; $r = -0.90**$	$R^2 = 0.719$; $r = -0.85**$
Inpara-2	$y = -0.0156x + 95.68$	$y = -0.0026x + 7.52$	$y = -0.0123x + 25.95$
	$R^2 = 0.842$; $r = -0.92**$	$R^2 = 0.838$; $r = -0.92**$	$R^2 = 0.783$; $r = -0.89**$
IR64	$y = -0.0293x + 96.68$	$y = -0.0047x + 5.58$	$y = -0.0183x + 22.21$
	$R^2 = 0.791$; $r = -0.89**$	$R^2 = 0.700$; $r = -0.84**$	$R^2 = 0.700$; $r = -0.84**$
Ciherang	$y = -0.0261x + 102.90$	$y = -0.0040x + 5.27$	$y = -0.0175x + 22.21$
	$R^2 = 0.759$; $r = -0.87**$	$R^2 = 0.701$; $r = -0.84**$	$R^2 = 0.703$; $r = -0.84**$

(*Continued*)

Table 5. (Cont.)

Varieties	Plant height	Root dry weight	Straw dry weight
Cibogo	y = -0.0244x + 101.83	y = -0.0039x + 5.43	y = -0.0169x + 23.65
	R^2 = 0.819; r = -0.91**	R^2 = 0.739; r = -0.86**	R^2 = 0.707; r = -0.84**
Ciapus	y = -0.0244x + 109.73	y = -0.0041x + 5.53	y = -0.0169x + 23.36
	R^2 = 0.708; r = -0.84**	R^2 = 0.702; r = -0.84**	R^2 = 0.712; r = -0.84**
Inpari-1	y = -0.024x + 99.86	y = -0.0040x + 5.69	y = -0.0167x + 22.42
	R^2 = 0.825; r = -0.91**	R^2 = 0.757; r = -0.87**	R^2 = 0.726; r = -0.85**
Inpari- 6	y = -0.0248x + 100.05	y = -0.0040x + 5.30	y = -0.0178x + 22.65
	R^2 = 0.782; r = -0.89**	R^2 = 0.720; r = -0.85**	R^2 = 0.702; r = -0.84**
Inpari-10	y = -0.0228x + 98.84	y = -0.0040x + 5.75	y = -0.0171x + 22.79
	R^2 = 0.783; r = -0.89**	R^2 = 0.879; r = -0.93**	R^2 = 0.725; r = -0.85**

brown (IRRI, 1996). Results of Mehraban *et al.* (2008) research on sand culture showed that increased concentrations of Fe solution from 50 mg L^{-1} to 500 mg L^{-1} linearly decreased roots and straw dry weight and showed a significantly high negative correlation. Ishizuka and Tanaka (1969) *in* Yoshida (1981) also showed the linear decrease in plant dry weight with increasing Fe concentration in culture solution. The Fe concentration in the culture solution were 2, 75, 150, and 300 mg L^{-1} Fe.

3.2 Grain yield and harvest index

For all varieties, the Fe concentration of acid sulfate soils negatively correlated with grain yield and harvest index (Table 6 and 7). Grain yield of varieties varied with Fe concentrations, and the highest per hill at low soil Fe concentration was shown by Inpara-1 which was not significantly different from Inpara-2, Mendawak, and Indragiri. At high Fe concentrations (1277.50 mg kg^{-1}), the highest grain yield was obtained from Inpara-1, whereas the lowest was shown by IR64 which was not substantially different from Batanghari and all irrigated rice varieties. However, an increase in Fe concentrations from 229.44 to 236.34 mg kg^{-1} resulted in a decrease in the grain yield of all varieties. Similarly, increasing the Fe concentration from 236.34 to 564.15 mg kg^{-1} and 564.15 to 1277.50 mg kg^{-1} caused the grain yield of all varieties to decrease (Table 6).

The varieties showed differences in harvest index (HI) at different soil Fe concentrations. At 229.44 mg kg^{-1} Fe, the largest harvest index was shown by Inpara-1, Margasari, Batanghari, Ciherang, and Inpari-6. Furthermore, at 1277.50 mg kg^{-1} Fe, the largest harvest index was consistently shown by Inpara-1 and Inpara-2 (Table 7). This data suggests that Inpara-1 was the most tolerant to Fe toxicity.

The relationship between Fe concentration and the grain yield per hill of Inpari-1, Inpara-2, and IR64 can be observed in Figure 2 which clearly

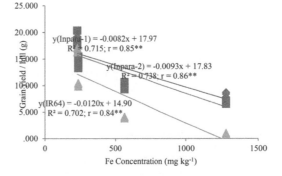

Figure 2. Relationship between Fe concentration and grain yield per hill of Inpara-1, Inpara-2, and IR64.

shows the negative correlation between both variables. Therefore a higher Fe concentration of acid sulfate soils leads to a lower grain yield. Inpara-1 had the highest grain yield followed by Inpara-2, whereas IR64 had the lowest. For this reason, IR64 is identified as a susceptible variety to iron toxicity (Table 6). Several researchers have reported that a decrease in grain yield due to iron toxicity mainly occurs in susceptible varieties (Audebert, 2006, Nozoe et al., 2008; Aboa and Dogbe, 2006).

There was a negative correlation between Fe concentration and harvest index. Therefore, a higher Fe concentration of acid sulfate soil results in lower harvest index of varieties. Inpara-1 had the largest harvest index followed by Inpara-2, while IR64 had the smallest (Table 7). Harvest index is described as the proportion of grain yields with plant growth, in this case the dry weight of straw (Yoshida, 1981). Large harvest index indicates more photosynthate allocation transported to grains rather than plant growth. It means that Inpara-1 and Inpara-2 were more efficient in utilizing all the resources for growth and development, and for this reason, their grain yields were high.

Table 6. Grain yield per hill (g) of 15 rice varieties at four Fe concentrations.

Varieties	Fe Concentration (mg kg^{-1})								Mean
	229.44		236.34		564.15		1277.50		
Margasari	16.66	cd	10.08	i	7.65	jk	3.61	mn	9.50
Lambur	18.51	ab	14.23	g	8.29	j	5.90	l	11.73
Batanghari	16.39	cd	10.49	i	4.41	m	1.68	op	8.24
Punggur	17.50	bc	12.27	h	7.22	jk	4.12	m	10.28
Indragiri	18.25	ab	14.57	fg	7.50	jk	5.68	l	11.50
Mendawak	18.74	a	12.39	h	7.26	jk	4.14	m	10.63
Inpara-1	19.12	a	14.80	efg	10.61	i	8.30	j	13.21
Inpara-2	18.87	a	14.30	g	9.88	i	6.83	kl	12.47
IR64	16.82	cd	10.20	i	3.61	mn	0.94	p	8.00
Ciherang	16.19	d	11.87	h	4.02	m	1.79	op	8.37
Cibogo	16.23	d	12.32	h	4.30	m	2.52	no	8.84
Ciapus	16.30	cd	12.22	h	4.19	m	2.17	op	8.72
Inpari-1	16.83	cd	11.87	h	4.58	m	2.43	o	8.93
Inpari- 6	15.56	def	11.86	h	3.72	m	1.45	op	8.15
Inpari-10	15.76	de	12.14	h	4.17	m	2.08	op	8.54
Mean	17.18		12.37		6.09		3.58		(+)

Number followed by the same letter in a column or a row was not significantly different (DMRT 5%).

Table 7. Harvest index of 15 rice varieties at four Fe concentration.

Varieties	Fe Concentration (mg kg^{-1})								Mean
	229.44		236.34		564.15		1277.50		
Margasari	0.53	c-f	0.44	lm	0.38	p	0.28	t	0.41
Lambur	0.58	a	0.50	f-i	0.43	lm	0.37	pq	0.47
Batanghari	0.54	b-e	0.48	ijk	0.34	qr	0.23	vw	0.39
Punggur	0.57	ab	0.49	g-k	0.43	lm	0.27	tu	0.44
Indragiri	0.57	ab	0.52	e-h	0.48	ijk	0.39	op	0.49
Mendawak	0.58	a	0.49	g-k	0.43	lm	0.33	r	0.46
Inpara-1	0.60	a	0.56	a-d	0.53	c-f	0.50	f-i	0.55
Inpara-2	0.59	a	0.53	c-f	0.49	g-k	0.41	mno	0.50
IR64	0.59	a	0.40	nop	0.31	rs	0.13	x	0.36
Ciherang	0.54	b-e	0.46	kl	0.38	p	0.24	uvw	0.41
Cibogo	0.56	a-d	0.49	g-k	0.39	op	0.26	tuv	0.42
Ciapus	0.57	ab	0.48	ijk	0.34	qr	0.21	w	0.40
Inpari-1	0.58	a	0.52	e-h	0.40	nop	0.29	st	0.45
Inpari- 6	0.53	c-f	0.48	ijk	0.32	rs	0.23	vw	0.39
Inpari-10	0.57	ab	0.50	f-i	0.37	pq	0.27	tu	0.43
Mean	0.57		0.49		0.40		0.29		(+)

Number followed by the same letter in a column or a row was not significantly different (DMRT 5%).

3.3 *Iron toxicity symptom*

Increasing the Fe concentrations increases iron toxicity symptoms. Severe symptoms at Fe concentration 229.44 mg kg^{-1} were observed in Inpari-1 and Inpari-6. However, at high Fe concentrations (1277.50 mg kg^{-1}), the most severe iron toxicity symptoms were shown by IR64, whereas the lightest was exhibited by Inpara-1 and Inpara 2 (Figure 3).

Furthermore, increasing Fe concentrations from 229.44 to 236.34 mg kg^{-1} led to an increase in iron toxicity symptoms in most of the varieties, except Inpara-1 and Margasari. However, an increase from 236.34 to 564.15 mg kg^{-1} and 564.15 to 1277.50 mg kg^{-1} increased symptoms in all the varieties (Figure 3). The smallest increase in the symptoms was exhibited by Inpara-1 followed by Inpara-2, whereas the most severe was shown by IR64. The relationship between Fe concentration and iron toxicity symptoms of Inpara-1, Inpara-2, and IR64 is illustrated in Figure 4. It has been observed that there is a positive correlation between both variables. This indicates that a higher Fe concentration in acid sulfate soils gives rise to a more severe iron toxicity symptom. Inpara-1 showed the lowest increase in the symptoms followed by Inpara-2, whereas IR64 exhibited the most significant increase. This phenomenon indicated that Inpara-1 was most tolerant

to iron toxicity followed by Inpara-2, while IR64 was the most susceptible.

4 CONCLUSION

Increased Fe concentrations in acid sulfate soils decreased rice agronomic traits (maximum tiller number, plant height, dry weight of roots and straw, grain yield and harvest index), but increased iron toxicity symptoms. Varieties Inpara-1 and Inpara-2 showed the smallest decrease in rice agronomic traits, while IR64 had the greatest decrease. Furthermore, iron toxicity symptoms of Inpara-1 and Inpara-2 were the lightest, while IR64 exhibited the most severe symptoms. In conclusion, Inpara-1 and Inpara-2 identified as the varieties tolerant to iron toxicity in the acid sulfate soils of tidal swampland.

ACKNOWLEDGEMENT

This experiment was funded by Indonesian Agency for Agricultural Research and Development (IAARD) through Agricultural Research Partnership with University Project. The authors would like to show gratitude to the Director of ISARI, Banjarbaru, for his assistance.

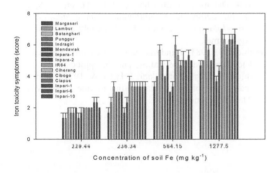

Figure 3. Iron toxicity symptom of 15 rice varieties at four Fe concentrations (score: 1 resistant; 9 susceptible). Vertical bars indicate s.e.

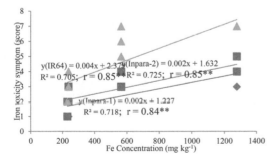

Figure 4. Relationship between Fe concentration and iron toxicity symptoms of Inpara-1, Inpara-2, and IR64.

REFERENCES

Aboa, K. and S. Y. Dogbe. 2006. Effect of iron toxicity on rice yield in the Amou-Oblo lowland in Togo. *In*: Audebert A, L.T. Narteh, P. Klepe, D. Millar, and B. Beks (eds.). Iron Toxicity in Rice-based Systems in West Africa. Africa Rice Center WARDA, Cotonou, Benin. 175 p.

Audebert, A and K. L. Sahrawat. 2000. Mechanism for iron toxicity tolerance in lowland rice. J. Plant Nutr. 23, 1877-1885.

Audebert, A. 2006. Iron toxicity in rice – environmental conditions and symptoms. *In:* Audebert A, L.T. Narteh, P. Klepe, D. Millar, and B. Beks (eds.). Iron Toxicity in Rice-based Systems in West Africa. Africa Rice Center WARDA, Cotonou, Benin. 175 p.

Benckiser, G., Santiago, S., Neue, H.U., Watanabe, I. & Ottow, J.C.G. 2005. Effect of Fertilization on Exudation, Dehydrogenase Activity, Population and Fe^{2+} Formatin in the Rhizosphere of Rice (Oryza sativa L.) in Relation to Iron Toxicity: 305-316. Netherlands: Springer.

Cai, M. Z., A. C. Luo, X. Y. Lin, and Y. S. Zhang. 2003. Nutrient uptake and partitioning in rice plant under excessive Fe^{2+} stress. J. Zhejiang Univ (Agric & Life Sci.), 29(3):305-310.

De Datta, S, K., H. U. Neue, D. Senadhira, and C. Quijano. 1994. Success in rice improvement for poor soils. *In*: Proc. of. The Workshop on Adaptation of Plants to Soil Stresses. 1-4 August 1993. University of Nebraska, Lincoln, nebraska. INTSORMIL Publication No. 94-2. Lincoln Nebraska, USA. Pp 248-268.

De Ikehashi, H and F. N. Ponnamperuma. 1978. Varieties tolerance of rice for adverse soils. *In*: Soil and Rice. International Rice Research Institute, Los Banos, Philippines.

IRRI. 1996. Standard evaluation system for rice. Int. Ric Res Inst. Los Banos, The Philippines.

Ismunadji, M,. W. S. Ardjasa, and H. R. Von Uexkull. 1989. Increasing productivity of lowland rice grown on iron toxic soil. Paper presented at International Symposium on Rice Production on Acid Soils of Tropics, June 26-30, 1989. Kandy, Srilanka.

Ismunadji, M., L. N. Hakim, I. Zulkarnain, and F. Yazawa. 1973. Physiological disease of rice in Cihea. Contr. Cent. Res. Inst. Agric. Pusat Penelitian dan Pengembangan Tanaman Pangan, Bogor. 4:10.

KEPAS. 1985. Tidal Swamp Agroecosystem of Southern Kalimantan: Workshop report on the sustainable intensification of tidal swamplands in Indonesia. Kelompok Peneliti Agroekosistem. Agency for Agricultural Research and Development, Jakarta.

Maas, A., Afandie, R. & Suryanto. 1992. Potensi dan kendala reklamasi lahan pasang surut. *Makalah pada Pertemuan Nasional Pengembangan Pertanian Lahan Pasang Surut, Cisarua 3-4 Maret 1992*.

Majerus, V., P. Bertin, and S. Lutts. 2007. Effects of iron toxicity on osmotic potential, osmolytes and Polyamines concentrations in the African rice (*Oryza glaberrima* Steud). Plant Science 173:96-105.

Mehraban, P., A.A. Zadeh, and H.R. Sadeghipour. 2008. Iron toxicity in rice (*Oryza sativa* L.) under different potassium nutrition. Asian Juornal of Plant Sciences, 2208: 1-9.

Noorsyamsi H and M. Sarwani. 1989. Management of the tidal for food crops: Southern Kalimantan experiences. IARD Journal (11): 18-24.

Nozoe, T., R. Agbisit, Y. Fukuta, R. Rodriguez, and S. Yanagihara. 2008. Characteristics of iron tolerantt rice lines developed at IRRI under field condition. JARQ 42(3):187-192.

Ottow, J. C. G., G. Benckiser and I. Watanabe. 1982. Iron toxicity of rice as a multHIle nutrition soil stress. *In:*. Proc. of Symposium on Tropical Agricultural Research.

Trop. Agric. Res. Series No. 15. Trop.l Agric. Res. Center. Ministry of Agric. Forestry and Fisheries, Japan. p. 167-179.

Ottow, J.C.G., Prade, K., Bertenbreiter, W. & Jack, V.A. 1989. Strategies to alleviate iron toxicity of wetland rice on acid sulphate soils. In: Deturk, P & F. N. Ponnamperuma (eds.) *Rice Production on Acid Soils of the Tropics. Proceeding of International Symposium, Institute of Fundamental Study, Kandy, Sri Lanka, 26-30 June 1989*.

Ponnamperuma, F.N. 1977. Screening rice for tolerance to mineral stresses. IRRI Research Paper Series, No. 6. Int. Rice Res. Ins., Los Banos, The Philippines.

Ponnamperuma, F.N., Bradfield, R. & Peech, M. 1955. Physiological disease of rice attributable to iron toxicity. *Nature* 175: 265.

Puslitbangtan. 1991. Sumber pertumbuhan produksi padi dan kedelai. Pusat Penelitian dan Pengembangan Tanaman Pangan, Bogor.

Tanaka, A., & Yoshida, S. 1970. Nutritional disorders of the rice plant in Asia. Int. Rice Res. Inst. Tech. Bull. 10. 51p. Int. Rice Res Inst., Los Banos, The Philippines.

Van Breemen, N. & Moormann, F.R. 1978. Iron toxic soils. In: *Soils and Rice: 781-799. Int. Rice Res. Inst., Los Banos, The Philippines*.

Virmani, S. S. 1977. Varietal tolerance of rice to iron toxicity in Liberia. Int. Ric.Res. Newsl. 2(1):4-5.

Widjaja-Adhi, I P.G., Suriadikarta, D.A., Sutriadi, M.T., Subiksa, I G.M. & Suastika, I W. 2000. Pengelolaan, pemanfaatan, dan pengembangan lahan rawa. hal. 127-164. Dalam: Adimihardja, A., L.I., Amien, F. Agus & D., Jaenuddin (eds.). *Sumberdaya Lahan Indonesia dan Pengelolaannya*. Bogor: Pusat Penelitian Tanah dan Agroklimat.

Yoshida, S. 1981. Fundamentals of rice crop science. Int. Rice Res. Inst., Los Banos, The Philippines.

Yoshida, S., D. A. Forno, and J. H Cook. 1971. Laboratory manual for physiological studies of rice. IRRI, Manila, The Philippines.

The efficacy of using climate data for developing food crops in wetlands: A case study from Kalimantan Island

W. Estiningtyas, S. Erni & Susanti
Indonesian Agroclimate and Hydrology Research Institute, Bogor, Indonesia

ABSTRACT: This paper presents the data and information of climate resources on Kalimantan Island which was one of the centers of food crops in the wetlands. Based on the analysis of Oldeman climate types on Kalimantan Island, it was dominated by the B1 Climate Type (30.4%). Most districts in Kalimantan have a constant trend of flooding and drought. However, there were 10 districts that demonstrate an increasing flood trend and 8 exhibiting an increasing drought trend. Based on the Katam analysis, it was suggested that the first planting time of WS 2018/2019 start September 3 to January 1-2 and the second planting January 3 to March 1-2. Distribution of paddy fields were in areas with wet climate types that experienced constant trends in flooding and drought. Climate information can help provide choices about where and what types of commodities to plant based on water availability and possible risks.

1 INTRODUCTION

The efforts of the Indonesian Government to gain self-sufficiency in food and the determination to become a world food basket in 2045, wetlands, especially swampland, become a reliable choice to realize this goal due to the extensive wetland area.

Wetlands are important for the carbon cycle, water balance, wildlife, biodiversity, and human food production (Neue et al., 1997). However, they can also affect climate regionally through land-atmosphere exchanges (Modathir et al., 2013). Wetlands consist of swamp and non-swamp land and Indonesia has 191,1 million hectares of land and about 46,6 million hectares is wetland (24,4%) (Nursyamsi, 2018). It has an area of swampland of about 34 million hectares. About 7.5 million hectares of swampland is potentially available for agricultural intensification. In accordance with its natural water abundance, two-thirds of the available land is suitable for a paddy rice system. However, lowland rice fields are normally used for rice-rice, rice-maize, or rice-soybean rotation, meaning that they become producers of secondary food crops such as maize and soybean.

Swampland is not an ideal land resource for agriculture due to its low fertility. However, it can be still managed profitably and sustainably. There are some advantages to utilizing swampland. For example, the land is relatively more available, there are fewer tenurial conflicts, and the abundant water availability can produce longer cropping periods in the dry season. The key management practices for it are water management, land management, high yielding varieties, amelioration, and fertilization.

Indonesian agriculture is challenged by the shrinking of land area due to relatively rapid population growth, demands in terms of quantity and quality, land conversion and extreme weather conditions, as well as aging farmers as working in food crop agriculture lack social-economic incentives.

The impact of climate change on the agricultural sector is quite significant. The impacts most felt are a result of changes in air temperature, changes in rainfall, changes in sea level, and increasingly extreme climate events. The relative impact of climate change on food security differs between regions (Gutman et al., 2000; FAO, 2005), both in tropical and subtropical regions. However, the impact in the tropics is greater because there is a considerable variation in rainfall (Slingo et al., 2005), which in turn disrupts the stability of the agricultural system (Koesmaryono et al., 2018). According to Boer (2017), the threat of climate change is at the center of agricultural production, especially in the southern part of the equator. This is indicated by a longer dry season and a shorter rainy period. The results of Foerster's research in Syahbuddin et al. (2015) stated that 1-2 meter sea level rise caused a total loss of harvested area in North Sumatra, West Sumatra, Lampung, Banten, West Java, Central Java, East Java and South Sulawesi of around 74,000-165.00 hectares which is the equivalent to a loss of production of between 238,650 -532,125 tons of rice. Drought events that occurred in the period 1844 to 1998 reported as many as 43 times but only six drought events were not related to the El-Nino phenomenon (Allan, 2000; Boer and Subbiah, 2005). These conditions had a significant impact on cultural strategies and agricultural production, especially food crops (Hamada et al., 2002; Haylock and McBride, 2001; IPCC, 2001; IPCC, 2007; Porter and Semenov, 2005; Betts, 2005; Osborne, 2005).

The FAO (2005) study shows that variability and climate change affect 11% of agricultural land in developing countries that reduce food production and reduce Gross Domestic Product (GDP) to 16%. Meanwhile, the impact of variability and climate change also reduce food crop production (cereals) in Southeast Asia between 2.5% and 7.8% (Fischer et al., 2002). According to Handoko et al 2008, variability and climate change with all its impacts have the potential to cause significant loss of food crop production at the following levels: 20.6% for rice, 13.6% corn, and 12.4% soybeans. Yet food needs, especially rice, continue to increase in line with population growth. It is estimated that in 2025 the population will reach 262 million with consumption of 134 kg of rice per capita, thus the national rice demand reaches 35.1 million tons or 65.9 million tons of GKG (Budianto, 2002). According to Cline (2007), by 2080 the decline in productivity of agricultural commodities in Indonesia due to global warming range from 15-25%. If CO_2 enrichment in the atmosphere considered, the decline in productivity could range from 5-15% of current productivity.

Developing wetlands as an of the potential support for the provision of planting land and national food providers must be supported by data and information related to land, climate and water resources so that they can be optimally utilized. One of the wetlands that have the potential to be developed and utilized is swampland. On the commemoration of the 38th World Food Day (HPS) in South Kalimantan in October 2018, the representative of the Food and Agriculture Organization (FAO) for Indonesia and Timor Leste, Stephen Rudgard, gave awards to the Indonesian Government for efforts to maintain the national food security carried out by Ministry of Agriculture. According to Rudgard (2018), the use of swamp land is very important for the cultivation of food for Indonesia's growing population. But it is even more important to have a sustainable agriculture approach. The determination of the government to maintain the raw land area of the national rice fields is realized through the Ministry of Agrarian Affairs and Spatial Planning which is currently drafting a Presidential Regulation (Perpres) regarding eternal rice fields. The Presidential Regulation is necessary to maintain the national rice fields. At present, the area of paddy fields has decreased due to the conversion of land for other purposes such as housing, industry, and gas stations.

The magnitude of the impact of climate change and extreme climate events shows that if efforts are not made to anticipate, adapt and mitigate climate change, then damage and losses will be ensured. Therefore, data and information related to climate are needed to minimize the impact of extreme climate change or events.

Data on the national distribution of swampland in Indonesia shows there are 4 large islands that have swampland, namely Sumatra 10.9 million hectares, Kalimantan 10.6 million hectares, Sulawesi 1.4 million hectares and Papua 10.5 million hectares (Nursyamsi, 2018). Kalimantan as one of the islands with a large area of wetland is the chosen study location. Kalimantan Island has mostly mountainous/hilly areas (39.69%), land (35.08%), coastal/tidal coastal plains (11.73%), alluvial plains (12.47%) and other (0.93%). Kalimantan has a tropical climate. The warmest month of the year is May with an average temperature of 27.4 ° C. In January, the average temperature is 26.7 ° C. During the year, the average temperature varies around 1.5 °C. Rainfall mostly occurs throughout the year. The average annual rainfall is 2992 mm. The driest month is July, with rainfall of 140 mm and a monthly average of 416 mm. The difference in precipitation between the driest month and the wettest month is 276 mm. The resources that already exist on Kalimantan Island can be studied and developed in order to produce further data and information that can be used to increase food production, especially rice.

The purpose of this paper is to present the climate data and information to support the development of wetlands on Kalimantan island.

2 METHODS

The climate information presented in this paper is in the form of the distribution of climate types, flood and drought trends, and integrated planting calendars by taking examples of cases on the island of Kalimantan.

Climate type is analyzed according to the Oldeman classification method. Climate classification is very useful for practical purposes, especially in the classification of agricultural crops in Indonesia. This climate classification is directed at food crops such as rice and secondary crops. Compared to other methods, the Oldeman classification method is more advanced because it takes into account other weather elements such as solar radiation associated with plant water requirements.

Oldeman creates a new system in climate classification associated with agriculture by using elements of the rainy seasons which are classified into climatic types based on the criteria of Wet Month (WM), Humid Month (HM) and Dry Month (DM) consecutively and limits attention to the chance of rain, effective rain and plant water needs. The concept is not complicated. A rice paddy requires an average of 145 mm of water per month in the rainy season, secondary crops need 50 mm of water per month on average in the dry season. The monthly rain can be expected to have a 75 percent chance of occurrence and equals 0.82 times the average rainfall. The monthly average is then reduced by 30. The effective rainfall for paddy fields is 100%, and the rainfall effective for crops with tightly closed crop canopy is 75%. Monthly rainfall can be calculated for rice or secondary crops (X) using long-term data:

Rice Paddy:	Secondary Crops (Palawija):
145= 1,0 (0,82 X -30)	50 = 0,75 (0,82 X - 30)
X = 213 mm/month	X = 118 mm/month

The values of 213 and 118 are rounded to 200 and 100 mm/month respectively, which are used as the wet and dry month determination limits. Therefore, the Wet Month (BB) is the month with an average rainfall of more than 200 mm. HM is the month with an average rainfall of 100-200 mm and DM is the month with an average rainfall of less than 100 mm. Furthermore, in determining the Oldeman climate classification uses the provisions of the length of the wet month period and successive dry months.

The main types of the Oldeman classification are divided into 5 types based on the number of consecutive wet months, while the sub-division is divided into 4 based on the number of consecutive dry months. Based on the 5 main types and 4 subdivisions, the climate types can be grouped into 17 Oldeman agro-climates ranging from A1 to E4 as presented with the Oldeman's Triangle (Figure 1). Oldeman issued a description of each type of agro-climate as in Table 1. The Oldeman classification results can be used to carry out agricultural activities, such as determining the beginning of the planting period, determining the planting pattern and intensity of planting.

Flood and drought trends were analyzed using trend analysis to establish the trends through the coefficient of the formed equation and its significance (p value). Trend determination is grouped into three outcomes. If the coefficient is positive and significant, the trend is up. If the coefficient is negative and significant, the trend is down. If the coefficient is positive or negative and not significant, it remains. The data used

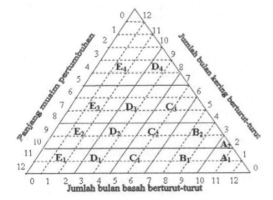

Figure 1. Oldeman's triangle.

Table 1. Description of Oldeman's agro-climate types.

Climate Type	Description
A1, A2	Suitable for continuous rice but less production because generally the solar radiation flux density is low throughout the year
B1	Suitable for continuous rice with good early planting season planning. High production if the dry season harvests
B2	Can plant rice twice a year with short age varieties and a short dry season sufficient for crops
C1	Rice planting can be once and crops twice a year
C2, C3, C4	Plant rice once and crops twice a year, but planting second crops must be careful not to fall on dry months
D1	Short-lived one-time rice planting and production can usually be high due to high radiation flux density. Time to palawija plant
D2, D3, D4	Only one rice or one crop per year may depend on the availability of irrigation water
E	This area is generally too dry, it may only be able to plant one crop once it depends on the rain

is on the area of rice fields affected by floods and droughts in the period 1989-2017, which were sourced from the directorate of food crop protection.

Integrated Cropping Calendar (ICC) Information System is prepared based on the approach of water availability through rainfall and its predictions for Wet Season (WS) and Dry Season (DS). In this paper the information presented is for WS2018/ 2019. The rainfall prediction used comes from the Meteorology, Climatology and Geophysical Agency (BMKG) which is analyzed to represent the rainfall of each sub-district. The results of the ICC analysis have been provided through web www.katam.lit bang.pertanian.go.id. Examples of information presented are planting time and cropping patterns of food crops.

3 RESULT AND DISCUSSION

Based on the analysis of the Oldeman climate type (Table 2), Kalimantan Island is dominated by the B1 Climate Type, which is equal to 30.4% of 56 districts/cities. Climate type B1 has 7 wet months and no dry months. Other types of climate distribution are type A (19.6%), C2 (12.5%), D1 (10.7%), C1 (8.9%), E1 (5.4%), E2 (3.6%), B2, C3, D2 and D3 1.8% respectively. The drier areas are mostly in East Kalimantan Province and a small area in North Kalimantan. Wetter areas are primarily in West Kalimantan and North Kalimantan bordering Malaysia and parts of Central Kalimantan. Districts with climate type E include Singkawang, North Hulu Sungai, Kutai Kartanegara, East Kutai, Mahakam Hulu and Nunukan. According to Oldeman, generally the area with climate type E is very dry and depends on rainfall. Areas with climate type D (D1, D2 D3, D4) include Sambas Regency, Ketapang, Katingan, Kota Baru, Banjar, Hulu Sungai Selatan, Balangan and Samarinda. Generally, areas with type D climate can use rice cultivation one time and another crop one time. Climate type C areas can plant rice once and other crops twice. This is linked with water availability in the second crop. Type C encompasses almost a quarter of Kalimantan, such as Paser, Pulang Pisau, Kubu Raya, Kapuas, and others. In climate B1 rice can be planted continuously. Production in the dry season will be higher because the level of resistance is lower so that the intensity of solar radiation reaching the plants is higher. The effect is higher yield. This can be different from the climate type A. In type A, rice can be planted continuously during the year, but due to cloudy and rainy days, the radiation that reaches the plants is lessened so the production is lower.

Oldeman's analysis is based on the amount of rainfall, along with the occurrence of climate change, the results of the analysis can experience changes from the Oldeman classification produced previously. The complete distribution of the Kalimantan Island climate type is presented in Figure 2.

The trends in flooding and droughts on Kalimantan Island were analyzed based on data in the area of rice fields affected by floods and droughts sourced from the Directorate of Food Crop Protection in 1989 to 2017 (29 years). The results of the data show that Kalimantan Island mostly has a constant trend of flooding and drought. Most of districts and cities in Kalimantan have not experienced a change in trends; however, there are 10 districts that have increased flood trends. Those districts are Singkawang City, Kotawaringin Timur Regency, South Barito, Barito Utara, Seruyan, Tanah Bumbu, Balangan, Paser, Penajam Paser Utara and Mahakam Hulu. Increased flood trends can occur due to several factors, such as climate

change which results in changes in rainfall patterns, watershed conditions that have decreased function, carrying capacity of land such as deforestation. Further research is needed to find out what factor is most influential.

Besides increasing flood trends in several districts, the results of the analysis also showed an increase in the drought trends. Increasing drought trends or more intense droughts from the meteorological side can occur due to increased air temperature and evaporation (Yang et al., 2012). Increasing trends in drought can have an impact on agricultural production. This is in agreement with Surmaini's results (2016) which states that drought in agriculture can have a significant impact on decreasing agricultural production. Drought is also quite difficult to predict and monitor. On Kalimantan Island, there are 8 districts with increasing drought trends: Palangkaraya City and Banjarbaru City, Lamandau District, Pulang Pisau, Tanah Bumbu, Balangan, East Kutai and Penajam Paser Utara. The increasing trend of drought in an area certainly requires treatment so it no longer intensifies. Surmaini (2016) states to significantly reduce the potential impact of drought, an agricultural drought early warning system is needed.

Based on this research, there are three districts that have an upward trend in both floods and droughts. Those districts are Tanah Bumbu and Balangan Regencies in South Kalimantan Province and North Paser Penajem Regency in East Kalimantan Province. In districts that have an upward trend of both floods and droughts, it is necessary to increase anticipation/adaptation. Given these conditions can have implications for the decline in production and a decrease in food security, it ultimately affects income (Ndamani and Watanabe, 2017). The impact of the decline in yields due to flooding and drought on economic conditions was also conveyed by De Silva and Kawasaki (2018).

Based on the results of the analysis, most districts in Kalimantan Island have a constant trend of flooding and drought and only 2 districts have a downward trend in drought - North Barito Regency in Central Kalimantan Province and Balikpapan City in East Kalimantan Province. One of the causes of the decline in drought is rainfall and the available water needs of plants. The decline in the drought trend can also occur due to several efforts, including adjusting the planting time with conditions of water availability or water adequacy, infrastructure improvements including the provision of facilities to anticipate droughts, in planting using early varieties - especially during the dry season, local government policies and regulations that support cultivation. The distribution of flood and drought trends is presented in Figures 3 and 4.

The challenge of rice production is getting difficult with increasingly uncertain climate conditions, particularly where they are closely related to the availability of water, the main determinant of rice production and other food crops. This climate

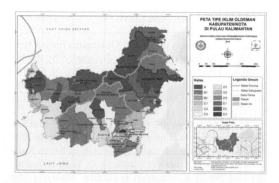

Figure 2. Oldeman climate type map of Kalimantan Island.

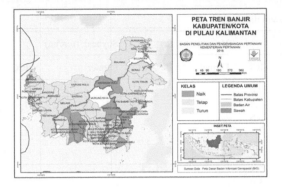

Figure 3. Distribution of flood trends in districts/cities on Kalimantan Island.

Figure 4. Distribution of drought trends in districts/cities on Kalimantan Island.

uncertainty often causes droughts, floods, attacks by plant pest organisms (OPT), as well as the advances or retreat of the rainy season which causes a shift in the planting season. Extreme climate events such as El Nino and La Nina correlate with decreases in rice production (Surmaini and Syahbuddin, 2016). For this reason, the Agricultural Research and Development Agency has made adaptation efforts by compiling a dynamic ICC that focuses on season and rainfall predictions from the Meteorological Climatology and Geophysics Agency that compiles time and cropping patterns of food crops (rice, corn and soybeans).

ICC is a tool for farmers and extension agents to make decisions in determining planting time, preparing seeds (varieties), processing land (fertilizer needs), labor requirements, and regulating the use of machine tools for land processing and harvesting (Surmaini and Syahbuddin, 2016). ICC also provides information on planting time for swamplands. The ICC recommends planting time based on the availability and needs of available water for rice, corn and soybeans. Rice requires more water than corn. While soybeans require relatively less water than corn. To optimize growth and increase production, the ICC also recommends types and doses of fertilizers, seed varieties and agricultural machinery.

Information contained in ICC can be used as a guide to increase productivity. This is supported by the results of the catastrophe validation that has been carried out in several provinces central to the production of rice in Indonesia, which shows that the productivity produced using information, cultivation technology, recommendation based on information from ICC is higher than farmers' intuition (Table 2).

Information on ICC of Kalimantan Island consists of rice fields on irrigated, rainfed and swamp fields. Based on the results of ICC analysis, the recommended planting time for rice for the 1st and 2nd planting paddy fields in Wet Season 2018/2019 is from September 3 to January 1-2 for the 1st and January3 planting to March 1-2 for the 2nd planting.

In general, the first planting time of the earliest MH was in West Kalimantan Province on Sep 3-Oct 1, followed by Central Kalimantan, South Kalimantan, East Kalimantan and North Kalimantan on January 1-2. While the earliest second plantings in the Provinces of West Kalimantan and North Kalimantan are January 3-February 1, followed by the Provinces of Central Kalimantan, South and East (Table 3). Dominan, planting in Kalimantan Island is only once a year since most of the island is land that relies exclusively on water from rainfall. Therefore, most farmers in Kalimantan plant local rice that is long-lived (8 months).

The verification of information results on planting time from 16 provinces show that planting during the Dry Season (DS) is better than during the Wet Season (WS) and generally the actual planting time is forward/backward 3 dasarians (1 month) (Yulianti et al., 2016). If the planting time at WS reverses 3 dasarians from ICC recommendations, it is safe because the recommended initial planting is to plant at that time, but if advanced 3 dasarians, it is not recommended because water/rainfall is not enough for rice plants. Yulianti et al., 2016, showed the results of validation that 7 from 8 provinces had higher productivity when applying ICC informat her higher productivity when applying ICC information such as planting time, fertilization and varieties compared to applying farmers' habits.

ICC information is a manifestation of climate change adaptation efforts which, if implemented, can increase national productivity and minimize risks due to extreme climate events such as floods higher productivity when applying ICC information such as plants. Yulianti et al., 2016, showed the results of validation that 7 from 8 provinces had higher productivity when applying ICC information such as planting time, fertilization and varieties compared to applying farmers' habits.

ICC information is a manifestation of climate change adaptation efforts which, if implemented, can increase national productivity and minimize risks due to extreme climate events such as floods and droughts and plant pest attacks. Kalimantan Island, which has swampy swamp land, tidal swamps and other extensive swamps is a potential

Table 2.   The results of ICC's validation in several province and districts.

Province	District	Paddy (ton/ha)		Maize (ton/ha)		Soybean (ton/ha)	
		Farmer	ICC	Farmer	ICC	Farmer	ICC
Nort Sumatera	Langkat	5.3	7.0	-	-	-	-
	Sergai	5.0	7.2	-	-	1.7	1.9
	Tapanuli Tengah	5.7	8.8	-	-	-	-
West Sulawesi	Mamuju	5.0	7.7	-	-	-	-
	Polewali Mandar	5.6	9.0	4	4.7	-	-
Central of Java	Batang	7.3	8.9	-	-	-	-
	Temanggung	6.6	7.4	-	-	-	-
	Kendal	5.8	5.9	-	-	-	-
Yogyakarta	Gunung Kidul	4.0	6.5	-	-	1.7	2.1
West Java	Kuningan	4.0	6.5	-	-	-	-
Central of Sulawesi	Konawe	4.0	6.5	-	-	-	-
Jambi	Kota Sungai Penuh	6.5	8.0	-	-	-	-
East Nusa Tenggara	Kupang	5.0	8.0	-	-	-	-
South Sumatera	Ogan Ilir	6.8	8.2	-	-	-	-
		5.6	7.7	-	-	-	-
South Sulawesi	Maros	6.2	7.4	-	-	-	-
	Pangkep	6.5	7.8	-	-	-	-

Source: Badan Penelitian dan Pengembangan Pertanian 2018

land for food supply. The results of Pramudia's analysis (2018) show that the potential for a tidal swamp planting area occurs mostly in WS, but the wider swamp planting area during the DS. In WS 2018/2019, the potential for planting on Kalimantan was Oct 2-3 to Jan 3-Feb 1 (284,000 ha) and fresh water swampland (lebak) 0 ha, while in DS 2019, the potential for planting fresh water swampland (lebak) was May 1-2 until July 3-August 1 (49,478 ha) and tidal swamp (6,645 ha). On tidal swamps and others in the peak rainy season planting occurs on Jan 1-3, while in the dry season on May 3-Jun 1 (Table 4).

Fresh water swampland is flooded throughout the rainy season so it cannot be planted. From table 4, it can be seen that fresh water swampland can be utilized when rainfall has decreased in the dry season. There is an additional planting area of around 49,478 ha when fresh water swampland can be planted in the dry season. Information on the amount of rainfall and its prediction of at least one season ahead is very important to be able to obtain an overview of whether there is an additional planting area or not and how wide it is.

Oldeman's climate type distribution illustrates the availability of water through rainfall which can be used to plant food crops based on the amount of rain and time period so that users can adjust the time and cropping pattern. Trends in floods and droughts provide a warning against the tendency for floods or droughts at the district level while ICC provides more detailed information on the use of climate resources for determining the start of planting and other recommendations related to crop cultivation

such as fertilizer, varieties and agricultural mechanization.

This climate information can be utilized by adjusting location and information needs, especially on Kalimantan Island, which is mostly dominated by non-irrigated paddy fields relying heavily on water availability through rainfall. Patterns and characteristics and magnitude of rainfall can be studied and delivered to users through climate data, therefore all impacts and losses can expect to be minimized.

Another important section of climate information is the prediction of rain. Different rain prediction methods have been developed to ensure the objectives are achieved (Estiningtyas et al., 2008). The use of global indices such as Nino 3.4 sea surface temperature as predictors of rainfall also further strengthens the interaction of the atmosphere with the occurrence of rainfall on the earth's surface (Estiningtyas, 2007). Global interactions and processes in the atmosphere are reflected in the phenomenon of climate change where the impact is significant on the agricultural sector. The results of the Estiningtyas and Muhammad (2017) shows that climate change is indicated by changes in air temperature and CO_2 concentration to a certain extent, increasing rice yields in rainfed areas during certain months, and the remainder has an impact on reducing production. To minimize the impacts and losses due to climate change, an early warning system has been developed. Some early warning information that can be utilized include early warning systems for flood, drought and developments in predicting El-Nino and La-Nina cycles. The seriousness of the Government, especially the Ministry of Agriculture, is exhibited

Table 3. Potential rice planting time by province in Kalimantan Island based on the Integrated Cropping Calendar (ICC).

Province	Adm Index	Vast of Rice Paddy Field (Ha)	Potential Rice Corp							
			West Season 2018/2019				Dry Season 2019			
			(OCTOBER 2018 - MARCH 2019)				(APRIL - SEPTEMBER 2019)			
			First Crop		Second Crop		First Crop		Second Crop	
			Dominant Start Planting	Area (Ha)	Dominant Start Planting	Area (Ha)	Dominant Start Planting	Area (Ha)	Dominant Start Planting	Area (Ha)
West Kalimantan	61	326,684	SEP III-OKT I (144,368 ha). DES II-III (84,068 ha). OKT II-III (55,767 ha)	300,969	JAN III-FEB I (144,736 ha). FEB II-III (56,408 ha). MAR I-II (16,855 ha).	0	APR II-III (84,305 ha)	84,305	-	0
Central Kalimantan	62	215,545	NOV I-II (159,946 ha). OKT II-III (3,302 ha). SEP III-OKT I (53 ha)	163,301	MAR I-II (121,637 ha). Feb II-III (1,024 ha). JAN III-FEB I (9 ha)	0	MAR III-APR I (40,087 ha)	0	-	0
South Kalimantan	63	431,437	NOV I-II (289,789 ha). NOV III-DES I (39,923 ha). OKT II-III (783 ha)	330,495	MAR I-II (180,902 ha). FEB II-III (652 ha)	0	MAR III-APR I (40,087 ha)	40,087	-	0
East Kalimantan	64	55,485	NOV I-II (31,515 ha). OKT II-III (8,242 ha). SEP III-OKT I (6,454 ha). OTHERS (1,905 ha)	48,116	MAR I-II (8,319 ha). FEB II-III (2,843 ha). JAN III-FEB I (1,659 ha)	0	APR II-III (438 ha). MEI I-II (123 ha)	561	-	0
North Kalimantan	65	29,558	JAN I-II (13,014 ha). SEP III-OKT I (10,704 ha). DES II-III (2,198 ha). OTHERS (780 ha)	26,696	JAN III-FEB I (2,714 ha). MAR I-II (197 ha)	0	MEI I-II (4,024 ha). APR II-III (558 ha)	4,582	-	0
Total		1,058,709		869,577		0		129,535		0

Sumber: Badan Penelitian dan Pengembangan Pertanian 2018

167

Table 4. Potential rice planting area per province in Kalimantan Island swamps.

POTENCY OF RICE PLANTING AREA IN TIDAL SWAMP AND OTHERS

WET SEASON 2018/2019

PROVINCE	SEP III-OKT I	OKT II-III	NOV I-II	NOV III-DES I	DES II-III	JAN I-III	JAN III-FEB I	FEB II-III	MAR I-II	TOTAL
Kalimantan Barat	-	14,186	-	24,401	13,318	657	-	-	-	52,561
Kalimantan Tengah	-	-	-	-	-	57,009	-	-	-	57,009
Kalimantan Selatan	-	-	-	-	-	151,781	15,043	-	-	166,824
Kalimantan Timur	-	15	-	615	84	1,425	-	-	-	2139
Kalimantan Utara	-	60	-	3879	-	1492	-	-	-	5481
TOTAL	-	14,261	-	28,895	-	212,364	15,043	-	-	284,014

POTENCY OF RICE PLANTING AREA IN TIDAL SWAMP AND OTHERS

DRY SEASON 2019

PROVINCE	MAR III- APR I	APR II-III	MEI I-II	MEI III-JUN I	JUN II-III	JUL I-II	JUL III-AGS I	AGS II-III	SEP I-II	TOTAL
Kalimantan Barat	-	1,287	-	1,980	-	63	-	-	-	3924
Kalimantan Tengah	-	-	-	-	-	332	-	-	-	332
Kalimantan Selatan	418	-	-	-	-	116	-	-	-	534
Kalimantan Timur	-	15	-	396	22	100	-	-	-	533
Kalimantan Utara	-	15	-	495	-	99	-	-	-	1322
TOTAL	418	1,317	-	2,871	-	710	-	-	-	6,645

(Continued)

Table 4. *(continued)*

POTENCY OF RICE PLANTING AREA IN FRESH WATER SWAMPLAND (LEBAK) AND OTHERS WET SEASON 2018/2019

PROVINCE	SEP III- OKT I	OKT II- III	NOV I-II	NOV III- DES I	DES II- III	JAN I- III	JAN III- FEB I	FEB II- III	MAR I- II	TOTAL
Kalimantan Barat	-	-	-	-	-	-	-	-	-	-
Kalimantan Tengah	-	-	-	-	-	-	-	-	-	-
Kalimantan Selatan	-	-	-	-	-	-	-	-	-	-
Kalimantan Timur	-	-	-	-	-	-	-	-	-	-
Kalimantan Utara	-	-	-	-	-	-	-	-	-	-
TOTAL	-	-	-	-	-	-	-	-	-	-

POTENCY OF RICE PLANTING AREA IN FRESH WATER SWAMPLAND (LEBAK) AND OTHERS DRY SEASON 2019

PROVINCE	MAR III- APR I	APR II- III	MEI I- II	MEI III- JUN I	JUN II- III	JUL I- II	JUL III- AGS I	AGS II- III	SEP I-II	TOTAL
Kalimantan Barat	-	-	-	-	-	-	-	-	-	3,515
Kalimantan Tengah	-	-	750	-	670	2,095	-	-	-	-
Kalimantan Selatan	-	-	25,681	-	17,792	-	708	-	-	45,181
Kalimantan Timur	-	-	-	-	782	-	-	-	-	782
Kalimantan Utara	-	-	-	-	-	-	-	-	-	-
TOTAL	-	-	27,431	-	-	2,095	708	-	-	49,478

by the compilation of the Minister of Agriculture Regulation of the Republic of Indonesia Number 39/PERMENTAN/HM.130/8/2018 concerning the Early Warning System and Handling the Impact of Climate Change in the Agriculture Sector.

To facilitate access to climate information, a reporting system is prepared that can be connected to the internet, therefore, users can get relatively fast and easy access to the information. For example, data about the Integrated Planting Calendar can be accessed at katam.litbang.pertanian.go.id. Due to today's rapid technological advancements, it is increasingly easier for users to access various climate-related data and information. The development of food crops in wetlands can be optimized by utilizing climate information so the development area is in accordance with available resources and minimizes various risks and impacts that are likely to occur.

4 CONCLUSIONS

Kalimantan Island is dominated by Climate Type B1 (30.4%) from 56 regencies/cities, and the distribution of other climate types is type A (19.6%), C2 (12.5%), D1 (10.7%), C1 (8.9%), E1 (5.4%), E2 (3.6%), B2, C3, D2 and D3 1.8% each. The drier regions of the climate are mostly located in East Kalimantan Province and only a small area in North Kalimantan. Wetter areas are mostly in West Kalimantan and North Kalimantan bordering Malaysia and parts of Central Kalimantan.

There are still several districts/cities on the island of Kalimantan that have a rising trend of floods or drought. Districts that have a flood trend are increasing. Those districts are Singkawang City, East Kotawaringin Regency, South Barito, North Barito, Seruyan, Tanah Bumbu, Balangan, Paser, Penajam Paser Utara and Mahakam Hulu. The districts where the trend of drought increased are the City of Palangkaraya and Kota Banjarbaru, Kabupaten Lamandau, Pulang Pisau, Tanah Bumbu, Balangan, East Kutai and Penajam Paser Utara. However, there are 3 districts that have an upward trend in both floods and droughts: Tanah Bumbu and Balangan Regencies in South Kalimantan Province and North Paser Penajem Regency in East Kalimantan Province.

Based on the results of the Katam analysis, the first planting time at WS 2018/2019 can be started from September 3 to January 1-2 and the second time on January 3 until March 1-2.

ACKNOWLEDGEMENTS

Thank you to the Head of the Agricultural Research and Development Agency for providing funding for this research through the KP4S mechanism. Our thanks also go to Atika Gustini and Dariin Firda from Indonesian Agriculture and Hydrology Research Institute who have helped in the preparation of data and mapping in this study.

REFERENCES

Allan R. 2000. ENSO and climatic variability in the past 150 year. In Diaz, H. & V., Markgraf (eds) *ENSO: Multiscale Variability and Global and Regional Impacts: pp. 3-55.* Cambridge: Cambridge Univ. Press.

Badan Penelitian dan Pengembangan Pertanian. 2018. Kalender Tanam Terpadu, Viewed 6 November 2018, http://katam.litbang.pertanian.go.id/main.aspx

Betts, R. 2005. Integrated approaches to climate–crop modelling: needs and challenges. *Phil. Trans. R. Soc. B* 360: 2049–2065. (doi:10.1098/rstb.2005.1739.)

Boer, R. 2017. *Implementasi NDC dan Integrasi Adaptasi dan Mitigasi Menuju Kedaulatan Pangan. Bahan tayang Lokakarya dan Seminar Nasional Adaptasi dan Mitigasi Perubahan Iklim. Bogor 13-14 September 2017.*

Boer, R., & Subbiah. 2005. Agriculture drought in Indonesia. In Boken, V.J., A.P., Cracknell & R.L., Heathcote (eds). *Monitoring and predicting agriculture drought: A global study:* p:330-344. New York: Oxford University Press.

Budianto, J. 2002. Tantangan dan peluang penelitian padi dalam perspektif agribisnis. Dalam: B. Suprihatno et al. (Eds.). *Kebijakan perberasan dan inovasi teknologi. Puslitbang Tanaman Pangan. Bogor.* p. 1-17.

Cline, W.R. 2007. *Global Warming and Agriculture: Impact Estimates by Country (Washington: Center for Global Development and Peterson Institute for International Economics).*

De Silva M.M.G.T. & Kawasaki A. 2018. Socioeconomic vulnerability to disaster risk: a case study of flood and drought impact in a Rural Sri Lankan Community. *Ecological Economics* 152: 131-140.

Direktorat Perlindungan Tanaman Pangan. 2017. *Data luas sawah terkenan banjir dan kekeringan. Direktorat Perlindungan Tanaman, Direktur Jenderal Tanaman Pangan, Kementerian Pertanian.*

Dukung Rawa untuk Lahan Sawah Pemerintah Susun Perpres 2018, Republika, Viewed 6 November 2018, http://republika.co.id/r/phrf3n453

Estiningtyas, W., Elza, S. & Kharmila, S.H. 2008. Penyusunan skenario masa tanam berdasarkan prakiraan curah hujan di sentra produksi pangan. *Jurnal Meteorologi dan Geofisika* 9(1).

Estiningtyas, W. 2007. Pengaruh tenggang waktu (time lag) antara curah hujan dengan suhu permukaan laut nino 3.4 terhadap performa model prediksi hujan. *Jurnal Meteorologi dan Geofisika* 8(1).

Estiningtyas, W. & Muhammad, S. 2017. Pengaruh perubahan iklim terhadap produksi padi di lahan tadah hujan. *Jurnal Meteorologi dan Geofisika* 18(2).

FAO. 2005. Impact of Climate Change and Diseases on Food Security and Proverty Reduction. *Special event background document for the 31st session of the committee on world food security. Rome, 23-26 May 2005.*

Fischer, G., Shah, M., Velthuizen, H.V. 2002. *Climate Change and Agricultural Vulnerability.* Luxemburg, Austria: IIASA.

Gutman, G.I., Csiszar & Romanov, P. 2000. Using NOAA/AVHRR products to monitor El Niño impacts: focus on

Indonesia in 1997-98. *Bull. Amer. Meteor. Soc.*, 81: 1189–1205.

Handoko I, Sugiarto Y, Syaukat Y. 2008. Keterkaitan Perubahan Iklim dan Produksi Pangan Strategis. Telaah kebijakan independen bidang perdagangan dan pembangunan oleh Kemitraan/Partnership Indonesia. *SEAMEO BIOTROP.* Bogor.

Hamada, J., Yamanaka, M.D., Matsumoto, J., Fukao, S., Winarso, P.A. & Sribimawati, T. 2002. Spatial and temporal variations of the rainy season over Indonesia and their link to ENSO. *J Meteorol Soc Jpn* 80:285–310.

Haylock, M. & McBride, J. 2001. Spatial coherence and predictability of Indonesian wet season rainfall. *J. Climate* 14" 3882-3887.

[IPCC] Intergovernmental Panel on Climate Change. 2001. *Climate Change 2001: The Scientific Basis*, eds Houghton JT, Ding Y, Griggs DJ, Noguer M, van der Linden PJ, Dai X, Maskell K, Johnson CA (Cambridge Univ Press, Cambridge, UK.

[IPCC] Intergovernmental Panel on Climate Change. 2007. *Contribution of Working Groups I, II and III to the Fourth Assessment Report of the Intergovernmental Panel on Climate Change Core Writing Team*, Pachauri, R.K. and Reisinger, A. (Eds.) IPCC, Geneva, Switzerland. pp 104.

Koesmaryono, Y., Las, I., Aldrian, E., Runtunuwu, E., Syahbuddin, H., Apriyana, Y., Ramadhani, F. & Trinugroho, W. 2018. Sensitivity and dynamics of paddy planting calendars against parameters of ENSO (El-Nino Southern Oscillation) and IOD (Indian Ocean Dipole) in the monsoon and equatorial areas. Project Report. Agricultural Research and Development Agency.

Modathir, A.H., Zaroug, M.B., Sylla, F., Giorgi, Elfatih, A. B., Eltahir, Pradeep, K. & Aggarwal. 2013. A sensitivity study on the role of the swamps of southern Sudan in the summer climate of North Africa using a regional climate model. *Theor Appl Climatol* 113:63.

Ndamani, F. & Watanabe T. 2017. Developing indicators for adaptation decision-making under climate change in agriculture: a proposed evaluation model. *Ecological Indicators* 76: 366-375.

Ncuc, H.U., Gaunt, J.L., Wang, Z.P., Becker-Heidmann. P. & Quijano, C. 1997. Carbon in tropical wetlands. *Geoderma* 79: 163–185.

Nursyamsi, D. 2018. *Tri Kelola Plus for Increasing Production and Productivity in Wetland Development in Agriculture.* Presentation materials in the International Workshop on Tropical Wetlands "Innovation in Mapping and Management for Sustainable Agriculture. Banjarmasin, South Kalimantan, Indonesia 19-21 October 2018.

Oldeman, L.R., Irsal & Muladi. 1980. *Contribution: The Agroclimatic Maps of Kalimantan, Maluku, Irian Jaya and Bali, West and East Nusa Tenggara.* Bogor: Central Research Institute for Agriculture.

Osborne, T.M. 2005. *Towards an integrated approach to simulating crop–climate inter-actions.* Ph.D. thesis, University of Reading.

Porter, J.R. & Semenov, M.A. 2005. Crop responses to climatic variation. *Phil. Trans. R. Soc.* B 360, 2021-2035. (doi:10.1098/rstb.2005.1752).

Pramudia, A. 2018. *Sistem Informasi Kalender Tanam Terpadu Katam Sawah dan Katam Rawa MH 2018/2019 disampaikan pada Bimtek Pertanian Modern untuk Pengembangan Lahan rawa Menjadi Lumbung Pangan.* Banjarbaru, 18-19 October 2018. Badan Penelitian dan Pengembangan Pertanian, Kementerian Pertanian.

Rudgard, S. 2018. HPS Kalsel, FAO Puji Terobosan Baru dari Kementan. URL: https://www.jpnn.com/news/hps-kalsel-fao-puji-terobosan-baru-dari-kementan

Slingo, J.M., Challinor, A.J., Hoskins, B.J. & Wheeler, T. R. 2005. Introduction: food crops in a changing climate. *Phil. Trans. R. Soc.* B 360: 1983-1989. (doi:10.1098/rstb.2005.1755)

Surmaini, E. 2016. Pemantauan dan peringatan dini kekeringan pertanian di Indonesia. *Jurnal Sumberdaya Lahan* 10(1): 37-50.

Surmaini, E. & Syahbuddin, H. 2016. Kriteria Awal Musim Tanam: Tinjauan Prediksi Waktu Tanam Padi di Indonesia. *J. Litbang Pert.* 35(2): 47-56.

Syahbuddin, H., Surmaini, E. & Estiningtyas, W. 2015. *Pengembangan Pertanian Berbasis Ekoregion dari Perspektif Keragaman Iklim.* Buku Pembangunan Pertanian Berbasis Ekoregion. IAARD Press.

Yang, C., Yu, Z., Hao, Z., Zhang, J. & Zhu, J. 2012. Impact of climate change on flood and drought events in Huaihe River Basin, China. *Hydrology Research* 43.1-2: 1-9. [https://iwaponline.com/hr/article-pdf/43/1-2/14/371254/14.pdf]. [6 November 2018].

Yulianti, A., Sirnawanti, E. & Ulpah, A. 2016. Introduction Technology of Cropping Calendar-Information System (CC-IS) for Rice Farming as A Climate Change Adaptation in Indonesia. *International Journal on Advance Science Engineering Information Technology* 6(1). ISSN: 2088-5334.

Identification and characterization of chilli and shallot developement regions based on soil, climate, and water resources in Tapin District, South Kalimantan

P. Rejekiningrum & S. Ritung

Indonesian Center for Agricultural Land Resources Research and Development, Bogor, Indonesia

ABSTRACT: The productivity of chilli and shallot in Indonesia is low, but the country has great opportunities to increase this either through intensification or extensification. Therefore, the objectives of this research were to compile and develop land resources data including climate and hydrology information, compose maps, and build recommendations on land, climate, and water management for chilli and shallot development regions. The study site was located at Hiyung Village of Central Tapin District characterized with swamps and water managed with bedded and furrow system (surjan) at 1 m height to avoid 0.5-1 m flood while the potential cultivation area was found in the upland of Sukamaja Village, Binuang District. Moreover, the region developed for chilli covered 185 ha in Central Tapin Sub District (BE.3) and the only form of irrigation was rainwater. The productivity in this sub-district increased with a low value of approximately 12-15 tons/ha. In addition, a potential a total area of 50,327 h for extensification with a medium development potential (P.2) was distributed throughout the sub-districts in Tapin District and the largest area for cultivation in the North Candi Laras Sub District covered 33,276 ha. Furthermore, the development region of shallots was found to be covering approximately 169 ha at Bungur and South Tapin Sub District with productivity reaching around 8-10 tons/ha, which means the increase in productivity was high. In Harapanmasa Village of South Tapin Sub District, shallot was cultivated 1-3 times after rice cultivation on the paddy fields and during the dry season water irrigation was manually supplied from a shallow dipped well of 2-3 m while the area is usually flooded during the rainy season. At the same time, the plant is widely developed in the paddy fields of the Shaba village in Bungur Sub District during the dry season using the same shallow dipped well water irrigation.

1 INTRODUCTION

Indonesia is an agricultural country with several natural resources and many choices of commodities developed to be dispersed and managed on a small scale. On the contrary, these commodities are partially exploited, and mono-commodity is the main practice, thereby, making farming inefficient, improvement difficult, and resource utilization less optimal with very little impact on the welfare of farmers. Therefore, the Ministry of Agriculture modified the agricultural development management policy by introducing a ministerial regulation (Permentan), Nr. 50, 2012, concerning **Guidelines for Agriculture Regions development** and revised to be Permentan Nr. 56, 2016.

The intention was to conduct holistic and integrated agricultural development activities on a focused location with attention on the commodity at an economical scale. It was projected to make the use of resources more optimal, more efficient funding, and expected to have a more significant impact on increasing production, productivity, and competitiveness of commodities to increase the welfare of farmers.

However, the implementation of these policies through a regional approach requires a spatial-based land resources data and information on a more detailed scale (1:50,000) to develop appropriate recommendations and directives. On this scale, land resource data provides information on potential, biophysical constraints, area, and distribution in a region for the development of agricultural commodities such as chilli and shallots. This is important because land management recommendations are needed to support commodity development in each region.

Furthermore, chilli and shallot are one of the strategic horticultural commodities in the national economy, and their high consumption rate must be counterbalanced with availability. Therefore, increasing their production is mandatory to minimize price volatility and control inflation of food commodities. According to the data obtained from Indonesian Statistic Agency (BPS), in 2016, the country was able to produce up to 1,045,587 tons of large chilli, cayenne pepper at 915,988 tons and shallots up to 1,446,860 tons.

The average productivity of large chilli was found to be 8.47 tons/ha, cayenne pepper was 6.7 tons/ha, and shallot was 9.67 tons/ha. This is, however, lower than the expected potential of 12-20 tons/ha for chilli and above 20 tons/ha for shallot.

This led to several efforts made by the country to increase productivity through both intensification and extensification, as observed in the use of region-based development. This is expected to cover every aspect of agribusiness system starting from the upstream to downstream.

Therefore, the objectives of this study were to compile and develop land resources data and information, production potential and institutions, compose maps 1: 50,000 scale, and build recommendations on land, climate, and water management for chilli and shallot development regions.

2 MATERIALS AND METHOD

Spatial identification and characterization of chilli and shallot development region were conducted at 1: 50,000 map scale. Desk work was used to analyze the available data, followed by field verification. The direction for the development was first based on the land suitability classes followed by existing land use and commodities production centers while the final consideration was the status of the forest region. However, the area should be a non-forest area (APL). The direction for developing the chilli and shallot commodity region was presented as tabular and spatial/map by district/city.

2.1 Time and place

The preparation of the 1:50,000 scale of chilli and shallot development region map in Tapin District, South Kalimantan Province was conducted in the 2017 fiscal year with most of the analysis conducted using desk work, and subsequently followed by field verification.

2.2 Materials and equipment

The materials used include Semi-detailed Soil Map at 1:50,000 scale of Tapin District updated version (BBSDLP, 2016), Topographical Map of Indonesia (RBI) at 1:50,000 scale (Bakosurtanal), District/city Administrative Boundaries Map (BPS, 2013), Map of Forest Area Status (KLHK, 2013), Indonesian Rice Field Map (Ministry of Agriculture, 2013), Land Use Map (BPN, 2012), Land Tenure Map (Ministry of Land Management and Spatial Planning/ATR, 2015), Tapin District land characteristics (LC) data, and Land Suitability Criteria for Agricultural Commodities (Ritung *et al.*, 2011).

2.3 Method

The chilli and shallot development region map at 1:50,000 map scale was divided into 7 stages of activities, and they include (1) preparation, (2) preparation of land evaluation units, (3) land evaluation analysis for land suitability maps, (4) map preparation for chilli and shallot region (draft), (5) field verification, (6) preparation of maps of chilli and shallot region (final), and (7) reporting. The flow chart is as presented in Figure 1.

2.3.1 Preparation

The preparation phase covered the compilation of materials required for analysis and they include Indonesian Topographical Map of Indonesia (RBI), Administrative Boundary Maps (BPS), forest status maps from KLHK, rice field map, Ministry of Agriculture, land use maps from BPN, Land Tenure Map from the ATR Ministry, and land suitability criteria for chilli and shallot.

Also, a literature study was also conducted to identify the conditions of the region (district/city), including climate, chilli, and shallot farming, statistical data on existing planting area, productivity and others.

2.3.2 Preparation of land evaluation unit

This is a unit of land used for suitability assessment, and it consists of characteristics such as soil, climate, and topography, and the unit number refers to the number in the land unit map. Furthermore, the soil characteristics data observed include drainage, texture, coarse material, soil depth, peat thickness and ripening, soil CEC, base saturation, pH, H_2O, C, organic content, N total, P_2O_5 and K_2O extract, HCl 25%, salinity, sulfidic depth, height and length of inundation, outcrops and rocks on the surface; climatic characteristics such as air temperature, annual rainfall, wet, and dry months (Oldeman et al., 1978; Schmidt and Fergusson, 1951; Puslitbangtanak, 2003), air humidity; and environmental characteristics such as elevation, and slope steepness.

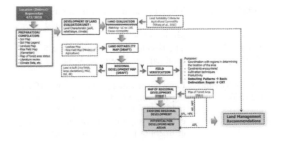

Figure 1. Flow chart of procedure in composing development region map for chilli and shallot.

2.3.3 *Land evaluation and preparation of land suitability map*

Land evaluation is the assessment of land-based on its performance when used for certain purposes (FAO, 1976). It was conducted for chilli and shallot by matching land quality/characteristics with use requirements. It also produced land suitability classes and limiting factors used as the basis to prepare land management recommendations. However, the suitability assessment criteria for the agricultural commodities was conducted according to Ritung *et al.* (2011).

Furthermore, the land suitability map was evaluated up to the sub-class level where each class was differentiated based on the lowest limiting factors following the Minimum Liebig Law. However, according to FAO (1976), land suitability classification hierarchy follows the pattern of Order-Class–Sub Class-Unit.

The Order describes the level of land suitability in general as presented by S for suitable and N for not suitable. Class is another level under the Order, and the difference is represented by suitable (S1), moderately suitable (S2), and marginally suitable (S3). However, denoting 'not suitable' with N is the same for all classes. Table 1 shows the distribution of land suitability classes. Moreover, sub-class is found under the Class based on the quality and characteristics of the land, and it has been observed to be the heaviest limiting factor. The next division is the Unit, which is another level in the sub-class based on additional characteristics of influencing management. However, it is important to state that all units in one sub-class have the same level in the class and have the same type of boundary at the sub-class level. One unit is different from the others in the additional properties or aspects of management required, and it is a detailed differentiator of the limiting factors.

Furthermore, Table 2 shows the quality and characteristics of the land used for the assessment and 13 quality and 25 characteristics were observed.

The results obtained form the draft of land suitability maps representing only the physical class. Furthermore, the draft was overlaid and updated with Rice Field Map (2013) and the Land Use Map of BPN (2012) and was used to prepare the agricultural region development maps.

On the contrary, the land suitability analysis conducted in South Africa found dynamic evaluation guidelines can assess the suitability of crops in specific ranges and the information required to make land use decisions (Nethononda *et al.*, 2014).

Table 1. Distribution of land suitability classes.

Land Suitability Classes	Limiting factor
Suitable (S1)	The land has minor or no limiting factor with no significant effect on land productivity.
Moderately Suitable (S2)	The land has a limiting factor affecting productivity. It requires input but can be overcome by farmers.
Marginally Suitable (S3)	Land has heavy limiting factors affecting its productivity, requiring more input from the S2 class. Therefore, high capital is required mostly through government or private assistance or intervention.
Not Suitable (N)	This has a very heavy limiting factor, which is difficult to overcome.

Source: FAO (1976); Djaenuddin *et al.* (2003)

Table 2. Land quality and characteristics in land suitability assessment.

No	Land Quality	Land Characteristic	Data Source
1.	Temperature (tc)	Mean Annual Temperature (°C)	Local climate station/secondary data (BMKG)
2.	Water Availability (wa)	Rainfall (mm) Humidity (%)	Local climate station/secondary data (BMKG)
3.	Oxigen Availability (oa)	Drainage	Field observation
4.	Rooting media (rc)	Texture, coarse material (%), effective depth (cm), peat ripening, and peat thickness (cm)	Field observation
5.	Nutrient retention (nr)	Soil CEC (me/100 g), base Saturation (%), soil pH, and C organic (%)	Lab analysis (specifically pH was also conducted in the field)
6.	Nutrients availability (na)	N total (%), P_2O_5 (mg/100 g), K_2O (mg/100 g)	Laboratory analysis
7.	Toxicity (xc)	Salinity (mmhos/cm)	Laboratory analysis
8.	Sodicity (xn)	Alkalinity (%)	Calculation
9.	Sulfidic hazard (xs)	Sulfidic depth (cm)	Field observation
10.	Erosion hazard level (eh)	Slope (%), Erosion hazard (cm/year)	Field observation, Calculation
11.	Landslide hazard	Slope (%), and landslide hazard	Field observation
12.	Flooding hazard/inundation (fh)	Inundation (cm/month)	Field observation
13.	Land preparation (lp)	Surface rock (%), rock outcrop (%)	Field observation

Source: FAO (1976); Djaenudin *et al.* (2003), Ritung *et al.* (2011).

2.3.4 Preparation of chilli and shallot region development maps

The maps for the development of chilli and shallot region were prepared based on the results of land suitability assessment which include suitable, moderately suitable, and marginally suitable classes. The suitable class is located on APL land including upland or rainfed lowland with rice planting index of 1 planting/year (IP-100) and in the commodities production center. They have the potential for chilli and shallot development and are considered for expansion or extensification.

A chilli and shallot production center is only considered a development region if it has a minimum area of 50 ha and interconnected to another according to the established rules. Moreover, these regions are differentiated based on productivity improvements that might be achieved from existing conditions to optimal productivity of each commodity. Therefore, three classes were distinguished for shallot and large red chilli in Table 3 and cayenne pepper in Table 4.

However, the potential land for extensification of chilli and shallot was divided into 2 classes, including high development potential (P1) for suitable land (S1), and moderate development potential (P2) for moderately suitable (S2) and marginally suitable (S3) and when the limiting factors such as nutrient retention, nutrient and water availability are relatively easy to be improved.

2.3.5 Field verification

Field verification of map draft was prioritized on the development region, commodity production centers and the land's potential for development. This was conducted to assess the accuracy of the draft covering (i) chilli and shallot commodity centers, (ii) crop performance in the field, (iii) land use, (iv) land management, as well as people's preferences for the commodity.

(i) Plant Performance

Verification of crop accuracy was conducted by comparing the actual land suitability class with the performance of plants in the field. This is important because there is often a difference between the land suitability class results obtained from desk work and plant performance in the field. This is due to the lack of information on improvements on land quality such as the application of fertilizer and dolomite, drainage channels improvement, terrace construction, and others.

(ii) Land Use

High demands on land have led to variation in how it is used, however, this is not usually followed by adequate updating of data and information. This often leads to overestimation, especially on the extent of potential unutilized land. Similarly, existing agricultural land is being used for non-agricultural purposes. The lack of information on changes in land use leads to low accuracy of land suitability maps and development of compiled agricultural areas. Therefore, field verification is expected to improve map accuracy.

(iii) Land Management

In addition to verification of plant performance and land use, it is necessary to conduct discussions with policymakers in the regions, including the Department of Agriculture, to understand people's preferences on of chilli and shallot, their development patterns with other strategic commodities as well as the collection of supporting data.

Table 3. Class of increased productivity for shallot and large red chilli.

Productivity increase	Optimum productivity	Yield Gap
	ton/ha/season	
Low	20	< 3,8
Medium	20	3,8-7,8
High	20	> 7,8

Table 4. Class of increased productivity for cayenne pepper.

Productivity increase	Optimum productivity	Yield Gap
	ton/ha/season	
Low	15	< 2,9
Medium	15	2,9-5,9
High	15	> 5,9

3 RESULTS AND DISCUSSION

3.1 Land resources characteristics

3.1.1 Location characteristics

Tapin District is a district in South Kalimantan Province geographically located at 2 ° 32′43 "- 3 ° 00′43" S, and 114 ° 46′13 "- 115 ° 30′33 "E with an area of 217,495 hectares (BPS Tapin District, 2015). It is adjacent to South Hulu Sungai District in the north, Banjar District in the south, South Hulu Sungai District to the east, and to the west is Barito Kuala District. Most of its areas are lowland with an elevation of 0-7 m above sea level with only 1.2% of the district having an altitude >500 m (Figure 2).

Administratively, the district is divided into 12 sub-districts with 135 villages and the largest sub-district is North Laras Temple with an area of 681.40 km² which is 31.33% of the whole district while the smallest is North Tapin with an area of 32.34 km².

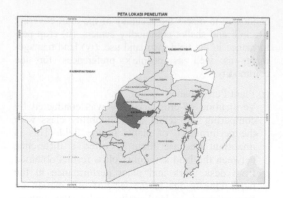

Figure 2. Map of location of Tapin District, South Kalimantan.

3.1.2 *Parent materials, landforms, and reliefs*

Based on Geology Map of Banjarmasin sheet (1713) and Amuntai (1712) (Center for Geological Research and Development, 1994) most of Tapin District consists of Alluvial Formation (Qa) which is composed of kaolinite clay deposits, dust with sand, peat, and porous inserts and loose chunks from river and swamp deposits. The parent materials found include fine alluvium deposits such as clay and dust, a little sand, and some in the form of marines and thin organic matter, sedimentary rocks, old volcanic rocks as well as rivers and swamps (Sikumbang and Heryanto, 1994).

Surface deposits consist of Alluvium (Qa) formation, which is classified as the youngest (quaternary) material originating from river and swamp, consisting of clay, sand, mud, silt and organic matters. It also spreads around floodplains, depression/basin or swampy swamps and along the Tapin River, Negara River, and other rivers. Furthermore, extensive distribution in the southern part of the research area includes Central and South Tapin Sub Districts (Rantau and Sungai Salak), North and South Candi Laras Sub District and Bakarangan Sub District.

Sedimentary rocks consist of limestone to quaternary rock formations. Tertiary-aged rocks are dominated by rocks and sandstones while quarterly rocks are dominated by quartz sand. It spreads on the plains to the hills and the mountains of Meratus and consists of Dahor, Warukin, Berai, Tanjung, and Pitap formations.

Furthermore, the Dahor Formation (Tqd) is composed of quartz sandstones found in transitional regions between the lowlands and hills and spreads to Binuang, Tatakan, Kalumpang, and South Tapin areas. The Warukni (Tmw) Formation is composed of sandstones found in undulating terrain to gentle slopes of Binuang District, Tatakan, South Tapin, Lokpaikat, and Puncak Harapan. The Berai (Tomb) Formation is composed of limestone inserts and claystone and these rocks form hills and karst plains in Salam Babaris and Binuang Sub Districts. Moreover, the Tanjung (Tet) Formation is an intermediary

between stone ponds, claystone, siltstone, limestone and conglomerates in hilly areas of Piani, Bungur, Mirawah and Padangbatung while the Pitap Formation (Ksp) is composed of flysch deposits in the form of spawning between rock-watering, claystone, siltstone, shale, flint, limestone, and basalt lava and distributed in hilly areas of Batuampar, Buniin Jaya, and Hangui.

Volcanic rocks in the form of a Haruyan (Kvh) formation are composed of molten basalt lava and distributed in the hilly areas and mountains of Hatungun, Piani, and G. Batupanggung. Moreover, breakthrough (intrusive) rock consists of granite rock combined with granodiorite and diorite and its distribution is not too wide in the eastern part of Tapin District, covering the sub-districts of Piani, G. Belayawan, and G. Kelatikan.

Based on the updated version of the Semi-Detail Land Map on a scale of 1:50,000 (BBSDLP, 2016), the Tapin District landform was divided into 6 landform groups, which are Alluvial, Fluvio-Marin, Peat, Karst, Tectonic, Vulcanic, and several others. Moreover, the reliefs were found to have varied from flat to mountainous with the 0-1% flat area found in alluvial and fluvio-marin land covering 146,686 ha in landforms. Furthermore, tectonic relief varied from flat, undulating slopes 3-8% covering an area of 12,836 ha, rolling slopes 8-15% covering an area of 13,228 ha, hillocky relief slopes 15-25% covering an area of 5,071 ha, and hilly slopes 25-40% covering an area of 16,013. Mountainous reliefs are found in old volcanic landforms and intrusions (slopes> 40%) covering an area of 10,980 ha.

3.1.3 *Soil condition*

The information obtained from the Semi-Detailed Soil Map of Tapin District of 1: 50,000 scale (BBSDLP, 2016) classified the soil, according to National Soil Classification (Subardja *et al.*, 2016), into 9 types and they include Organosol, Alluvial, Latosol, Cambisol, Gleisol, Nitosol, Podsolik, Mediteran, and Oxisol.

The peat consists of Organosol Saprik and Hemik which are equivalent to Terric Haplosaprists and Sapric Haplohemists according to Soil Taxonomy classification. Alluvial soils developed from the river sediments in the form of clay and silt and marine deposits in environmental conditions that are almost always inundated or much influenced by reduction and oxidation processes, which are classified into 3 types of soil, Alluvial Sulfidik, Alluvial Sulfik, and Alluvial Gleik. Moreover, latosol soil found developed from basaltic rock found in the hilly landforms and old volcanic mountains with only one soil type named Latosol Haplik. It is generally very deep, well-drained drainage, udic moisture regime (moist soil, never dry), colored dark brown or more red, hue 7.5 YR and chroma more than 4, or redder than 7.5 YR, fine texture, acidic soil reaction, soil CEC, and low base saturation.

Table 5. Suitability of chilli and shallot plants in Tapin District, South Kalimantan province.

No	District	Chilli			Shallot		
		S3	N	Total (Ha)	S3	N	Total (Ha)
1	Bakarangan	7.106		7.106	7.106		7.106
2	Binuang	10.607	4.371	14.978	10.607	4.371	14.978
3	Bungur	3.173	5.769	8.943	3.173	5.769	8.943
4	South Candi Laras	27.622		27.622	4.283	23.340	27.622
5	North Candi Laras	61.797		61.797	5.357	56.440	61.797
6	Hatungan	3.867	6.058	9.925	3.867	6.058	9.925
7	Lokpaikat	6.367	4.241	10.608	6.367	4.241	10.608
8	Piani	11.350	7.711	19.061	11.350	7.711	19.061
9	Salam Babaris	1.991	3.831	5.821	1.991	3.831	5.821
10	South Tapin	12.746	4.364	17.109	11.552	5.557	17.109
11	Central Tapin	30.086		30.086	28.710	1.375	30.086
12	North Tapin	2.829		2.829	2.829		2.829
Grand Total		**179.541**	**36.345**	**215.886**	**97.192**	**118.694**	**215.886**

Furthermore, cambisol soil types found have a low base saturation of <60%, acidic, moderately drained drainage, which is classified as Cambisol Distrik which is an equivalent of Typic Dystrudepts and Cambisol Gleik or Aquic Dystrudepts according to the Soil Taxonomy classification. Based on its characteristics, Gleisol observed is divided into 3 types and they include Gleisol Distrik, Gleisol Fluvik, and Gleisol Sulfik. Moreover, Nitosol found developed from acidic sedimentary rocks (claystone and sandstones) with distribution narrow in the tectonic plain and associated with the Podsolik Haplik dan Cambisol Distrik. However, only Nitosol Distrik was observed and it has a lower candic horizon, clay CEC <16 cmol (+)/kg clay, generally very deep, well-drained drainage, fine texture, acid, soil CEC, and very low KB. The Podsolik soil found decreased the Podsolik Haplik and Podsolik Kandik.

Mediteran is a type of soil developed further with the arrangement of the A-Bt-C horizon and characterized by the presence of a clay silicate illuviation horizon that meets the requirements of argillic or candic and has a high base saturation (>50%). The soil developed from limestone parent material in the environment of the udic soil moisture regime and spread narrowly to the karst hills and only two types were generated, Mediteran Litik and Mediteran Ortoksik which is an equivalent of Lithic Hapludalfs and Typic Kanhapludalfs.

Oxisol developed from volcanic material, which was found in volcanic landforms with slopes of 8-25% and generated Oksisol Kandik and Oksisol Haplik which are equivalent to Typic Kandiudox and Typic Hapludox according to the classification of Soil Taxonomy (Soil Survey Staff, 2014).

3.2 Potential of land resources

The map of chilli and shallot development region was prepared based on the suitability of land for the two commodities and the lands recommended for the development and expansion are classified as suitable, (S1), moderately suitable (S2), and marginally suitable (S3) if the limiting factors are relatively easy to be improved.

The results of the land suitability assessment are presented in Table 6 and the marginally suitable (S3) area for chilli was found to be 179,541 ha, while not suitable area (N) was 36,345 ha. The marginally suitable area spread across all sub-districts with the highest found in Candi Laras Utara Sub District to be 61,797 ha. However, the dominant limiting factor was low rainfall, poorly drained drainage, and acid to very acid pH with an area of 98,759 ha.

In addition, non-suitable lands for chili were found in seven sub-districts with the highest observed in Piani Sub District covering 7,711 ha and the dominant limiting factor was a slope of more than 15% with an area of 36,345 ha.

Furthermore, land suitability assessment for marginally suitable class (S3) for shallot has a total area of 97,192 ha and not suitable class (N) of 118,694 ha. The most marginal land area was observed in Central Tapin District with an area of 28,710 ha while the dominant limiting factor includes poorly drained drainage of depression swamp, and acid to very acid soil reaction (pH), with an area of 44,232 ha.

However, non-suitable lands for shallot were found almost in all sub-districts except in the Districts of Bakarangan and North Tapin. The highest was observed in North Candi Laras sub-district to be covering 56,440 ha while the dominant limiting factors were very poorly drained drainage, very acidic soil reaction (pH), and low rainfall covering an area of 82,473 ha.

3.3 Climate and water resources characteristics

Tapin District has tropical climates, most of which are swamp and paddy fields and the average dry months (<60 mm) of 3-4 months, and wet month (>100 mm) of 8-9 months. The data obtained from local stations on the rainfall from the last 4 years showed the average annual rainfall to be 1871-1927 mm with a monthly average of 156-161 mm. Therefore, the district is categorized as a wet climate and the highest monthly rainfall of 284 mm occurred in December at Central Tapin station and 251 mm at South Tapin as shown in Figures 3 and 4. The annual rainfall in Tapin District from 2013 to 2016 fluctuated with the highest at 2397 mm found in 2013 at Middle Tapin and the highest in South Tapin was 2104 mm in 2014 as shown in Figures 5 and 6.

According to the Schmidt and Ferguson climate types, this region is included in type C with 8-9 wet months and 3-4 dry months. The air temperature in

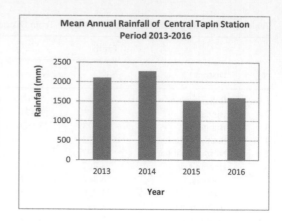

Figure 5. Mean annual rainfall of Central Tapin station from 2013 to 2016.

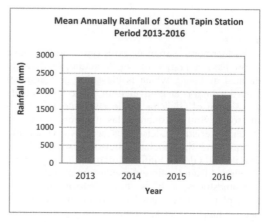

Figure 6. Mean annual rainfall of South Tapin station from 2013 to 2016.

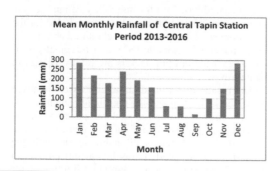

Figure 3. Mean monthly rainfall of Central Tapin station for the period of 2013-2016.

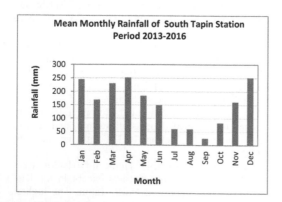

Figure 4. Mean monthly rainfall of South Tapin station for the period of 2013-2016.

the period of 2007-2016 ranged from 26.4°C to 27.3°C with the highest average temperature found in November and the lowest in January as shown in Figure 7. Furthermore, the shallot areas were found in South Tapin and Bungur. In the South Tapin Sub-district, shallot was discovered to be planted 1-3 times after rice on the paddy fields in Harapanmasa Village, using water sources from the 2-3 m depth wells dipped in the dry season and the area is usually flooded in the rainy season. However, in Bungur, the plant is developed in the Shaba village rice fields in the dry season, and the source of water was also from the same wells.

The potential of shallot was found on the upland of Asam Randah Village, Hatungan District and the water sources rely solely on rainwater. Moreover, the chilli area was found to be located in Central Tapin District in Hiyung Village, developed on the

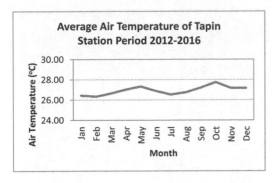

Figure 7. The average air temperature fluctuation of Tapin District.

Table 7. Legend of chilli development region map of Tapin District, South Kalimantan Province.

Symbol	Description	Area	
		Ha	%
Development region			
BE.3	Low productivity increase	185	0,37
Subtotal		*185*	*0,37*
Development potential			
P.2	Medium development potential	50.327	99,63
Subtotal		*50.327*	*99,63*
Total		**50.512**	**100,00**

swampy swampland, managing water with a 1 m surjan system to avoid puddles of 0.5-1 m. While the potential for chilli was found in the upland fields of Sukamaja Village, Binuang Sub District and the water sources rely solely on rainwater.

3.4 Chilli and shallot region

Chilli development region in Tapin District is presented in Figure 8 and Table 7. Only one development region was discovered in an area with a low productivity increase (BE.3) covering 185 ha and located in Central Tapin Sub District. The productivity of chilli in this region was found to be around 12-15 tons/ha. Furthermore, it is a shallow inundate swamp area with land suitability found to be marginally suitable (S3) with dominant limiting factors observed to be poorly drained drainage, acidic to very acidic soil reaction (pH), low N nutrient, and low to very low P_2O_5.

In addition, approximately 50,327 ha of land was found to have a moderate potential for extensification (P.2) spreading across all sub-districts. The largest area was Candi Laras Utara District covering an

area of 33,276 ha with marginal suitability. The dominant limiting factors were poorly drained drainage and very acid soil reaction (pH) and the suitable area was 27,795 ha.

The shallot development area in Tapin Regency is presented in Figure 9 and Table 8. Only one development region was discovered in an area with a high productivity increase (BA.1) covering 169 ha and located in Bungur and South Tapin Sub Districts.

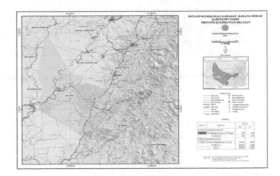

Figure 9. Map of the shallot development region in Tapin District.

Table 8. Legend map of shallot development region in Tapin District, South Kalimantan province.

Symbol	Description	Area	
		Ha	%
Development region			
BA.1	Low productivity increase	169	0,36
Subtotal		*185*	*169*
Development potential			
P.2	Medium development potential	46.510	99,64
Subtotal		*50.327*	*46.510*
Total		**50.512**	**46.679**

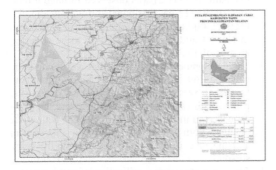

Figure 8. Map of chilli development region in Tapin Distric.

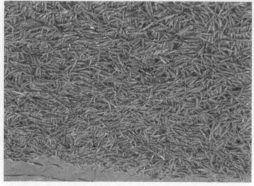

Figure 10. Performance of chilli plants on surjan of shallow swamp land by in Hiyung Village, Central Tapin Sub District.

Figure 11. Performance of shallot on bedded alluvial land in Bungur and South Tapin Sub District.

The productivity has high increase to around 8-10 tons/ha and all the lands in BA.1 are classified as marginal (S3). However, the dominant limiting factor includes acidic to very acidic soil reaction (pH), and low to very low N and P_2O_5.

For the extensification of shallot in this area, a total area of 46,510 ha land with moderate development potential (P.2) was discovered to spread across all sub-districts with the highest area located in Central Tapin Sub District covering an area of 10,783 ha considered moderately suitable. The dominant limiting factors were poorly drained drainage and acidic soil reaction (pH) covering 22,008 ha.

Moreover, Figure 10 shows the performance of chilli plants on surjan of shallow swamp land in Hiyung Village, Central Tapin Sub District while Figure 11 indicates those of shallot on bedded alluvial land in Bungur and South Tapin Sub District.

4 CONCLUSIONS

The conclusions drawn from this study are as follows.

1. Chili development region is developed in swamps of Central Tapin District in Hiyung Village covering 185 ha (BE.3), and the water was managed by surjan system at 1 m height to avoid 0.5-1 m flood. The potential area for its development was found in the dry fields of Sukamaja Village, Binuang District with water source being rainwater.

2. A potential land with medium development for extensification (P.2) was observed to have a total area of 50,327 ha distributed throughout the districts of Tapin District, and the largest was found in North Candi Laras Sub District to be covering an area of 33,276 ha.

3. The development region for shallot was found to be covering 169 ha in Bungur and South Tapin Sub District. It was planted 1-3 times after rice on the paddy fields of Harapanmasa Village and the highest observed in Shaba village of Bungur with water obtained from 2-3 m well during the dry season, and the area is usually flooded during the rainy season.

4. There was no optimal utilization of water sources for irrigation. Therefore, it is necessary to explore and exploit other potential water resources,

including deep groundwater. Other water harvest system such as building reservoirs, channel reservoir and long storage are recommended for supplementary irrigation.

5. Moreover, technology should be implemented to determine planting time with an integrated planting calendar in order to achieve optimal growth and yields.

REFERENCES

Balai Besar Penelitian dan Pengembangan Sumberdaya Lahan Pertanian (BBSDLP). 2016. *Atlas Peta Tanah Semidetail Kabupaten Tapin* Skala 1:50.000. Bogor: Balai Besar Penelitian dan Pengembangan Sumberdaya Lahan Pertanian, Badan Penelitain dan Pengembangan Pertanian.

BPS Kabupaten Tapin. 2015. *Kabupaten Tapin Dalam Angka*. Jakarta: Badan Pusat Statistik.

BPN (Badan Pertanahan Nasional). 2012. *Peta Penggunaan Lahan Indonesia*. Jakarta: BPN.

Djaenudin, D., Hendrisman, M., Hidayat, A. & Subagyo, H. 2003. *Petunjuk Teknis Evaluasi Lahan untuk Komoditas Pertanian*. Bogor: Balitanah.

FAO. 1976. A framework for land evaluation. FAO Soil Bulletin 32. FAO, Rome. ISBN 92-5-100111-1. 65 p.

http://id.climate-data.org/continent/asia/. Climate-data.org.

http://www.bakosurtanal.go.id/bakosurtanal/peta-rbi/

Kementerian Pertanian. 2013. Peta Sawah pada Kawasan Pertanian Kabupaten Tapin, Kalimantan Selatan. Kementerian Pertanian.

KLHK (Kementerian Lingkungan Hidup dan Kehutanan). 2013. *Peta Status Kawasan Hutan Indonesia*. Jakarta: KLHK.

Nethononda, L.O., Odhiambo J.J.O & Paterson, D.G. 2014. Land suitability for specific crop ranges using dynamic land suitability evaluation guidelines for small-scale communal irrigation schemes. *Bulgarian Journal of Agricultural Science* 20(6): 1349-1360. Agricultural Academy.

Oldeman, L.R., Las, I. & Muladi. 1980. *An Agroclimatic map of Kalimantan*, scale 1:3,000,000. Bogor: Central Research Institute for Agriculture.

Puslitbangtanak. 2003. *Atlas Sumberdaya Iklim Pertanian Indonesia* skala 1:1.000.000. Bogor: Pusat Penelitian dan Pengembangan Tanah dan Agroklimat, Badan Penelitian dan Pengembangan Pertanian.

Ritung, S., Nugroho, K., Mulyani, A. & Suryani, E. 2011. *Petunjuk Teknis Evaluasi Lahan untuk Komoditas Pertanian. Edisi Revisi 2011*. Bogor: Balai Besar Penelitian dan Pengembangan Sumberdaya Lahan Pertanian, 166pp.

Schmidt, F.H. & Ferguson, J.H.A. 1951. *Verhandelingen No.42 Rainfall Types Based On Wet And Dry Period Rations For Indonesia With Western New Guinee*. Jakarta: Kementrian Perhubungan Djawatan Meteorologi dan Geofisika.

Sikumbang, N. & Heryanto, R. 1994. *Peta Geologi Lembar Amuntai (1712) Kalimantan*, skala 1:250.000. Bandung: Pusat Penelitian dan Pengembangan Geologi.

Soil Survey Staff. 2014. *Keys to Soil Taxonomy. 12nd ed*. Washington DC: USDA Natural Resources Conservation Service.

Subardja, D.S., Ritung, S., Anda, M., Sukarman, Suryani, E. & Subandiono, R.E. 2016. *Petunjuk Teknis Klasifikasi Tanah Nasional. Edisi 2/2016*. Bogor: Balai Besar Penelitian dan Pengembangan Sumberdaya Lahan Pertanian (BBSDLP), Badan Penelitian dan Pengembangan Pertanian, Kementerian Pertanian, 53 pp.

Potential of land resources for agriculture in Asmat region, Papua Province, Indonesia

R.A. Gani & Sukarman

Indonesia for Center Agricultural Land Resources Research and Development of the Ministry of Agriculture (ICALRRD), Bogor, Indonesia

ABSTRACT: Approximately 94% of the Asmat region located on the southern coast of Papua Province is lowland swamp. In order to explore the potentials of this land for agriculture development, mapping, land resource correlation, and suitability assessment at the 1:50,000 scale was conducted. This study aimed to identify the potential and direction of the land resources through land suitability maps for agriculture, especially food crops on a semi-detail map. The research was conducted through the use of standardized methods developed by the Indonesia Center for Agricultural Land Resources Research and Development (ICALRRD). Land suitability assessment was conducted using Land Suitability Assessment System version 2.01 (SPKL v.2.01) software in accordance with the Guidelines for Land Suitability Assessment for Strategic Agricultural Commodities Semi-Detail Level Scale 1:50,000 (ICALRRD, 2017b). 2,517,204 hectares of the land area was mapped, and only 47,774 ha could be optimized for agricultural activity. The soil mapping exercise found six types of soils, which are Alluvial, Organosol, Regosol, Kambisol, Gleisol, and Podsolik. Furthermore, the land suitability map recommended 14 strategic commodities including paddy rice, rainfed rice plants, upland rice crops, swampy rice paddy, corn, sweet potatoes, cassava, soybeans, red chili, shallots, oil palm, white sugar cane, and elephant grass feed plants. However, Protected Forests, National Parks and Limited Production Forests covering 2,142,205 hectares were not recommended for agricultural purposes. Therefore, to plan for the expansion or extensification of land for agriculture in the region, only 47,774 hectares were recommended, with approximately 210,080 hectares not suitable, and some 117,346 hectares covered by settlements and water bodies.

1 INTRODUCTION

After 15 years of establishment, the Region of Asmat in Papua Province continues to grow to this day. The vast land resources and huge potentials make it possible for the region to become a future agricultural base, especially in the eastern part of Indonesia. One of the supporting forces to make this happen is the existence of information on land resources and their potentials.

The increasingly sharp growth in demand for renewable food and bio-energy continually decreases land availability. Therefore, it is necessary to expand to other regions based on the inventory of potential and suitable land resources for sustainable agriculture in the face of climate change (Sulaiman et al., 2019; Zabel et al., 2014).

Information is very important, effective and efficient in the development of land resources, both in terms of its potential and limitations (Gong et al., 2011; Mendas and Delali 2012; Mohsen et al., 2017).

Operational-scale soil mapping (scale 1:50,000) is information support that can be used as a reference basis for the preparation of other thematic maps needed for land resources development such as land suitability maps. This is because it has more detailed information about the properties of the soil, its extent, as well as distribution in an area. The land suitability map at a scale of 1:50,000 plays an important role in providing information on the suitability of various agricultural commodities or identifying limiting factors of growth, extent, and distribution (Hikmatullah, 2014).

Soil Mapping at a scale of 1:50,000 is another basic reference in preparing management recommendations to increase land productivity. For local governments, a map of 1:50,000 can be used in preparing or revising Regional Spatial Plans (RTRW) to allocate more appropriate space according to its potentials (ICALRRD, 2017a).

This paper aimed to identify and make appropriate recommendations on land resources based on information obtained from soil maps, land suitability maps for agriculture, especially food crops, and directives of agricultural commodities on a semi-detail scale of 1:50,000 on Asmat region in the Papua Province.

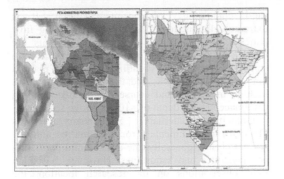

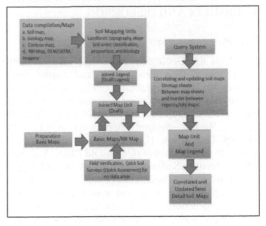

Figure 1. Maps of location and administration in the Asmat region, Papua Province (BPS, 2016).

2 METHODS

The study was conducted in May 2017 in the Asmat region, Papua Province to correlate and update the soil map on a semi-detailed scale. The soil data was based on legacy (ICALRRD, 2008), the Agro Ecology Zone (AEZ) soil mapping data (BPTP Papua, 2012) and recent data obtained from the correlation of soil maps and updating of the semi-detailed scale soil maps of 2017. These were used as a baseline for the land characteristics while land suitability map and direction of the agricultural commodities were freshly conducted. The commodity land map was included in the land management recommendations for the optimization and development of strategic agricultural commodities (ICALRRD, 2017c).

2.1 Soil map preparation from the correlation and updating scale of 1:50,000

Soil maps were prepared on a scale of 1:50,000 from the correlation and updating using a data and map compilation and validation approach. Data and maps were obtained from the results of soil mapping surveys conducted in previous years (ICALRRD, 2008). The soil map has been correlated with others from regions directly adjacent to Asmat on an equal scale. The steps involved are shown in Figure 2.

2.2 Preparation of land suitability map scale of 1:50,000

The land suitability maps on a scale of 1:50,000 were prepared based on semi-detailed soil maps on a scale of 1:50,000 obtained from the correlation and updated soil maps. The map used in this paper is for rice food crops such as field rice irrigation, rainfed lowland rice, tidal field rice, swamp field rice, and upland field rice, and tubers like sweet potato and cassava. The stages involved in preparing semi-detailed maps for land suitability and commodity direction are shown in Figure 3.

The results of these stages produced a land evaluation unit containing complete land characteristics data according to the standards processed using the System

Figure 2. Flowchart of procedure for correlation and updating of soil map in the Asmat region, Papua Province (ICALRRD, 2017a).

Land Suitability Assessment (SPKL 3.0) software and land typology (Bachri S. et al. 2016).

2.3 Direction of agricultural commodities

The results obtained from the preparation of the land suitability map were overlaid with the land use and the area status map. This produced a map of development in the form of commodity directives which can be used to provide information on potential improvements through existing field rice land or intensification (I), diversification on dry land and mixed plantations (D), alley crops on plantation land (C), and new openings/ extensification (E). The land directed towards the development of strategic commodities is located in Other Use Areas (APL), Conversion Production Forests (HPK), and Production Forests (HP). Lands outside these areas such as Protected Forest areas, National Parks, and water bodies will not be included in the expansion plan (ICALRRD, 2017c). The steps involved are shown in Figure 3.

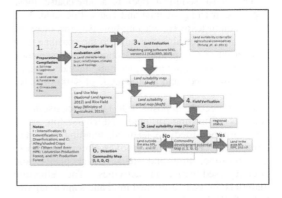

Figure 3. Flow chart for the preparation of semi-detailed maps of land suitability and direction of agricultural commodities in the Asmat region, Papua Province (ICALRRD, 2017c).

3 RESULTS AND DISCUSSION

3.1 *Parent materials*

Based on the Geological Map scale of 1: 250,000 Mapi Sheets (Suwarna, 1995), and Asmat Sheet (Heriyanto and Panggabean, 1995), the Region of Asmat has eight geological formations which are Recent River Deposits (Qr.1), Old River Deposits (Qr.2), Recent Swamp Deposition (Qs.1), Old Swamp Deposition (Qs.2), Coast Sediment (Qc), Young Beach Sediment (Qc.1), Old Beach Sediment (Qc.2), Tertiary Sedimentary Rocks-Carter (TmQp).

The river recent sediment (Qr. 1) is a loose deposit consisting of sand, mud, and gravel. It is floodplain sediment, generally, with a very fine texture, winding flow pattern with a horseshoe shape, locally seen river steps, and almost parallel rivers and straightness, the cover layer is quite thick, the precipitation is active and occupies an altitude of <50 m above sea level. The alluvium materials observed are sand, mud, and gravel.

The old river sediment (Qr.2) is a deposit composed of mud, sand, and gravel, as a result of a rather compact floodplain sediment, which is generally with a smooth texture, visible straightness, and a cover layer thick enough, and occupying an altitude of <20 m above sea level. The alluvium materials observed are mud, sand, and gravel.

The swamp recent deposition (Qs.1) is a very fine deposit composed of clay, silt, and fine sand. It contains carbonaceous material, medium fine texture, flowing-elbow flow pattern, visible straightness, a tightly closed but thin layer, with active settling. The alluvium materials include clay, silt, and fine sand at an altitude of <250 m above sea level.

The old swamp sediment (Qs.2) is also a very fine deposit consisting of mud, and fine carbonaceous sand and peat, coarse-pile textured, smooth-elbow-patterned flow pattern, straightness, and a tight cover plant. The alluvium materials found are mud and fine sand of carbon and peat at an altitude of <250 m above sea level.

The coast sediment (Qc) is a fine-coarse loose deposit consisting of mud and coarse-fine sand. It is fine-medium textured, winding flow patterns, there are forms of grooves almost parallel to the coastline, thin and sparse cover plants, and ongoing sedimentation activities. The alluvium materials include coarse-grained mud and sand at an altitude of <100 m above sea level.

Coast recent deposits (Qc.1) are coarse-loose deposits consisting of mud and coarse-fine sand. They are fine-medium textured, winding flow patterns, there are forms of grooves almost parallel to the coastline, thin and sparse cover plants, and ongoing sedimentation activities. The alluvium materials are coarse-grained mud and sand at an altitude of <100 m above sea level.

The Old Coast Deposition (Qc.2) is a rather solid clastic deposit consisting of mud, silt and coarse-fine sand containing carbonaceous material. It is extremely fine-textured, high-density winding flow patterns, borders with young beach deposits are marked by the "v" shape and rotating, and there are steps not far from the coastline. The cover plant is not so tight and thick, and the alluvium materials include mud, silt, and coarse-fine sand containing carbonaceous at a depth of <100 m above sea level.

Quaternary-Quaternary Sedimentary Rocks (TmuQp) are sedimentary rocks of marine and terrestrial clastic, generally fine-grained and consisting of claystones and sandstones. However, based on field observations on the research location, the parent soil material in Asmat Region consisted of (a) clay and sand deposits, (b) sand deposits (c) clay deposits (d) organic matter (e) silt (f) claystones and sandstones (g) marine sediment (clay) (h) marine sediment (sand) (i) marine sediment (clay and sand).

3.2 *Group of landforms*

Soil mapping could be grouped into 6 (six) landforms, which are Alluvial (A), Marine (M), Fluvio Marine (B), Peat (G), Tectonics (T) and Mileneous (X) (Ritung et al., 2017). It was discovered that Asmat Region was dominated by the Alluvial Group (A) covering an area of 1,452,865 ha or 57.72%, followed by the Fluvio-Marine Group (B) about 222,540 ha or 8.84% while the least was the Tectonic Group with 39,251 ha or 1.56%. The distribution of landform units is shown in Table 1.

The Alluvial landforms group consist of floodplain sub-groups, meander lanes, back swamps, sandbanks, former old river flows, alluvial plains, flow paths, fan heads, fan center, fan feet, and colluvial plains. The Marine is divided into coastal sand, back and coastal basins, the tidal plains of mud, and the tidal swamp behind. The Fluvio-marine is divided into estuarine deltas, estuarine plains along the river, and fluvio-marine plains. The Peat is further categorized into topogen freshwater peat and tidal topogen peat. The Tectonic is divided into tectonic plains, tectonic hills, and parallel ridges while the Mileneaouse consists of settlements, water bodies, and small islands.

3.3 *Soil classification of correlation and updating soil map*

Soil resource data is very important in supporting government programs in the regions. These data can be obtained through research and mapping of land resources conducted systematically in stages (Waas et al., 2014). The data will be very beneficial to local governments in the development of their regions, especially those that support the availability of agricultural land.

Soil can be classified according to the National Soil Classification System (ICALRRD, 2016) and the USDA Soil Taxonomy System (Soil Survey Staff, 2014). It is, however, important to state that correlation and update of soil map with semi-detailed scale have

Table 1. Group of landforms and soils in Asmat Region, Papua Province (Ritung et al., 2017).

Landforms	Soils	Area Hectares	%
Alluvial Groups		**1,452,865**	57,72
1. Flood Plain	Aluvial Gleik (Typic Fluvaquents)	121,973	4,85
2. Meander Belt	Gleisol Eutrik (Aeric Endoaquepts)	31,797	1,26
3. Back Swamp	Aluvial Gleik (Typic Fluvaquents)	510,553	20,28
4. Sand Bar	Regosol Gleik (Aquic Udipsamments)	2,236	0,09
5. Old River Channel	Gleisol Fluvik (Fluvaquentic Endoaquepts)	45,149	1.79
6. Alluvial Plain	Gleisol Distrik (Typic Endoaquepts)	340,157	13.51
7. Stream Belt	Gleisol Eutrik (Aeric Endoaquepts)	66,300	2.63
8. Fan Head	Kambisol Eutrik (Typic Eutrudepts)	52,916	2.10
9. Fan Middle	Kambisol Eutrik (Typic Eutrudepts)	69,851	2.77
10. Fan Base	Kambisol Gleik (Typic Dystrudepts)	115,842	4.69
11. Colluvial Plain	Regosol Eutrik (Typic Quartzip samments)	96,091	3.82
Marine Groups		**211,705**	8.41
1. Sand Beach	Regosol Distrik (Typic Udipsamments)	16,796	0.67
2. Beach Ridges and Swales Recent	Aluvial Sulfidik (Typic Sulfaquents)	7,332	0.29
3. Tidal Mud Flat	Gleisol Sulfik (Sulfic Endoaquents)	86,279	3.43
4. Tidal Back Swamp	Aluvial Sulfidik (Typic Sulfaquents)	101,298	4.02
Fluvio Marine Groups		**222,540**	8.84
1. Estuarine Delta	Aluvial Sulfidik (Typic Sulfaquents)	1,073	0.04
2. Estuarine Flat Along Major Rivers	Gleisol Sulfik (Sulfic En doaquepts)	80,087	3.18
3. Fluvio-Marine Flat	Aluvial Sulfidik (Typic Sulfaquents)	141,380	5.62
Peat Groups		**495,211**	19.67
1. Fresh Water Topogenous Peat	Oranosol Saprik (Typic Haplosaprists)	430,105	17.09
2. Tidal Topogenous Peat	Oranosol Sulfidik (Typic Sulfihemists)	65,106	2.59
Tectonic Groups		**39,251**	1.56
1. Tectonic Plain	Kambisol Distrik (Typic Dystrudepts)	13,228	0.53
2. Hilly Tectonic	Kambisol Distrik (Typic Dystrudepts)	17,526	0.70
3. Parallel Ridges and Hills	Podsolik Haplik (Typic Hapludults)	8,497	0.34
Miscellaneous Groups		**95,632**	3.80
1. Settlement	-	803	0.03
2. Water Body (lakes, rivers)	-	92,397	3.67
3. Small Islands	-	2,432	3.67
Total		2,517,204	100.00

been previously conducted in the region and classified into 34 units of soil mapping (SPT) consisting of six main dominant groups and eighteen subgroups.

The six dominant groups include Alluvial, Organosol, Regosol, Kambisol, Gleisol, and Podsolik while the eighteen subgroups are made up of Organosol Saprik, Organosol Fibrik, Organosol Hemik, Organosol Sulfidik, Organosol Sulfurik, Alluvial Gleik, Alluvial Sulfidik, Regosol Eutrik, Regosol Distrik, Regosol Gleik, Kambisol Gleik, Kambisol Eutric, Kambisol Region, Gleisol Fluvik, Gleisol Distrik, Gleisol Eutrik, Gleisol Sulfik, and Podsolik Haplik. These soils developed from parent materials of organic matter, as well as sediments of clay and organic matters, sand, clay and sand, marine clay, colluvial, silt, silt and clay, and sandstone and mudstone.

In the Alluvial group, the soils found were Alluvial Gleik (Typic Fluvaquents), Gleisol Eutrik (Aeric Endoaquepts), Regosol Eutrik (Typic Quarzipsamments), Regosol Gleik (Aquic Udipsamments), Gleisol Fluvik (Fluvaquentic Endoaquepts), Gleisol Distrik (Aeric Endoaquepts), Kambisol Eutrik (Typic Eutrudepts), Kambisol Gleik (Aquic Dystrudepts). In Marine, they include Regosol Distrik (Typic

185

Table 2. Soils at the Asmat region based on National Soil Classification (Subardja et al., 2016) and the equivalent with USDA (Soil Survey Staff, 2014).

Groups (ICALRRD)	Sub Groups (ICALRRD, 2016)	Sub Groups (Soil Survey Staff, 2014)
Organosol	Organosol Fibrik	Hemic Haplofibirists
	Organosol Hemik	Typic Haplohemists
		Sapric Haplohemists
	Organosol Saprik	Terric Haplosaprists
		Typic Haplosaprists
	Organosol Sulfidik	Typic Sulfihemists
	Organosol Sulfidik	Typic Sulfohemists
Aluvial	Aluvial Gleik	Typic Fluvaquents
	Aluvial Humik	Humaqueptic Endoaquents
	Aluvial Sulfid ik	Typic Sulfaquents
Regosol	Regosol Gleik	Typic Psammaquents
		Aquic Udip samments
	Regosol Distrik	Typic Udip samments
Kambisol	Kambisol Gleik	Aquic Dystrudepts
		Aquic Eutrudepts
	Kambisol Distrik	Typic Dystrudepts
	Kambisol Eutrik	Typic Eutrudepts
Gleisol	Gleisol Sulfik	Sulfic End oaquepts
	Gleisol Fluvik	Fluvaquentic Endoaquepts
	Gleisol Distrik	Typic Endoaquepts
	Gleisol Eutrik	Typic Endoaquepts
Podsolik	Pods solik Hap lik	Typic Hapludults

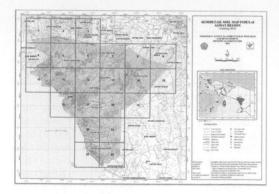

Figure 4. The index of semi detail soil map scale 1:50.000 in Asmat region, Papua Province.

Udipsamments), Alluvial Sulfidik (Typic Sulfaquents), and Gleisol Sulfik (Sulfic Endoaquepts), and for the Fluvio-marine, Alluvial Sulfidik (Typic Sulfaquents) and Gleisol Sulfik (Sulfic Endoaquepts).

The Peat groups were found to include Organosol Saprik (Typic Haplosaprists) and Organosol Sulfidik (Typic Sulfihemists). The Tectonic include Kambisol Distrik (Typic Dystrudepts) and Podsolik Haplik (Typic Hapludults). The distribution is as shown in Table 1, Table 2 and Figure 4 while the legend of soil map is presented in Table 4 of Appendix 1.

3.4 *Land suitability for agriculture commodity*

Land suitability for agricultural commodities in this paper covers food crops such as paddy field irrigation rice, paddy swamp lowland field rice, paddy tidal lowland field rice, paddy rainfed lowland rice, paddy upland rice, sweet potatoes, and cassava; horticultural crops such as red chili and red onions; plantation crops such as white sugar cane, cocoa, and oil palms; and animal

feed forage like elephant grass. From this research, it was discovered that food crops are the most suitable for the 2,517,205 ha of land mapped, followed by horticultural and plantation commodities. The distribution area of land suitability for each commodity is as shown in Figure 5.

Paddy field rice irrigation was found to cover the highest suitable land with 1,325,393 ha moderately suitable (S2) and 236,626 ha marginally suitable (S3). It was discovered that 837,386 ha was not suitable (N) and the remaining (Pusdatin, 2014).

Red chilli and red onions have the same broad yield on land suitability assessment. The suitable lands for red chili or red onion were found to be 1,459,638 ha with 1,297,396 ha moderately suitable (S2) and 939,767 ha marginally suitable (S3). The remaining 162,242 ha were not suitable (N) and 117,800 ha not rated (TD).

Cocoa plants were found to have the highest suitability for plantation crops by covering 1,475,449 ha with 215,128 ha moderately suitable (S2) and 1,260,321 ha marginally suitable (S3). The remaining 923,955 ha were not suitable (N) and 117,800 ha not rated (TD).

The dominant limiting factors observed for the land suitability of food crops include soil drainage, moderate soil texture, acidic soil pH value, the maturity of peat classified as hemist, sulfidic material at a depth of 40-75 cm, P and K nutrient contents of moderate to low, and 5-8% slope.

In order to find a lasting solution to the drainage problem, it is necessary to make channels to improve soil aeration, which is essential for the growth of plant roots.

Medium to slightly coarse soil texture is another limiting factor that must be addressed. Organic matter must be added to the soil to improve the organic content, soil texture, water-retaining ability, and nutrients of the soil.

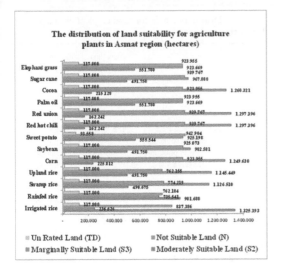

Figure 5. Distribution of lands suitability for agriculture in the Asmat region of Papua Province.

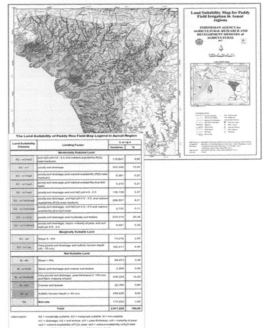

Figure 6. Map and legend of land suitability for paddy rice field in the Asmat region, Papua Province.

The maturity level of the peat was categorized as sapric and hemic. Therefore, the drainage system of the land must be regulated. The depth of sulfidic material was also found to be between 40-75cm for food crop cultivation. A groundwater management system must be employed to raise the groundwater level above the sulfidic material layer.

The soil pH was generally acidic for lands considered moderately suitable for food crops. To improve this condition, agricultural lime such as dolomite must be applied to improve the pH of the soil as well as the availability of alkaline elements such as Ca and Mg.

The total P and K content (25% HCl extraction) was classified as low to moderate on moderately suitable lands. This can be rectified through the timely and appropriate application of P and K fertilizers.

Another limiting factor observed was 5-8% slope (eh) on marginal lands. To overcome this, conservation techniques through the creation of terraces was applied to avoid erosion hazards on the land.

Example of the map for land suitability for agriculture commodities scale 1:50.000 in Asmat region, Papua Province.

3.5 Direction for agricultural commodities and land management recommendations

A Map of Agricultural Commodity Direction was produced from the overlay of land suitability with land use maps and equipped with a legend. The agricultural commodities directed include bio-physically suitable agricultural land (class S1, S2, and S3) located in Other Land use Area (APL), Production

Forests (HP, and Production and Conversion Forests (HPK) areas with a minimum of 5 ha. The Landuse Direction is shown in Figure 8.

The figure shows that agricultural commodities directed include rice such as field rice irrigation and swamp field rice, upland rice, corns, soybeans, sweet potatoes, and cassava. They were further divided into 13 units of direction in a pattern to develop new expansion area or extensification (I) by planting superior strategic commodities on new openings of 47,774 ha previously occupied by shrubs or swamp shrubs, open land or grasslands. These lands were located in Akat, Atsy, Fayit, Kasuari Beaches, Sawa Erma, Suator, and Suru-suru region. The ones not recommended covered 2,142,005 ha and include Protected Forests, National Parks, and Limited Production Forests. Those classified as not suitable covered 210,080 ha while settlement and water body (X) were found on 117,346 ha. The distributions are as shown in Figures 7 and 8.

3.6 Potential for agricultural commodity development in dry land in the Asmat region

Table 3 shows the distribution of drylands that can be developed for food crops in the Asmat region through the Extensification (E) pattern to be 5,627 ha. It was

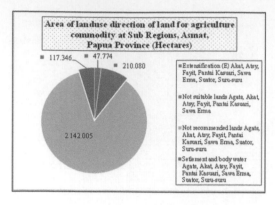

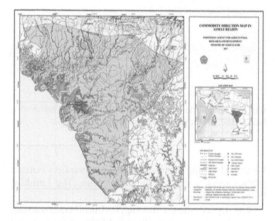

Figure 7. The distribution area of land use direction for agriculture commodity development in subregions, Asmat region, Papua Province.

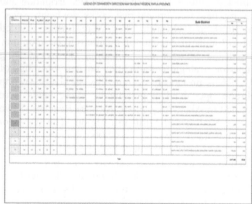

Figure 8. Map and legend of agriculture commodity direction Asmat region, Papua Province.

discovered that Other Use Areas (APL) covered an area of 135 ha, Conversion Production Forest (HPK) 58 ha and Production Forest (HP) 5,435 ha. Food crops such as upland rice, corn, soybeans, sweet potatoes, and cassava can be developed on these lands.

Table 3. Distribution area for food crops at dry land in Asmat region, Papua Province.

Landuse Types	Commodities	Area (ha)			
		APL	HPK	HP	TOTAL
Diversification/ Mix crops (D)	-	-	-	-	-
Extensification (E)	Upland rice, coms soybeans, sweet potatoes, cassava	135	58	5.435	5.627
Estate(C)	-	-	-	-	-
TOTAL		135	58	5.435	5.627

Notes
APL: Other landuse area
HPK: Conversion Production Forest
HP: Production Forest

4 CONCLUSIONS

This study has mapped 2,517,204 hectares of land but only 47,774 hectares could be used for agriculture activities in Asmat. From the analysis, six main types of soils were found in the region, and they include Alluvial, Organosol, Regosol, Kambisol, Gleisol, and Podsolik. The land suitability map also recommended 14 strategic commodities which are paddy rice, rainfed rice plants, upland rice crops, swampy rice paddy, corn, sweet potatoes, cassava, soybeans, red chili, shallots, oil palm white sugar cane, and elephant grass feed plants. Furthermore, Protected Forests, National Parks and Limited Production Forests covering 2,142,205 hectares were not recommended for agricultural activities while 47,774 ha were suggested for new area expansion or extensification of agriculture. Approximately 201,080 ha were observed not to be suitable, and 117,346 ha were settlement and water body.

ACKNOWLEDGEMENT

The first author shows appreciation to the ICALRRD for project updating and correlation soil mapping. The authors also show appreciation to the people of Asmat, Papua that assist them during the soil map correlation and mapping.

REFERENCES

Bachri S. Y. Sulaeman, Ropik S., H. Hidayat, dan A. Mulyani. 2016 Petunjuk Pengoperasian Sistem Penilaian Kesesuaian Lahan (SPKL) v. 2.01. Balai Besar Sumberdaya Lahan Pertanian. Bogor.

BPS Kabupaten Asmat. 2016. Asmat in Figure 2016. Bereau of Statistic Asmat Region. catalog No. 1102001.9401.

BPTP Papua. 2012. Laporan Pemetaan AEZ Kabupaten Asmat, Provinsi Papua. Balai Pengkajian Teknologi Pertanian Papua. Merauke.

Gong, J., Liu, Y. & Chen, W. 2011. Land suitability evaluation for development using a Matter-Element Model: a case study in Zengcheng, Guangzhou, China. Land Use Policy 29(2): 464-472.

Heriyanto, R. & Panggabean, H. 1995. Map of Geology, Asmat Sheet, Irian Jaya, scale 1:250.000. Bandung: Pusat Penelitian dan Pengembangan Geologi.

Hikmatullah, S. Ritung, Sukarman dan K. Nugroho. 2014. Pe-tunjuk Teknis Survei dan Pemetaan Tanah Tingkat Semi Detail Skala 1:50.000. ICALRRD, Indonesia Agency for Agricultural Research and Development, Ministry of Agri-culture. Bogor.

ICALRRD, 2016. National Soil Classification System 2nd editions. 60 pages. Bogor: ICALRRD. Indonesia Agency for Agricultural Research and Development, Ministry of Agriculture.

ICALRRD, 2017a. Atlas Peta Tanah Semi Detail Kabupaten Asmat (Skala 1:50.000). Bogor: ICALRRD, Indonesia Agency for Agricultural Research and Development, Minis-try of Agriculture.

ICALRRD, 2017b. Peta Kesesuaian Lahan untuk Pertanian Kabupaten Asmat (Skala 1:50.000). Bogor: ICALRRD. In-donesia Agency for Agricultural Research and Development, Ministry of Agriculture.

ICALRRD, 2017c. Peta Arahan dan Rekomendasi Pengelolaan Lahan Kabupaten Asmat (Skala 1:50.000). Bogor: Indonesia Agency for Agricultural Research and Develop-ment, Ministry of Agriculture.

ICALRRD. 2008. Semi Detail Soil Mapping at Asmat Region (scale 1:50.000). Report. Indonesia Agency for Agricultural Research and Development (ICCALRD), Ministry of Agri-culture. Bogor.

Mohsen, B., Kaveh Madani, M., Hossein Hashemi & Azadi, P. 2017. Iran's Land Suitability for Agriculture. www.nature.com/scientificreports. https://doi.org/10.1038/s41598-017-08066-y

Mendas A. & Delali, A. 2012. Integration of multi criteria dcision analysis in GIS to develop land suitability for agri-culture: application to durum wheat cultivation in the re-gion of Mleta in Algeria. Journal Computers and Electron-ics in Agriculture. https://doi.org/10.1016/j.compag. 2012.02.03.

Pusdatin. 2014. Statistik Lahan Pertanian Tahun 2009-2013. Sekretariat Jenderal Kementerian Pertanian. Jakarta.

Ritung S, Suparto, E. Suryani, K. Nugroho, dan C. Tafakresnanto. 2017. Petunjuk Teknis Pedoman Klasifikasi Landform untuk Pemetaan Tanah di Indonesia. Indonesia Central for Agricultural Land Resource Research and Development (ICALRRD). Ministry of Agriculture Bogor. 49 p.

Soil Survey Staff. 2014. Keys to Soil Taxonomy. Twelfth Edi-tion, 2014: 362 p. Washington DC: Natural Resources Conservation Service-United States Department of Agricultural.

Subardja, D., Ritung, S., Anda, A., Sukarman, Suryani, E. dan Subandiono, R.E. 2016. Petunjuk Teknis Klasifikasi Tanah Nasional. Edisi ke-2: 60 pages. Bogor: ICALRRD, Indonesia Agency for Agricultural Research and Development, Ministry of Agriculture.

Sulaiman, A.A., Sulaeman, Y. & Minasny, B., 2019. A Framework for the Development of Wetland for Agri-cultural Use in Indonesia. Resources, 8(1), p.34.

Suwarna, N. 1995. Map of Geology, Mapii Sheet, Irian Jaya, scale 1:250.000. Bandung: Pusat Penelitian dan Pengem-bangan Geologi.

Waas, E.D., Ayal, J. & Sheny Kaihatu. 2014. Evaluasi dan penentuan jenis tanah di Kabupaten Seram Bagian Barat. Agros 16(2): 336-348.

Zabel, F., Brigitta Putzenlechner & Wolfram Mauser. 2014. Global Agricultural Land Resources - A High Resolution Suitability Evaluation and Its Perspectives until 2100 under Climate Change Conditions. http://journals.plos.org/plosone/article?id=10.1371. pone.0107522#abstract0#.

Appendix 1

Semi Detailed Soil Map Legend of Asmat Region

No. Soil MU.	Soil Unit	Proportion	Landform	Parent material	Relief (% slope)	Large Ha	%
1	Aluvial Gleik, deep, poor drainage, fine texture, slightly acid, medium cation exchangeable capacity, and medium base saturation (*Typic Fluvaquents*)	D	Flood plain	Mud sediments	Flat (<-1)	121,973	4.85
	Gleisol Fluvik, deep, poor drainage, fine texture, slightly acid, medium cation exchangeable capacity, and medium base saturation (*Fluvaquentic Endoaquepts*)	F					
2	Aluvial Gleik, deep, poor drainage, slightly coarse texture, slightly acid, low cation exchangeable capacity, and medium base saturation (*Typic Fluvaquents*)	D	Back swamp	Clay and organic matter sediments	Flat (<-1)	510,553	20.28
	Gleisol Fluvik, deep, poor drainage, fine texture, slightly acid, medium cation exchangeable capacity, and medium base saturation (*Fluvaquentic Endoaquepts*)	F					
	Organosol Saprik, shallow, poor drainage, sapric, acid, high cation exchangeable capacity, and very low base saturation (*Terric Haplohemists*)	M					
3	**Regosol Gleik**, deep, poor drainage, coarse texture, slightly acid, low cation exchangeable capacity, and medium base saturation *(Aquic Udipsamments)*	D	Sand bar	Sand sediments	Nearly flat (1-3)	2,236	0.09
	Aluvial Gleik, deep, poor drainage, slightly coarse texture, slightly acid, low cation exchangeable capacity, and medium base saturation (*Typic Fluvaquents*)	F					
4	**Gleisol Fluvik**, deep, poor drainage, fine texture, slightly acid, medium cation exchangeable capacity, and medium base saturation (*Fluvaquentic Endoaquepts*)	D	Former old river flows	Clay and sand sediments	Flat (<-1)	45,149	1.79
	Aluvial Gleik, deep, poor drainage, slightly coarse texture, slightly acid, low cation exchangeable capacity, and medium base saturation (*Typic Fluvaquents*)	F					
5	**Gleisol Eutrik**, deep, poor drainage, fine texture, slightly acid, medium cation exchangeable capacity, and medium base saturation (*Aeric Endoaquepts*)	D	Meander belt	Clay and sand sediments	Flat (<-1)	31,797	1.26
	Aluvial Gleik, deep, poor drainage, coarse slightly texture, slightly acid, low cation exchangeable capacity, and medium base saturation (*Typic Fluvaquents*)	F					

(Continued)

Semi Detailed Soil Map Legend of Asmat Region

No. Soil MU.	Soil Unit	Propor-tion	Land-form	Parent material	Relief (% slope)	Large Ha	%
6	**Gleisol Eutrik**, deep, poor drainage, fine texture, slightly acid, medium cation	D	Stream belt	Clay and sand sediments	Nearly flat (1-3)	66,300	2.63
	exchangeable capacity, and medium base saturation (*Aeric Endoaquepts*)						
	Gleisol Fluvik, deep, poor drainage, fine texture, slightly acid, medium cation	F					
	exchangeable capacity, and medium base saturation (*Fluva-quentic Endoaquepts*)						
7	**Gleisol Distrik**, deep, poor drainage, fine slightly texture, acid, medium cation	D	Alluvial plain	Clay and sand sediments	Flat (<-1)	133,540	5.31
	exchangeable capacity, and medium base saturation (*Typic Endoaquepts*)						
	Kambisol Gleik, deep, moderately drainage, fine slightly texture, acid, medium	F					
	cation exchangeable capacity, and medium base saturation (*Aquic Eutrudepts*)						
8	**Kambisol Distrik**, deep, well drainage, fine slightly tex-ture, acid, low cation	D	Alluvial plain	Clay and sand sediments	Nearly flat (1-3)	206,617	8.21
	exchangeable capacity, and low base saturation (*Typic Distrudepts*)						
	Kambisol Gleik, deep, moderately drainage, fine slightly texture, acid, low cation	F					
	exchangeable capacity, and low base saturation (*Aquic Dystrudepts*)						
9	**Kambisol Eutrik**, deep, well drainage, fine slightly tex-ture, slightly acid, medium	D	Fan head	Colluvium	Rolling (8-15)	52,916	2.10
	cation exchangeable capacity, and high base saturation (*Typic Eutrudepts*)						
	Kambisol Gleik, deep, moderately drainage, fine slightly texture, acid, low cation	F					
	exchangeable capacity, and high base saturation (*Aquic Dystrudepts*)						
10	**Kambisol Eutrik**, deep, well drainage, fine slightly tex-ture, slightly acid, medium	D	Fan middle	Colluvium	Undulating (3-8)	69,851	2.77
	cation exchangeable capacity, and high base saturation (*Typic Eutrudepts*)						
	Kambisol Gleik, deep, moderately drainage, fine slightly texture, acid, low cation	F					
	exchangeable capacity, and low base saturation (*Aquic Dystrudepts*)						

(*Continued*)

Semi Detailed Soil Map Legend of Asmat Region

No. Soil MU.	Soil Unit	Proportion	Land-form	Parent material	Relief (% slope)	Large Ha	Large %
11	**Kambisol Gleik**, deep, moderately drainage, fine slightly texture, acid, low cation exchangeable capacity, and low base saturation *(Aquic Dystrudepts)*	D	Fan base	Colluvium	Nearly flat (1-3)	115,842	4.60
	Kambisol Eutrik, deep, well drainage, fine slightly texture, slightly acid, medium cation exchangeable capacity, and high base saturation *(Typic Eutrudepts)*	F					
12	Kambisol Eutrik, deep, well drainage, fine slightly texture, slightly acid, medium cation exchangeable capacity, and high base saturation *(Typic Eutrudepts)*	D	Colluvila plain	Colluvium	Flat (<1)	92,476	3.67
	Kambisol Gleik, deep, moderately drainage, fine slightly texture, acid, medium cation exchangeable capacity, and low base saturation *(Aquic Dystrudepts)*	F					
13	Kambisol Eutrik, deep, well drainage, fine slightly texture, slightly acid, medium cation exchangeable capacity, and low base saturation *(Typic Quartzipsamments)*	D	Colluvila plain	Colluvium	early flat (1-3)	3,615	0.14
	Regosol Gleik, deep, poor drainage, tekstur slightly kasar, acid, medium cation exchangeable capacity, and low base saturation *(Typic Psammaquents)*	F					
14	Regosol Distrik, deep, drainase slight quickly, coarse texture, acid, medium cation exchangeable capacity, and medium base saturation *(Typic Udipsamments)*	D	Sand beach	Sand sediments	Nearly flat (1-3)	16,796	0.67
	Gleisol Sulfik, deep, very poor drainage, very fine texture, acid, medium cation exchangeable capacity, and medium base saturation *(Sulfic Endoaquepts)*	F					
15	Aluvial Sulfidik, deep, very poor drainage, very fine texture, acid, medium cation exchangeable capacity, and medium base saturation *(Typic Sulfaquents)*	D	Beach ridges and swales recent	Clay marine sediments	Flat (<1)	7,332	0.29
	Gleisol Sulfik, deep, very poor drainage, very fine texture, acid, medium cation exchangeable capacity, and medium base saturation *(Sulfic Endoaquepts)*	F					

(Continued)

						Large	
No. Soil MU.	Soil Unit	Propor-tion	Land-form	Parent material	Relief (% slope)	Ha	%
16	Gleisol Sulfik, deep, very poor drainage, very fine texture, acid, medium cation	D	Tidal mud flat	Mud and marine clay sediments	Flat (<-1)	86,279	3.43
	exchangeable capacity, and medium base saturation *(Sulfic Endoaquepts)*						
	Aluvial Sulfidik, deep, very poor drainage, very fine texture, acid, medium cation	F					
	exchangeable capacity, and medium base saturation *(Typic Sulfaquents)*						
	Organosol Saprik, deep, poor drainage, sapric, acid, high cation exchangeable	M					
	capacity and low base saturation *(Typic Endoaquept*s)						
17	Aluvial Sulfidik, deep, very poor drainage, very fine texture, acid, medium cation	D	Tidal back swamp	Marine clay sediments	Flat (<-1)	101,298	4.02
	exchangeable capacity, and medium base saturation *(Typic Sulfaquents)*						
	Gleisol Sulfik, deep, very poor drainage, very fine texture, acid, medium cation	F					
	exchangeable capacity, and medium base saturation *(Sulfic Endoaquepts)*						
18	Aluvial Sulfidik, deep, very poor drainage, very fine texture, acid, medium cation	D	Estuar-ine delta	Clay and sand marine sediments	Flat (<-1)	1,073	0.04
	exchangeable capacity, and medium base saturation *(Typic Sulfaquents)*						
	Gleisol Sulfik, deep, very poor drainage, very fine texture, acid, medium cation	F					
	exchangeable capacity, and medium base saturation *(Sulfic Endoaquepts)*						
19	Gleisol Sulfik, deep, very poor drainage, very fine texture, acid, medium cation	D	Estuar-ine flat along major rivers	Sand sediments	lat (<-1)	80,087	3.18
	exchangeable capacity, and medium base saturation *(Sulfic Endoaquepts)*						
	Aluvial Sulfidik, deep, very poor drainage, very fine texture, acid, medium cation	F					
	exchangeable capacity, and medium base saturation *(Typic Sulfaquents)*						
20	Aluvial Sulfidik, deep, very poor drainage, very fine texture, acid, medium cation	P	Fluvio-marine flat	Mud sediments	Flat (<-1)	53,788	2.14
	exchangeable capacity, and medium base saturation (Typic Sulfaquents)						

Semi Detailed Soil Map Legend of Asmat Region

(Continued)

Semi Detailed Soil Map Legend of Asmat Region

No. Soil MU.	Soil Unit	Proportion	Landform	Parent material	Relief (% slope)	Large Ha	Large %
21	Aluvial Sulfidik, deep, very poor drainage, very fine texture, acid, medium cation	D	Fluvio-marine flat	Mud sediments	Flat (<-1)	87,592	3.48
	exchangeable capacity, and medium base saturation *(Typic Sulfaquents)*						
	Gleisol Sulfik, deep, very poor drainage, very fine texture, acid, medium cation	F					
	exchangeable capacity, and medium base saturation (*Sulfic Endoaquepts*)						
22	Organosol Saprik, deep, very poor drainage, sapric, slightly acid, high cation	D	Fresh water topogenous peat (shallow)	Organic matter	Flat (<-1)	259,614	10.31
	exchangeable capacity, and very low base saturation (*Typic Haplosaprists*)						
	Aluvial Humik, deep, poor drainage, fine texture, slightly acid, medium cation	F					
	exchangeable capacity, and medium base saturation *(Humaqueptic Endoaquents)*						
23	Organosol Hemik, deep, very poor drainage, hemic, very acid, high cation	D	Fresh water topogenous peat (moderate)	Organic matter	Flat (<-1)	124,395	4.94
	exchangeable capacity, and very low base saturation *(Sapric Haplohemists)*						
	Organosol Saprik, deep, very poor drainage, sapric, very acid, high cation	F					
	exchangeable capacity, and very low base saturation (*Typic Haplosaprists*)						
24	Organosol Hemik, very poor drainage, hemic, very acid, high cation exchangeable	D	Fresh water topogenous peat (deep)	Organic matter	Flat (<-1)	46,096	1.83
	capacity, and very low base saturation (*Sapric Haplohemists*)						
	Organosol Fibrik, very poor drainage, sapric, very acid, high cation exchangeable	F					
	capacity, and very low base saturation (*Hemic Haplofibrists*)						
25	Organosol Sulfidik, very poor drainage, hemic, very acid, high cation exchangeable	D	Tidal Topogenous Peat (moderate)	Organic matter	Flat (<-1)	46,635	1.85
	capacity, and very low base saturation (*Typic Sulfihemists*)						
	Organosol Sulfurik, very poor drainage, hemic, very acid, high cation echangeable	F					
	capacity, and very low base saturation (*Typic Sulfohemists*)						

(Continued)

Semi Detailed Soil Map Legend of Asmat Region

No. Soil MU.	Soil Unit	Propor-tion	Land-form	Parent material	Relief (% slope)	Large Ha	Large %
26	Organosol Sulfidik, very poor drainage, hemic, very acid, high cation exchangeable	D	Tidal Topo-genous Peat (deep)	Organic matter	Flat (<-1)	18,471	0.73
	capacity, and very low base saturation (*Typic Sulfihemists*)						
	Organosol Sulfurik, very poor drainage, hemic, very acid, high cation echangeable	F					
	capacity, and very low base saturation (*Typic Sulfohemists*)						
27	Kambisol Distrik, deep, well drainage, fine slightly texture, acid, medium cation	D	Tectonic plain	Sandstone and mudstone	Nearly flat (1-3)	2,733	0.11
	exchangeable capacityse, and medium base saturation (*Typic Dystrudepts*)						
	Kambisol Gleik, deep, moderately drainage, fine texture, slightly acid, medium	F					
	cation exchangeable capacity, and high base saturation (*Aquic Dystrudepts*)						
28	Kambisol Distrik, deep, well drainage, fine slightly texture, acid, medium cation	D	Tectonic plain	Sandstone and mudstone	Undulating (3-8)	4,385	0.17
	exchangeable capacity, and medium base saturation (*Typic Dystrudepts*)						
	Kambisol Gleik, deep, moderately drainage, fine slightly texture, slightly acid,	F					
	medium cation exchangeable capacity, and low base saturation (*Aquic Dystrudepts*)						
29	**Podsolik Haplik,** deep, well drainage, fine texture, acid, medium cation	D	Tectonic plain	Sandstone and mudstone	Rolling (8-15)	6,110	0.24
	exchangeable capacity, and low base saturation (*Typic Hapludults*)						
	Kambisol Distrik, deep, well drainage, fine slightly texture, acid, medium cation	F					
	exchangeable capacity, and medium base saturation (*Typic Dystrudepts*)						
30	Kambisol Distrik, deep, well drainage, fine slightly texture, slightly acid, medium	D	Hilly tectonic	Sandstone and mudstone	Hillocky (15-25)	15,336	0.61
	cation exchangeable capacity, and medium base saturation (*Typic Dystrudepts*)						
	Podsolik Haplik, deep, well drainage, fine texture, acid, medium cation	F					
	exchangeable capacity, and low base saturation (Typic Hapludults)						
31	Kambisol Distrik, deep, well drainage, fine slightly texture, slightly acid, medium	P	Hilly tectonic	Sandstone and mudstone	Hilly (25-40)	2,190	0.09
	cation exchangeable capacity, and high base saturation (*Typic Dystrudepts*)						

(*Continued*)

Semi Detailed Soil Map Legend of Asmat Region

No. Soil MU.	Soil Unit	Proportion	Landform	Parent material	Relief (% slope)	Large Ha	%
32	Podsolik Haplik, deep, well drainage, fine slightly texture, slightly acid, medium	D	Parallel ridges and hills	Sandstone and mudstone	Rolling (8-15)	1,035	0.04
	cation exchangeable capacity, and low base saturation (*Typic Hapludults*)						
	Kambisol Distrik, deep, well drainage, fine slightly texture, acid, medium cation	F					
	exchangeable capacity, and medium base saturation (*Typic Dystrudepts*)						
33	Kambisol Eutrik, deep, well drainage, fine slightly texture, slightly acid, medium	D	Parallel ridges and hills	Sandstone and mudstone	Hillocky (15-25)	479	0.02
	cation exchangeable capacity, and high base saturation (*Typic Eutrudepts*)						
	Podsolik Haplik, deep, poor drainage, fine texture, acid, low cation exchangeable	F					
	capacity, and low base saturation (*Typic Hapludults*)						
34	Kambisol Eutrik, deep, poor drainage, fine slightly texture, acid, medium cation	D	Parallel ridges and hills	Sandstone and mudstone	Hilly (25-40)	6,983	0.28
	exchangeable capacity, and high base saturation (*Typic Eutrudepts*)						
	Podsolik Haplik, deep, poor drainage, fine slightly texture, acid, medium cation	F					
	exchangeable capacity, and high base saturation (*Typic Hapludults*)						
X1	Settlement					803	0.03
X3	Water body (lake, rivers)					92,397	3.67
X4	Mud bar					2,432	0.10
Total						**2,517,204**	**100.00**

Author Index